Over 70% of the population in industrialised nations live in cities; in the next decade so will most of the world's population. This volume examines the impact of urban living on human health and biology. Cities pose numerous and diverse social and biological challenges to human populations which bear little resemblance to the forces that molded human biology through millions of years of evolution. Urban populations in industrialised nations have distinctive patterns of behavior, social stratification, stress, infectious disease, diet, activity and exposure to pollutants from years of industrialisation. These features affect diverse aspects of human function including human nutrition, energy expenditure, growth and reproduction. This volume begins with an introduction to the recent history of urbanism and poverty, infectious disease, reproductive function, child health, nutrition, physical activity and psychosocial stress. The book will appeal to workers in urban planning, human biology, anthropology, preventive medicine, human ecology and related areas.

T0297131

SOCIETY FOR THE STUDY OF HUMAN BIOLOGY
SYMPOSIUM SERIES: 40

Urbanism, Health and Human Biology
in Industrialised Countries

PUBLISHED SYMPOSIA OF THE
SOCIETY FOR THE STUDY OF HUMAN BIOLOGY

*Numbers 1–9 were published by Pergamon Press, Headington Hill Hall, Headington. Oxford
OX3 0BY. Numbers 10–24 were published by Taylor & Francis Ltd, 10–14 Macklin Street,
London WC2B 5NF. Further details and prices of back-list numbers are available from the
Secretary of the Society for the Study of Human Biology.*

Urbanism, Health and Human Biology in Industrialised Countries

EDITED BY

L. M. SCHELL

Departments of Anthropology and Epidemiology
University at Albany, State University of New York

S. J. ULIJASZEK

Institute of Biological Anthropology
University of Oxford

CAMBRIDGE
UNIVERSITY PRESS

CAMBRIDGE UNIVERSITY PRESS
Cambridge, New York, Melbourne, Madrid, Cape Town, Singapore, São Paulo, Delhi

Cambridge University Press
The Edinburgh Building, Cambridge CB2 8RU, UK

Published in the United States of America by Cambridge University Press, New York

www.cambridge.org
Information on this title: www.cambridge.org/9780521117630

First published 1999
This digitally printed version 2009

A catalogue record for this publication is available from the British Library

Library of Congress Cataloguing in Publication data
Urbanism, health, and human biology in industrialised countries /
[edited by] Lawrence M. Schell and Stanley J. Ulijaszek.
   p.   cm. – (Society for the Study of Human Biology symposium series; 40)
ISBN 0 521 62097 X
1. Urban health. 2. Urbanization – Health aspects. 3. Cities and towns – Health
aspects. 4. Urban ecology – Health aspects. 5. City dwellers – Health aspects. I.
Schell, Lawrence M. II. Ulijaszek, Stanley J. III. Series.
RA566.7.U73 1999
306.4'61 – dc21 98-4450 CIP

ISBN 978-0-521-62097-0 hardback
ISBN 978-0-521-11763-0 paperback

Contents

Contributors

C. Behrman
Department of Anthropology
University Museum
University of Pennsylvania
Philadelphia, PA 19104-6398
USA

David Clark
School of Natural Environmental
Sciences
Coventry University
Priory Street
Coventry CV1 5FB
UK

Stefan Czerwinski
Department of Genetics
Southwest Foundation for Biomedical
Research
PO Box 760549
San Antonio, TX 78227
USA

George DiFerdinando
Division of Family and Local Health
State of New York, Department of
Health
Corning Tower, Empire State Plaza
Albany, NY 12237
USA

Elizabeth Dowler
Department Epidemiology and
Population Health
London School of Hygiene and
Tropical Medicine
2 Taviton Street
London WC1H 0BT
UK

Peter T. Ellison
Department of Anthropology
Harvard University
Cambridge, MA 02138
USA

Penny Gordon-Larsen
Center for Community Partnerships
University of Pennsylvania
Philadelphia, PA 10910-3246
USA

Rebecca Huss-Ashmore
Department of Anthropology
University Museum
University of Pennsylvania
Philadelphia, PA 19104-6398
USA

Francis E. Johnston
Department of Anthropology
University of Pennsylvania
Philadelphia, PA 19104-6398
USA

A. J. McMichael
Department of Epidemiology and
Population Health
London School of Tropical Medicine
2 Taviton Street
London, WC1H 0BT
UK

Melissa Parker
International Medical Anthropology
Programme
Brunel University
Uxbridge
Middlesex
UK

Jean Peters
Section of Public Health
School of Health and Related
Research
University of Sheffield
30 Regent Street
Sheffield S1 4DA
UK

Tessa M. Pollard
Department of Anthropology
University of Durham
43 Old Elvet
Durham, DH1 3HN
UK

Lawrence M. Schell
Department of Anthropology
University of Albany
Albany, NY 12222
USA

A. D. Stark
Bureau of Environmental and
Occupational Epidemiology
New York State Department of Health
Albany, NY 12237
USA

Stanley J. Ulijaszek
Institute of Biological Anthropology
University of Oxford
58 Banbury Road
Oxford OX2 6QS
UK

Acknowledgments

The Wenner Gren Foundation for Anthropological Research is gratefully acknowledged for its support of the symposium held by the Society for the Study of Human Biology, entitled Urbanism, Human Biology and Health in Industrialized Countries, on which this volume is based.

Part I
The urban environment

1 Urbanism, urbanisation, health and human biology: an introduction

LAWRENCE M. SCHELL AND STANLEY J. ULIJASZEK

> 'The city is a fact in nature, like a cave, a run of mackerel or an ant heap. But it is also a conscious work of art, and it holds within its communal framework many simpler and more personal forms of art. Mind *takes form* in the city; and in turn, urban forms condition the mind.'
>
> Lewis Mumford, *The Culture of Cities* (1938)

The world is becoming an increasingly urban place. Current estimates indicate that one half of the world's population will be living in urban centers by the year 2005, largely the result of but two centuries of rapid urbanisation, and there are no apparent social forces that strongly oppose it.

Urbanism involves the concentrated inhabitation of human populations in relatively small areas, while urbanisation is the process of becoming urban and includes population growth by migration, natural increase and the changing scale of economic activity associated with this change. Both influence human biology and behavior in ways distinct from rural lifeways. Humans, like all other living things, are affected by their environment, whether that environment is completely unaltered by human activity or made entirely from it. Evolution and adaptation continue to operate in human populations through differentials in morbidity, mortality and reproduction. While humans use culture and behavior to modify their environments, culture is only an incomplete buffer to a range of environmental stressors, and can add new stressors to the environment to which human populations may adapt and evolve (Schell 1997a,b; Ulijaszek and Huss-Ashmore 1997). For example, in urban environments, culture may impose and concentrate psychosocial stress, time-budget stress and the stresses associated with industrialisation, including environmental pollution. Urban environments may be conceptualised as a complex of stressors, often different in nature and degree from rural environments, and created by human culture. These environments may thus pose numerous diverse challenges to human adaptation and evolution (Cassel 1977; Eisenbud 1978; Hardoy and Satterthwaite 1997; Schell, Smith and Bilsborough 1993; Schwirian *et al.* 1995).

The evolutionary past of humans may not have adapted them for urban living. Urban environments probably differ in fundamental ways from the environments in which humans evolved since the morphological and physiological characteristics of hominids evolved over 5 million years or more and became established in the hominid line at least 50,000 years before cities arose. Virtually all of human evolution occurred in response to physical and social demands, many of which may be absent from current urban environments, and which in turn present new challenges and stressors that were never present in the evolutionary past. Exposure to pollutants in air, water and food has become routine, psychosocial stress is commonplace, and diet and activity patterns have altered.

It is not clear whether the match of paleolithic response patterns to modern urban stressors is an example of preadaptation or maladaptation. Indeed, the question is one that embraces a large body of work done in a variety of fields and is ideally answered by human biologists, focusing as they do on long-term adaptation, evolution and human biological variability. Whether the current state of human health and well-being in cities is an adaptive success or a failure, there can be little doubt that the change from pre-urban to urban living has had consequences for health, disease and human population biology generally.

The city as a human product

The most important characteristics of cities that pertain to health and biology are difficult to determine, and one of the intellectual challenges of analysing the effects of urbanism on human biology derives from the multiple dualities that characterise cities themselves. The quote from Mumford that begins this chapter refers to these dualities. In one duality, cities are built environments, expressing a concept of living space that is derived from human thought and culture, yet they are also cultural responses to features of the local physical environment and reflect the history of local land use and settlement. In one sense, cities are products of the interaction, at individual, household, neighborhood, group, and population level, between the expression of a mental groundplan held by those units of organisation, and the impression made by physical features of the environment.

In another duality, urban places are comprised of social and physical features. Many of the salient urban stressors are social ones; other stressors being consequences of the physical construction of the city itself, such as its climate and atmosphere, as well as its lighting and acoustic characteristics. Since cities are built environments, the physical and social environments are interactive and ongoing products of the duality of human expression and impression. Although quite similar dualities may be said to exist to a degree in every human settlement form, and in any human product, cities with their vast size and numerous opportunities for interaction between the mental groundplan and the historically based physical environment, between social organisation and environmental features, may be extraordinary windows into an essential human complexity.

Measurement and the urban–rural contrast

Despite the complexity of cities, there have been analyses of urbanism by philosophers, social theorists, writers and scientists from ancient to current times (Plato; Rousseau 1755; Hobbes 1651; Mumford 1961). Most of these investigations were conducted by comparing urban to rural people or places. Such comparisons made by natural historians and early physical anthropologists of the eighteenth century contrasted two vastly different societies and environments. Topinard in his *Anthropologie* (1878) credits the demise of the human species in urban environments to the process of domestication, it being thought that domestication itself resulted in animals that were weaker than their feral cousins. J. J. Rousseau (1755), among other social philosophers, railed against the insidious and effeminising effects of urban life. Rousseau's noble savage was as far from urbanised as a human being could be. When the social philosophers and naturalists first made urban–rural comparisons as analyses of the human predicament, urban and rural places were more distinct than they are today, and there were far fewer cities and less variation among them. The urban–rural contrast depends on a view of such places as existing without profound variation, such that any contrast can serve to stand for all such contrasts. It may be that such comparisons began when cities were less complex and varied than now; however, in contemporary studies, results of comparisons between urban and rural places or peoples can often be foretold by the definitions used to create the comparison groups.

Much of our view of urban environments and their effects on human populations is based on past observations of human interactions with qualitatively different types of city environment than those which most people experience now in industrialised nations. Observations of city life have a long history in Western thought, and, for the most part, cities have been viewed as unhealthy places. The unhealthiness of cities became enshrined in Western thought by social philosophers of the eighteenth century when large cities were characterised by poor quality and unhealthy housing, poor and contaminated food supplies, poor and often absent sanitation, very high mortality and very short human lifespans. When this traditional view is combined with modern scientific progessivism which has the aim of improving the lot of humankind, the result is an approach to the topic of health in cities which is negative in perception. Such negative bias can produce unobjective research on contemporary cities and their effects on human biology.

Research employing the urban–rural contrast may combine all cities into one negative stereotype. However, cities vary enormously depending on the local and regional contexts of their development. Variation among cities may be conceptualised not in terms of different qualities but of differences in the quantities of urban qualities. Greater understanding of urbanism requires the measurement of these characteristics as quantitatively and precisely as possible. Some basic measures of urbanism are possible, including those relating to the physical environment such as climate, temperature, altitude and pollutant levels. A number of basic social factors are also amenable to measurement, including population size, density, and social and economic gradients. The measurement of diet and energy expenditure patterns may be equally difficult in cities as in rural populations. Factors which are difficult to measure include psychosocial stress and information density. Recent developments in the non-invasive sampling and measurement of endocrine products has made the study of psychosocial stress and reproductive function more objectively based.

Many of the variables under consideration are neither exclusively social nor physical in nature. Because cities combine social and physical features, the categorisation of variables into one or the other is meaningful only for the simplest characteristics, or as a starting point for further investigation. To cite but two examples, pollutants are largely products of human activity. Their production depends on the social organisation of resources and work, yet they are measured as parts of the physical environment. Human physical energy expenditure patterns also depend on the physical and social landscape that cities present. Social, physical and biological factors are interwoven in urban environments, and a biocultural approach is needed that emphasises these links, rather than an approach from either purely social or biological disciplines.

Measurement of urban features is important because measurement itself is a quality of science. A dichotomy is a crude scale of measurement, but it reduces a continuum of difference into but two values distinguished usually by an arbitrary criterion. In the history of science, a new quality is measured imprecisely at first, but as more is learned from its crude measurement, more precise metrics are devised. In the history of study of urban environments, investigators have begun with a crude metric, the urban–rural contrast, but are now able to advance the study of urban human biology by employing more precise measuring tools. The dichotomous urban–rural contrast obfuscates the factors that influence human biology in populations which vary in degree and type of urbanisation, since the contrast involves a comparison of many factors combined, and can only detect the sum result of all such factors acting on any biological outcome of interest. When there is no difference between the urban and rural place in some biological outcome, such as growth, it means little. There could be many powerful factors in each environment that affect growth, but they may counterbalance one another. The urban environment may contain numerous factors that pertain to human biology, but they cannot be detected with the simple urban–rural contrast. Furthermore, since the sum difference depends on which individual factors are present in the contrast, each urban–rural comparison contrasts different constituent factors which makes generalisation across different studies meaningless. The opportunities for the accumulation of knowledge by the aggregation of results from different urban–rural contrast studies is thus very limited. Measurement of urban qualities should allow analysis and modeling of the urban biocultural system, and should permit identification of individual environmental factors that may influence biological outcomes, and, if measured properly, allow for comparison across studies.

The diminution of differences between urban and rural populations

In industrialised nations now, the difference between urban and rural communities has lessened substantially. Three hundred years ago and more, occupational specialisation could distinguish the two circumstances. Farming has been the hallmark of rural life for the majority of the world's populations since the onset of cities, and with few exceptions is not practiced extensively by urban populations. Urban–rural contrasts of the eighteenth century compared communities with farming economies to urban ones based on trade and banking. In the nineteenth century the influence of urban manufacturing was added to the contrast. Today, urban and rural places do not necessarily differ much qualitatively. One way in which difference has been reduced is by the spread of contemporary communications systems, which provide rural and urban people with the same information conveyed through national newspapers, radio, magazines and television. Through these means, urban and rural populations receive similar forms of passive entertainment that can both influence social mores and suggest patterns of appropriate behavior and values.

Radio and television news provides virtually the same information about national and international events to both rural and urban populations. Advertising may be more regional than news broadcasts, but frequently urban and rural populations receive the same message extolling the benefits of the same goods and services.

One of the clearest indicators of cultural difference is diet. In many parts of the world, urban and rural populations have very similar diets. However, until recently this was not so. Early urban populations relied on imports from rural food-producing areas, and the process resulted in differences in the quality and types of food available to urban residents. Urban importation of food lead to the development of one arm of public health service: the regulation of the national food supply. This now-traditional arena of public health regulation developed during the late nineteenth and early twentieth centuries when some foods were heavily adulterated (milk was diluted and then supplemented with chalk to restore bulk), and refrigeration was primitive. McKeown (1976) credits a change in dietary quality with the rise of population growth in industrialised countries, although public health regulation does not play a large part in his theory. However, regulation of food production was very important in the creation of safe urban food supplies; regulation of dairy herds and milk purity was one of the first areas of regulation achieved in the United States (Rosen 1958).

Currently, food quality is assured by the existence of standardised food products, which are important to rural populations as well as urban ones. The commoditisation of agriculture and the generation of standard food products and national systems of food distribution have contributed to this homogenisation of urban and rural diets. In many parts of the world, most rural and urban residents can now eat similar foods, within economic, social and cultural constraints. Few rural residents now grow their own food and most shop for produce, as well as packaged items from stores. The stores obtain their produce, meats and fish from distant areas without much regard to what is grown locally. Many of these stores are local representatives of national companies that serve both rural and urban areas. When urban and rural residents dine outside the home, they may eat at a local restaurant that is one of a national chain of restaurants. In the United States, a MacDonald's hamburger is the same whether purchased in a rural or urban community. In general, differences are not in the presence or absence of features but in how common and how concentrated urban factors are. In industrialised countries, studies of urban populations are thus relevant to rural ones because factors are largely the same in both and they differ more in quantity than in quality. Once quantification and measurement of specific urban qualities is employed, the roles of individual factors may be distinguished and compared among urban populations and across urban and rural ones as well.

Urban populations

Because urban society contains people from many different cultures that share the same city but act relatively independently of one another, urban populations do not experience the same urban features. Ethnic and cultural distinctions may represent differences in behavior, local physical environment, diet, activity pattern, buffering and exposure to risk factors for poor health, as well as possible genetic differences in phenotypic responses to stressors and disease exposure.

Urban places have migrant populations. In earlier times, such populations might be reasonably homogeneous, and be of the same or similar ethnicity to the majority urban population. However, in more recent times and with improved and cheaper transport, migrants can come from all parts of the world. For example, in England in the 1500s, urban migrants came predominantly from rural areas, and if from overseas, from other parts of Europe. In 1440, the migrant population, based on a tax survey, was 1% of the total population of the country (Coleman and Salt 1992) and perhaps 14% of the urban population. Of these migrants, over 80% came from Northern Europe, France, Ireland and Scotland (Thrupp 1957). In current urban populations, the vast majority of these migrants have vanished into the general population and cannot be discerned as distinct ethnic identities (Coleman and Salt 1992). Migrants from outside of Europe since that time, and before the 1939–45 war, are limited largely to about 10,000 Africans imported as slaves to become servants in London in the eighteenth century (Shyllon 1977). After 1945, the increasingly easy availability of cheap long-distance travel helped new patterns of international migration which, by the 1980s had resulted in an overwhelmingly urban non-European-origin population in Britain of some 2.6 million, comprising migrants from the Indian subcontinent, Africa, the Caribbean and Asia (Coleman and Salt 1992), and representing about 5% of the urban British population, which is 92% of the total British population (Department of Economic and Social Statistics 1992). By contrast, Malaysia, which is currently undergoing the sort of rapid urbanisation that Britain underwent in the eighteenth and nineteenth centuries, had 37% of its total population living in urban places in 1980. Of this, the majority were of Chinese ethnicity, followed by Malay, then Indian (Figure 1.1), the Chinese showing a decline in the proportion of their population as urban dwellers, and the Malays showing a doubling in urban population (Swee-Hock 1988). Across the period 1947–80, the proportion of the total population as urban

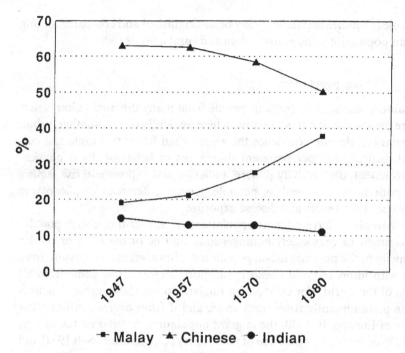

Figure 1.1   Distribution of the urban population of Malaysia by ethnicity, 1947–1980.
Source: From Swee-Hock (1988)

has doubled, as has the proportion of Malays that are urban dwellers. This urban mix shows considerable economic differentiation, although it is difficult to ascribe social differentiation across these categories except with respect to access to resources, control of business, trade and bureaucratic structures. Cultural differentiation on the basis of religion shows that, with the exception of Malays, these are not homogeneous groups. While 99% of Malays are Muslim, 83% of Indians are Hindu, and 56% and 38% of Chinese are Buddhist and Confucian, respectively; 7% and 5% of Indians are Christian and Muslim, respectively, while 3% of Chinese are Christian. Thus, the ethnic variation of cities can change with time, and although there may be some dominant cultural groups in cities, there are also substantial minorities often living side-by-side with other groups, whose health experience is different from their neighbours by virtue of occupying a different niche.

While it might be expected that ethnic identity might persist across generations, there is evidence to suggest that ethnic intermarriage frequency in urban centers may be influenced by residential distance, and educational similarity (Peach and Mitchell 1988). Thus, within any urban context, populations are constantly being defined in ethnic terms and becoming assimilated to varying degrees with the dominant cultural norms of the country and city. However, the dominant cultural norms of the city may also change, and the menu of human possibilities within the city is increased.

Urban society is also characterised by gradients of socioeconomic status which may be linked, if only loosely, to ethnic and cultural variation. Within the same city, there may be groups distinguished by socioeconomic characteristics such as income, education and occupation. Differences in these characteristics can greatly affect their risk to health. Such risks arise from occupational disease and disability (Syme and Berkman 1976), the physical and social environment of their neighbourhoods (Smart *et al.* 1994; Ennett *et al.* 1997; Crum *et al.* 1996), as well as diet and activity patterns (Marmot 1997; Kooiker and Christiansen 1995; Lip *et al.* 1996; Cox 1994). Socioeconomic characteristics pertain directly to economic power that can be mustered to avoid risks to health, buffer exposure to such risks and attend to adverse consequences from experiences of those risk factors (Thouez 1984; D'Arcy and Siddique 1985). For example, socioeconomic status may be closely related to pollutant exposure (Schell and Czerwinski 1998), dietary buffering of pollutant effects (Czerwinski this volume), and differences in health care utilisation (Breen and Figueroa 1996). Furthermore, gradation in socioeconomic status within cities may be steep, with the richest and poorest members of a nation living in virtually the same political space but with little overlap in physical space or physical contact, with some important exceptions (DiFerdinando, and Parker, both in this volume).

Stratification by socioeconomic characteristics also may be subtle, and the qualitative division of populations into richer and poorer sectors may hide important distinctions by which environmental exposure might be mediated. One example of this is housing. In developing countries, the urban poor may live in a range of housing types (Table 1.1), including settlements, shanty towns, tenements and public housing projects. The pattern of tenure and thus the hold that a family has on this local household environment varies, from legal ownership, rental, illegal possession or none at all. The stressors that populations living in these different types of housing experience will thus also vary enormously, and include, for many, the psychosocial stress of simply maintaining the limited tenure they have. Furthermore, the provision of basic amenities like sanitation and water will also vary with the tenure status of the settlements in general.

There may also be great variation in health across different low socioeconomic status urban populations. For example, in Karachi, Pakistan, health differentials between middle-class and slum areas are enor-

Table 1.1. *General typology of low-income settlements*

Type	Land acquisition	Tenure	Land and physical characteristics/ propensity for upgrading
(1) Irregular settlements (1a) Squatters	invasion of public or private land	*de facto* or *de jure* 'ownership'	Periphery where security of tenure then up-grading is likely. Otherwise static shantytown structures with little consolidation
(1b) Illegal sub-divisions	sale of private land	owners, although titles may be imperfect	Periphery. In various phases of consolidation and upgrading through 'self-help'
(2a) Shanty towns	Sale of customary land squatter or, more usually, renters	rental	Mostly down town and around city centre/small plots. Structures with few public amenities and little prospect of their provision. Little likelihood of self-help improvements because: (a) no security of tenure; (b) difficulty of creating an investment cash surplus for improvement (because of rental outgoings); (c) very small plot sizes
(2b) 'Street sleepers'	may have regular sleeping places	none	Down town/inner city. Minimal shelter, removed daily. Often associated with workplace

Table 1.1 *(cont.)*

Type	Land acquisition	Tenure	Land and physical characteristics/ propensity for upgrading
(3) Tenements	converted large houses or purpose-built tenements	rental	Mostly down town. An increasing proportion of new rental accommodation is located in older irregular settlements (1a, 1b). Single room per family and share services
(4) Public housing projects			
(4a) Complete units	Government purchase and sales	owner or rental Owner	High density, good amenity, but relatively expensive. Periphery. Basic services mostly installed but house construction varies. Self-help and mutual aid to improve external house structure
(4b) Incomplete units –site and services –'core' units	Government purchase and sale (sometimes interrelated agencies sponsorship)		

Source: WHO Environmental Health Division, unpublished data from Harpham, Lusty and Vaughan (1988).

mous, such that differences in infant mortality rates are on average over three times greater in the latter (Table 1.2). However, infant mortality rate differences between different slum areas are also large, with infant mortality rate varying 1.6 fold between slums with the highest and lowest infant mortality rates. These differences reflect variation in environmental quality based on the history of the slum, its legal status and infrastructure, and housing type and tenure, to a degree that crude social or economic indicators cannot discern. Furthermore, infant mortality within each slum varies enormously according to religion. In both slums where religion has been related to infant mortality rate, Hindus have about twice the death rate of Muslims, while in one slum where Christians live, they have infant mortality rates similar to those of Muslims. Religion is here a marker of cultural, but also economic, differences within settlements that are taken to be homogeneous with respect to infrastructure. Clearly, these religious groups are living in overlapping urban microenvironments, but with very

Table 1.2. *Intra-urban and intra-slum differentials in infant mortality rates in Karachi, Pakistan, 1985*

Area	Infant mortality rate	Deaths of under-fives as % of all deaths
Orangi[a]	110	51
Chanesar Goth[a]	95	50
Muslim	77	
Hindu	146	
Grax[a]	152	55
Christian	123	
Muslim	139	
Hindu	286	
Karimabad[b]	32	18

[a] A *katchi abadi* (slum).
[b] Middle-class area.
Source: Agha Khan University students, 1986 (unpublished), Harpham, Lusty and Vaughan (1988).

different health outcomes, because of cultural, social and economic differences at the household level. Such differences also exist among ethnically distinct populations in cities around the world, and human biologists need to identify the variables that create these micro-environmental differences that lead to important differences in health outcome.

The Society for the Study of Human Biology Symposium on Urbanism, Human Biology and Health, held on 10–11 April 1997 at the Department of Biological Anthropology, University of Cambridge, and Corpus Christi College, Cambridge considered ways in which human biology could focus in new ways on urban issues, given the increasingly urban nature of human populations. The great strength of this meeting was the multidisciplinary perspective taken by the speakers, who included epidemiologists, anthropologists, nutritionists as well as human biologists. Fundamental to understanding urban human biology is the historical and geographical nature of urban places. In this volume, the authors of the first two chapters, McMichael and Clark, consider separately the interrelated nature of historical and geographical perspectives of urbanism. McMichael describes the commonalities of economic type across urban centers and the ways in which it has changed across history. Before the nineteenth century, many large cities were built on trade economies, but many also grew as suppliers of energy and centers of manufacturing. This transition involved tremendous migration from the countryside, and marked increases in urban population size and density. There was a concomitant increase in infectious disease, undernutrition and general deterioration of human health and well-being. However, by the late nineteenth and early twentieth centuries, government legislation had played a crucial role in improving conditions of sanitation, housing, labour and food supply.

The health and biology of urban residents is directly influenced by urban factors such as transportation, violence, climate and activity. Furthermore, the effect of cities extends far beyond their municipal boundaries, and McMichael describes the way in which cities today, fed as they are with food and energy from very distant places, impress ecological footprints on non-urban people and lands. In his chapter, Clark describes the extensive and remarkable diversity in urban forms and distributions around the globe in the present day. He clearly shows why terms such as 'developed' and 'developing' are at best classificatory devices only, and do not explain variation in the distribution and type of urbanism across the world. Indeed, variation in distributions of urban centers is the result of the interplay of history, culture and resources. Today, tremendous variation exists among nations in the growth of urban populations, the proportion of urban residents and the spread of urban lifestyles. It is thus impossible to identify urbanism as a single, simple characteristic which has meaning in human biology.

In chapter 2, Huss-Ashmore and Behrman address the issue of diversity and character of urban places, using as study material the human populations of trailer parks. These defy classification in common categories of urban places, and by employing an ethnographic approach, the authors are able to show the complexity of the interrelationships among ideology, physical environment and human health and biology of populations living in these places. This serves as a useful model for the investigation of urban places using an adaptability framework.

The recent rise of infectious diseases in urban places of the industrialised world is addressed by DiFerdinando (chapter 5), who describes changes in the frequencies and distributions of two infections, tuberculosis (TB) and human inmmunodeficiency virus (HIV). Both diseases have been associated throughout their histories with urban populations, but during the emergence of HIV and the re-emergence of TB, each was originally associated with different social segments of urban populations. An important change occurred when interaction between the two segments increased. Urban populations inhabit overlapping microenvironments, and such overlap has changed the risk factors for these diseases, and changed the notion of their causes. One hundred years ago many urban residents were exposed to TB, and the presence of illness depended on variation in susceptibility rather than exposure. People defended themselves against infection and the disease was caused by susceptibility. Now, in HIV affected populations, TB is 'caused' by exposure to the infective agent since HIV increases susceptibility greatly.

The future of the species is dependent on our reproductive ecology, and Ellison considers how reproduction may be affected by features of urban life. Past ecological transitions, from foraging to agriculture, and from agriculture to industrialisation, are shown to have altered reproductive ecology dramatically. Both transitions involved changes in residential patterns that are part of urbanisation. The post-industrial city also can have an impact, and Ellison describes several mechanisms and pathways, termed "proximate determinants," by which urban factors can influence parameters of reproductive function, specifically ovarian function. Among these factors are urban patterns of maturation and aging, anxiety, dietary composition, exposure to pollutants, as well as levels of energy expenditure and energy balance.

Contemporary urban environments in industrialised countries contain features that impact on human biology and health. The idea of health risk is developed in the chapter by Schell and Stark, who describe the current asthma epidemic in industrialised countries as a way of illustrating how a characteristic of urban populations can be studied by seeking the relationships between the disorder and aspects of the urban physiosocial environment. Although pollution is often included among the new environmental features of cities, rarely are animal allergens considered, even though they are pollutants in the strict sense. The use of an alternative approach, the identification of a factor in the environment and tracing its effects in the population, is shown by the authors to be useful in studying the effects of low level lead pollution on child development. The illustrations in this chapter emphasise the importance of the measurement of environmental features, and the moving away methodologically from urban–rural comparisons. In the chapter by Peters, the focus is on changing patterns of urban health in industrialised Asia. Contrasting social and economic differences in health experience and lifestyles, Peters paints a vivid picture of urban complexity whose impact on human biology is largely positive, in many ways still unmeasured, and difficult to predict into the future.

Urban poverty and nutrition are considered by Dowler, who describes the ways in which socioeconomic factors impact on dietary availability in Britain. Urban poverty is clearly shown to be associated with low intakes of all nutrients, including energy. Among poor households, the poorest ones have the lowest intakes of nutrients and the least diverse food base. This is in contrast to the example given by Johnston and Gordon-Larsen, who in their chapter on poverty, nutrition and obesity in the United States, show poor Philadelphia children to have higher body mass index and intakes of energy, protein and other nutrients relative to United States norms. Johnston and Gordon-Larsen's use of a microlevel, biocultural approach shows that diet and nutrition of the urban poor do not generalise across the United Kingdom and the United States.

The subsequent chapter, by Czerwinski, also considers nutritional status and poverty, this time in relation to environmental toxicants. Many people are exposed to pollutants through the foods they eat, such that being well nourished can also mean having a higher pollutant burden. Using new data from the Albany Pregnancy Infancy Lead Study, the author shows that the urban poor experience greater exposure to toxicants such as lead, and their dietary patterns may expose them to more food-borne toxicants. Furthermore, dietary items that might moderate toxicant absorption may be less commonly consumed by the urban poor. Pollutants, as relatively new factors in urban environments, interact with one of the oldest, social stratification, to produce a greater health risk for the urban poor.

For centuries cities have been known for their crowding, noise, confusion, and general disorder. Pollard questions whether these features are genuinely stressful, and considers how to investigate their impact on human biology and health. Citing new work that employs non-invasive methods to measure endocrine and neurotransmitter correlates of psychosocial stress, Pollard shows that urban features in the areas of work, home, travel do indeed change neuroendocrine markers of stress. However, cultural and psychological constructs are important for understanding the stressfulness of urban features. On the one hand, the perception of stress turns out to be an inaccurate predictor of the neuroendocrine stress markers while the sense of control or its lack, over physical and social environmental features, has been identified as playing a prominent role in generating feelings of stress.

Ways in which physical activity is influenced by social, cultural, economic and geographical factors in urban settings are discussed in the chapter by Ulijaszek. Urban populations and the environments they inhabit are highly diverse, this being largely related to position in the economic status system, and reflected in energy expenditure levels. Cultural and demographic factors may have an underlying effect on the way in which the interplay between these factors varies in different urban centers. This chapter points to a relative shortage of information about the human energetics of urban populations in ecological contexts, and suggests that there is a great need to study urban compexity in relation to physical activity.

In chapter 14, Parker discusses the transmission of the human immunodeficiency virus by sexual contact, and the networks that exist in order to find such contact. Based on a study of HIV+ men in London,

Parker's observations show that sexual networks for potential HIV transmission in London cut across social class and across countries, notably the urban United States and Australia. The importance of the study of urban pathways is discussed in the final chapter by Ulijaszek and Schell, as one possible new way for the study of urban human biology. Other approaches include the use of epidemiological techniques with clearly defined urban variables within an adaptability framework, and the use of ethnography, and modernisation studies for the identification of new, urban characteristics that impact on human biology. Vital to the future of urban human biology is the reformulation of human ecology within an adaptability framework in which social and organisational constructs are regarded as components of the stress environment.

References

Baldassare, M. and Wilson, G. (1995). More trouble in paradise: urbanization and the decline in surburban quality-of-life ratings. *Urban Affairs Review*, **30**, 690–708.

Breen, N. and Figueroa, J. B. (1996). Stage of breast and cervical cancer diagnosis in disadvantaged neighborhoods: a prevention policy perspective. *American Journal of Preventive Medicine*, **12**, 319–26.

Cassel, J. (1977). The relation of the urban environment to health: towards a conceptual frame and a research strategy. In *The Effect of the Man-Made Environment on Health and Behavior*, ed. L. E. Hinkle, Jr. and W. C. Loring, pp. 129–42. Atlanta: US Department of Health, Education, and Welfare.

Coleman, D. and Salt, J. (1992). *The British Population*. Oxford: Oxford University Press.

Cox, R. H. (1994). Dietary risks of low income African-American and White women. *Family and Community Health*, **17**, 49–59.

Crum, R. M., Lillie-Blanton, M. and Anthony, J. C. (1996). Neighborhood environment and opportunity to use cocaine and other drugs in late childhhod and early adolescence. *Drug and Alcohol Dependence*, **43**, 155–61.

D'Arcy, C. and Siddique, M. C. (1985). Unemployment and health: an analysis of 'Canada Health Survey' data. *International Journal of Health Services*, **15**, 609–35.

Eisenbud, M. (1978). *Environment, Technology, and Health: Human Ecology in Historical Perspective*. New York: New York University Press.

Ennett, S. T., Flewelling, R. L., Lindrooth, R. C. and Norton, E. C. (1997). School and neighborhood characteristics associated with school rates of alcohol, cigarette, and marijuana use. *Journal of Health and Social Behavior*, **38**, 55–71.

Hardoy, J. E. and Satterthwaite, D. (1997). Health and environment and the urban poor. In *International Perspectives on the Environment, Development, and Health: Toward a Sustainable World*, ed. G. S. Shahi, B. S. Levi, A. Binger, T. Kjellstrom, and R. Lawrence, pp. 123–62. New York: Springer Publishing Company.

Harpham, T., Lusty, T. and Vaughan, P. (1988). *In the Shadow of the City*.

Community, Health and the Urban Poor. Oxford: Oxford University Press.

Hobbes, T. 1651 (1958). *Hobbes' Leviathan* (reprint from the 1651 edition and essay by W. G. Pogson Smith). Oxford: Clarendon Press.

Kooiker, S. and Christiansen, T. (1995). Inequalities in health: the interaction of circumstances and health-related behavior. *Sociology of Health and Illness*, 17, 495–524.

Lip, G. Y., Luscombe, C., Mccarry, M., Malik, I. and Beevers, G. (1996). Ethnic differences in public health awareness, health perceptions, and physical exercise: implications for heart disease prevention. *Ethnicity and Health*, 1, 47–53.

Lynch, J. W., Kaplan, G. A. and Salonen, J. T. (1997). Why do poor people behave poorly? Variation in adult health behaviors and psychosocial characteristics by stages of the socioeconomic lifecourse. *Social Science and Medicine*, 44, 809–19.

Marks, N. F. (1996). Socioeconomic status, gender, and health at midlife: evidence from the Wisconsin Longitudinal Study. *Research in the Sociology of Health Care*, 13(Part A), 135–52.

Marmot, M. (1997). Inequality, deprivation, and alcohol use. *Addiction*, 92 Supplement 1, s13–20.

McKeown, T. (1976). *The Modern Rise of Population.* London: Edward Arnold.

Mumford, L. (1961). *The City and History: Its Origins, its Transformations, and its Prospects.* New York: Harcourt, Brace, and World.

Mumford, L. (1970). *The Culture of Cities.* Westport, Conn.: Greenwood Press.

Peach, C. and Mitchell, J. C. (1988). Marriage distance and ethnicity. In *Human Mating Patterns*, ed. C. G. N. Mascie-Taylor and A. J. Boyce, pp. 31–45. Cambridge: Cambridge University Press.

Plato (1974). *The Republic.* Trans. and Introduction D. Lee. Baltimore: Penguin.

Rosen, G. (1958). *A History of Public Health.* New York: MD Publications.

Rousseau, J. J. (1992 [1755]). *Discourse on the Origins of Inequality*, vol. III. ed. C. Kelley, trans. J. Bush *et al.* Hanover, NH: for Dartmouth College by University Press of New England.

Schell, L.M. (1997a). Culture as a stressor: a revised model of biocultural interaction. *American Journal of Physical Anthropology*, 102, 67–77

Schell, L. M. (1997b) The evolution of human adaptability: society, funding and the conduct of science. In *Human Adaptability, Past, Present and Future*, ed. S. J. Ulijaszek and R. A. Huss-Ashmore, pp. 281–94. Oxford: Oxford University Press.

Schell, L. M. and Czerwinski, S. (1998). Environmental health, social inequality and biologic difference. In *Human Biology of Social Inequality*, ed. S. Strickland and P. Shetty. Cambridge: Cambridge University Press.

Schell, L. M., Smith, M. and Bilsborough, A. (eds.)(1993). *Urban Ecology and Health in the Third World.* Cambridge: Cambridge University Press.

Schwirian, K., Nelson, A., and Schwirian, P. (1995). Modeling urbanism: economic, social, and environmental stress in cities. *Social Indicators Research*, 35, 201–23.

Shaftel, N. (1978). A history of the purification of milk in New York, or, 'How now, brown cow'. In *Sickness and Health in America*. ed. J. W. Leavitt and R. L. Numbers, pp. 275–91. Madison, WI: University of Wisconsin Press.

Shyllon, F. O. (1977). *Black People in Britain 1555–1833.* London: Oxford University Press.

20   L. M. Schell and S. Ulijaszek

Smart, R. G., Adalf, E. M. and Walsh, G. W. (1994). Neighborhood socioeconomic factors in relation to student drug use and programs. *Journal of Child and Adolescent Substance Abuse*, **3**, 37–46.
Swee-Hock, S. (1988). *The Population of Peninsular Malaysia*. Singapore: Singapore University Press.
Syme, S. L. and Berkman, L. F. (1976). Social class, susceptibility and sickness. *Journal of Epidemiology*, **104**, 1–8.
Thouez, J. P. (1984). Cancer mortality differential by social milieu: the case of the Montreal Metropolitan Area, 1971. *Social Science and Medicine*, **18**, 73–81.
Thrupp, S. (1957). A survey of the alien populations of England in 1440. *Spectrum*, **32**, 262–73.
Topinard, P. (1878) *Anthropology*. London: Chapman and Hall.
Ulijaszek, S. J. and Huss-Ashmore, R. A. (eds.)(1997). *Human Adaptability, Past, Present, and Future*. Oxford: Oxford University Press.
Wister, A. V. (1996). The effects of socioeconomic status on exercise and smoking: age-related differences. *Journal of Aging and Health*, **1**, 467–88.

2 Urbanisation and urbanism in industrialised nations, 1850–present: implications for health

A. J. MCMICHAEL

Editors' introduction

When examined in detail, cities are so highly diverse and continually changing that grouping them for comparison to rural settlements requires their essential diversity to be ignored. Each city represents one possible position along a continuum of change and diversification. Although the current configuration of each city has developed from its own individual history, this chapter shows that certain commonalties have also existed. Before the nineteenth century many large cities were built on a trade economy, but, by the middle of that century, cities also grew as suppliers of energy, and as centers of manufacturing. The transition involved tremendous migration from the countryside, and marked increases in population sizes and densities of cities took place. There was a concomitant increase in infectious disease, undernutrition, and the deterioration of human health and well-being. However, by the late nineteenth and early twentieth centuries, government legislation had played a crucial role in improving conditions of sanitation, housing, labor and food supply. The role of government continues to be important in regulating urban hazards to health from such factors as pollutants and traffic. As McMichael shows, the health and biology of urban residents can be traced to direct influences from urban factors such as transportation, violence, climate and activity. Furthermore, the effect of cities extends far beyond their municipal boundaries. Cities today consume with food and energy from very distant places and impress an ecological footprint on non-urban people and lands. This satellite effect of cities should be considered when assessing the impact of urbanism on human biology and health in non-urban populations.

Introduction

Cities are an expression of the social nature of humans, of their collective approach to meeting basic needs and solving problems, and of their technological and cultural ingenuity. Nevertheless, only during the last one-twentieth of the lifespan of the modern human species *Homo sapiens* have humans begun to live in towns and cities, i.e., during the last approximately 8,000 years out of a total 150,000 years. The advent of systematic labor-intensive agriculture, with taxable and tradeable surpluses and an infrastructure of landholdings ('property'), led to the rise of centralised hierarchical societies with urban elites: rulers, priests, soldiers, bureaucrats and technicians. Over ensuing millennia, cities became the natural centres of government, learning, and artistic endeavor, and the nodes of commerce and trade; they have been an irresistible lure for many and a haven of hope – albeit often illusory – for the dispossessed and displaced.

Even so, until well into the twentieth century the great majority of humankind lived in non-urban settings. Indeed, by around 3,000 years ago there were still only an estimated 4 cities in the world with over 50,000 inhabitants (including Thebes and Memphis in Egypt), and, by 2,000 years ago, when world population approximated to 200 million people, there were only about 40 such cities (Schell 1988). In contrast, the urbanisation process has surged in recent decades, as the ratio of urban to rural dwellers within populations has accelerated upwards. City living – urbanism – is therefore a rather recent mass experience, and cannot have played a significant part in the long formative processes of human biological evolution.

The processes of urbanisation and urbanism should not be thought of as expressions of natural laws. A similar misplaced generalisation led us to believe that the "demographic transition" in population mortality and fertility rates, as widely observed in Western countries since the mid-nineteenth century, follows a preordained path. Today, however, we realise that the process may follow differing trajectories – and perhaps even stall. Further, the shift in population health profile associated with the demographic transition – i.e. the "epidemiological transition" from a profile dominated by infectious diseases to one dominated by non-infectious diseases – is not, as was previously thought, irreversible. Today, developed countries are experiencing new and resurgent infectious diseases: HIV/AIDS, drug-resistant tuberculosis, several new hepatitis viruses, and many more. These infectious diseases impinge most on the susceptible urban poor, in cities ranging from London to Moscow to New York. Meanwhile, in developing countries, the urban poor typically encounter both the persistence of infectious disease and the acquisition of non-infectious degenerative diseases.

Urban populations may vary in their susceptibility to particular health-endangering influences because of differences in the levels of other risk factors (e.g. nutritional status, immunological integrity and, less probably, genetic characteristics). Human populations also differ in aspects of their cultural histories, social relations, technological resources and behavioral responses. Hence, epidemiological studies of the mortality impact of air pollution within urban North American populations do not necessarily predict the impact of that same exposure in Jakarta or São Paulo.

We should therefore tread cautiously in seeking unifying theories of urbanism and human biology. Nevertheless, as Schell (1988) has pointed out: "The extreme difference between the environment of our early biological adaptation and the environment of today's populations has consequences for modern people in urban surroundings." We are thus challenged to seek a more systematic understanding of the biological and health consequences of urbanism. A further dimension of inquiry is also emerging as we perceive the issues of urbanism and health within an ecological framework: how do urban ecosystems function, affect the sustainability of human health, and impact upon the wider environment? What are the appropriate composite indices of urban ecosystem health (Guidotti 1995)?

Historical perspective

In 1800, only 5% of the world population lived in cities of over 10,000 people. By 2000, half of the world's 6 billion people will be urban dwellers. Humankind is becoming a predominantly urban species. Meanwhile, these urban populations are also becoming "older." Viewed against the usual pace of history, these are dramatic and rapid changes in human ecology.

The best-documented historical account on urbanisation, urbanism and human health comes from Great Britain, the first country to begin industrialising. In the 1850s, Britain was the world's most urbanised nation – although at that time almost two-thirds of the British population still lived in the countryside. During the latter decades of the nineteenth century, major changes in the demographic and economic circumstances influenced the urbanisation process in the industrialising Western nations. First, because of social progress in housing, food supplies and literacy and specific public health advances, especially in the amelioration and control of infectious diseases, life expectancy increased – from around 40 years in 1850 in England and Wales to around 50 years in the early 1900s (Table 2.1). With this declining mortality, especially in infancy, urban populations no longer needed to restock with rural immigrants. Indeed, the slower turnover of urban living space and jobs meant that most new arrivals now became part of slum-dwelling populations.

The second major change was in the economic role of cities within industrialising national economies. Cities ceased to be parasitic on the national economy and, instead, became the industrial engines of national

Table 2.1. *Improvement in life expectancy at birth in England and Wales over the past 160 years*

Period	1838-54	1871-80	1881-90	1891-1900	1901-10	1910-12	1920-22	1930-32	1950-52	1960-62	1970-72	1980-82	1990-92
Life expectancy at birth	39.9	41.4	43.7	44.1	48.5	51.5	55.6	58.7	66.4	68.1	69.0	71.	73.4

Source: Table 14, Office of Population Censuses and Surveys, 1994.

wealth. Following Britain's lead, new industrial towns developed widely across Europe, particularly around the coalfields in Belgium, France and Germany's Ruhr Valley.

During those same decades, cities began to sprawl. As industrialisation ensued and urban slums developed, the inner-city wealthy moved out to newer, more salubrious, suburbs, serviced by newly installed radial roads, bridges, horse-drawn street-cars and railways (McMichael 1993). This process of sprawl was largely unplanned and often based on speculative developments, and it frequently obliterated large swathes of countryside and market gardens. Towards the end of the nineteenth century, electrification appeared. So, too, the urban air pollution from coal-burning power plants increased. London had begun building its undergound railway in the 1860s, initially steam driven. Other European and North American cities followed suit around the turn of the century, and their electrified underground railways facilitated the rapid movement of large numbers of people. In the USA in 1890, around 80% of urban railways were horse-drawn; by 1900 this had plunged to around 1%.

Suburban sprawl continued in the early twentieth century, assisted by the extension of suburban railway systems. As factories relocated towards the periphery of cities, the non-residential city centre typically converted into a hub of financial and commercial activity. The residential flight to the suburbs continued, soon to be abetted by the advent of the private motor car and the provision of water and sewerage systems. The building of peripheral motorways around large cities created a new vantage point for industry and for large-scale shopping centres. Thus emerged a new, dispersed, functionally differentiated, pattern of living and working in early twentieth-century cities.

As both urbanisation and the absolute rate of population growth in industrialised countries continued to increase around the mid-twentieth century, so a process of coalescence of cities emerged. For example, such urban agglomeration has occurred in The Netherlands (the so-called *Randstad*), in Japan where there is now a continuous urbanised zone between Tokyo and Kobe, and the eastern seaboard of the USA, where one quarter of the national population reside on less than 2% of the nation's land. This large-scale urbanisation has yielded a world in which cities of 10–20 million are no longer a novelty.

Even so, a gradual decline in the demographic importance of cities in industrialised countries has recently emerged. In Britain, the proportion of the population living in cities has actually diminished since the 1960s. In many developed countries, much of the residential and industrial development is now occurring outside, but near, major cities. These developments draw on the cities' high-quality infrastructure and services: piped water, sewers, telephones, television, garbage collection, paved roads and public transport. A more clearcut "counter-urbanisation" has developed over the past two decades, with development reorienting towards smaller towns and rural areas. Meanwhile, the social and economic decline of many erstwhile industrial cities, such as Liverpool and Bilbao, has created new social and public health problems (especially in inner-city sectors) and political pressures (WHO 1996). Long-term unemployment, persistent poverty, a weakening of kinship and community relations, a contraction in social services, and the physical decay of urban infrastructure have all predisposed to an environment in which alienation, hopelessness and violence can arise. In New York and London, for example, various of these socially corrosive processes, amplified by the prevailing ideology of free markets, smaller government and reduced welfare provisions, have resulted in around 1% of the population sleeping on the streets or in emergency shelters each night (Burrows et al. 1997).

We seem now to be at a crossroads. The post-Renaissance rise of cities has been a consequence of, first, mercantilism and then, industrialisation. The modern concentration of living and working within large, crowded, urban environments has been predicated on fossil fuel combustion, electrification and mechanised transport. Modern electronic communications and the deregulation of trade and financial transactions have boosted the role of cities as the hub of globalised private enterprise. But, for many, city life has become more difficult, more stressful, less satisfying and less healthy. Meanwhile, there is a nascent recognition that cities are overloading the ecosystems upon which they necessarily depend: their "ecological footprints" have become very large (Rees 1996). Urban populations are increasingly living beyond the ecological-infrastructural means, externalising their environmental impacts in ways that displace or defer the true costs to others (Ashton 1992).

As we enter the twenty-first century, there is a growing recognition of the need to reappraise urban human ecology and its social and health consequences, and to seek ways of redesigning cities that recapture a sense of human scale and create the possibilities of human community (WHO 1996), while retaining the energy, variety and dynamism that has been the historical hallmark of cities.

Public health responses to the urban way of life

Public health ideas, broadly construed, have often played a reactive, corrective, role in the development of cities in the industrialised world. The initial urban public health crisis became apparent in England in the 1840s. In 1839, recognising the mortality toll attributable to overcrowding, unsanitary conditions and air pollution in London's most blighted boroughs, the Registrar of Births, Deaths and Marriages wrote in his Annual Report that "a Park in the East End of London... would probably add several years to the lives of the entire [East End] population." In response, Victoria Park was created. In 1843, the Health of Towns Commission was created to address the health consequences of chaotic and rapid industrialisation. This was the decade of Edwin Chadwick's "sanitary idea", which spawned the great public health legislation of that era (Acheson 1990). The resultant improvements in housing standards, sewer systems, food safety and drinking water supplies curbed the ravages of cholera, diarrhoeal diseases, tuberculosis and pneumonia. In the ensuing 100 years, the overall death rate in England declined markedly, predominantly because of a great reduction in infectious disease mortality.

Ashton (1996) has argued that this "sanitary idea" was an idea of its time, the natural product of a reforming sanitary engineer – but, he says, the idea was ultimately flawed because of its mechanistic implications. It assumed that man would impose control over nature and environment, would separate the miasmas that emanated from sewage and refuge from the citizens' food, water and housing. What began with privies, paving and piped water thus led eventually to "garden city" suburbs and motorways: that is, engineering triumphs, but social liabilities. Ashton sees in the sanitary idea the origins of a technical-fix mentality, of the idea of "functional" separation of work, residence and recreation, of social atomism, and of a loss of ecological sensitivity.

Meanwhile, this reforming health-related approach to urban planning was promoted in the idealism of Benjamin Ward Richardson, an English physician, sanitarian and ardent disciple of Edwin Chadwick. In 1875, Richardson's vision of a healthy city, "Hygeia", proposed in 1875, was influential in both Britain and North America. Historians regard the early twentieth century as a turning point in urban health in the United States, particularly in relation to sanitation, pest control, food safety and housing standards (Schell, 1988). At that same time, the Canadian government's Commission on Conservation sought to achieve not just beautiful, but healthy, cities; to give emphasis "to the prevention of diseases, to health and to the prolongation of life" (Hancock 1993).

In the early decades of this century, commitments to public health persisted in many cities of the industrialised world. However, this idealism dissipated during the 1930s recession and World War II, and was then overshadowed by the heady postwar era of prosperity, automated production, industrial expansion and mass consumerism. Again, a need arose for a reactive response by the public health system. The urban air periodically thickened with smoke and sulphur dioxide; great smogs blighted cities in

Belgium and Pennslyvania and then, famously in 1952, London (McMichael 1996). An extra 4,000 deaths occurred in Greater London in the space of a week or so. Governments were compelled to introduce clean air legislation during the third quarter of this century.

In recent years, again, we have been faced with several health-endangering pressures in modern city life – including the adverse health consequences of the private motor car (air pollution, physical trauma, noise, reduced physical activity and social fragmentation) and increases in violent crime, drug abuse, long-term unemployment and homelessness. The persistence of socioeconomic disadvantage, and the emergence of a socially excluded underclass with weakened welfare support, has potentiated the re-emergence of various poverty-linked infectious diseases in urban populations (WHO 1995).

The influence of urbanism on human biology

Urbanism could, in principle, influence human biology in three ways. First, prolonged exposure to unfamiliar, environmentally stressful, urban factors could induce genetic evolution in humans. Second, the environmental conditions of urban living can directly influence human biological functioning. And, third, the cumulative and physically displaced impacts of urban populations upon regional and global biophysical systems, i.e. the "ecological footprints" of the city, could eventually result in widespread impacts on human health.

(1) Is there any evidence that humans have *genetically adapted* to city life? In general, any such processes would have taken many hundreds, more often thousands, of years. The amelioration of the early "crowd infections" (such as measles, smallpox, diphtheria, etc.) that swept virulently through ancient, immunologically naive, urban populations must have entailed genetic adaptations in both human host and infectious agent. Over more than 5,000 years, the evolving agrarian-based diet, consumed by urban populations, has put many unfamiliar demands upon human metabolism, resulting in genetic adaptations that enabled efficient usage of the differing food energy sources, especially cereal grains, milk and sugars. Hence today's inter-population genetic differences in, respectively, gluten sensitivity, lactose tolerance and insulin insensitivity (Allen and Cheer 1996). Among the few urban-specific candidate possibilities, it has for example been speculated (but not established) that the high prevalence of Tay-Sachs disease in Jews may be the consequence of a genetic trait that conferred resistance to tuberculosis, a scourge of young adulthood in the urban ghetto (Karlen 1993).

Could changes in the conditions of city living in just the past 150 years have affected the human genome? Inevitably, a few advantageous muta-tions must have occurred, randomly imbuing some individuals with a marginally reduced susceptibility to such things as air pollutants, allergens and noise. However, even *if* there had been sufficient time for the selective spread of those mutations, the proportion of new-borns who survive to post-reproductive adulthood is now so high as to diminish the primary mode of operation of the natural selection process. (This may be partially offset by the additional "genetic fitness" achieved by the elderly by dint of assistance they give to younger first-degree relatives of reproductive or pre-reproductive age.)

(2) The second category, the direct health impact of urbanism, is the most tangible and best understood. There are many ways in which physical, chemical and microbiological aspects of the ambient urban en-vironment affect health. Noise, population density, traffic hazards, air pollution, lead-contaminated drinking water and crowd-transmitted infec-tions are all well recognised urban risks to health. Likewise, the complex social environment of the modern city can affect health: the fragmentation of neighborhoods, isolation of the single and the elderly, the stresses of urban transport, fear of strangers and insufficient autonomy or status at work.

In his well-known analysis of declining mortality rates in Britain over the past two centuries, Thomas McKeown (1976) points to the nineteenth-century pre-eminence of improvements in social factors, followed by en-vironmental factors and then behavioral factors. He argues that the main social factor was the provision of adequate nutrition, which strengthened human biological defences against the infectious diseases of early indus-trial city life. Improved housing quality, safe water supplies, increasing literacy (especially of women) and the idea of domestic hygiene gave further important protection to infants and children against infectious agents. Most other commentators have broadly concurred with McKeown (e.g. Powles 1973; McKinlay and McKinlay 1977; Guha 1994). Szreter (1988), however, has argued the greater importance of deliberate public health interventions. In France, for example, substantial improvements in life expectancy emerged sequentially in Lyon (beginning in 1850s), Paris (1860s–1870s, more protractedly) and then Marseille (around 1890) in direct association with improvements in public water supply and sanita-tion to each of those cities (World Bank 1992).

Subsequently, McKeown suggested that, in contrast to the inferred dominance of physical and social environmental factors in nineteenth-century health gains, by later in the twentieth century the configuration of determinants had reversed. In developed countries behavioral factors now dominate, followed by environmental and, least important, social influences (McKeown 1979). Prominent consumer-based risk behaviors are: tobacco smoking, alcohol consumption, dietary excesses, sedentarism and freer sexual behaviors.

McKeown's diminution of the importance of social change as a determinant of population health is misleading. Yes, those readily achievable social changes that laid the foundations of hygienic, liberal, literate, democratic societies have already been made. However, a great many social deficits and inequities remain which have profound influences upon the health of urban populations. Although health indices in rich countries suggest that the *average* level of health is better in large cities than in the countryside, these measures of overall urban health are misleadingly weighted by the superior health experience of the comfortable middle-class majority of the urban population. In The Netherlands, for example, the life expectancy for the urban poor is five years less than for the well-to-do (McMichael 1993). Indeed, in developed countries the poorer segments of urban communities often have worse health status than the rural poor. Socioeconomic disadvantage entails greater risks of hazardous environmental exposures (local air quality, traffic noise and hazards, poorer housing stock, occupational hazards, etc.), greater likelihood of unhealthy behaviors (smoking, dietary imbalances, unprotected sex, etc.), and the psychosocial consequences of a sense of relative deprivation combined with a lack of social cohesion (Wilkinson 1996). At its most extreme, the life expectancy of the homeless, in London, is almost three decades below the population average. However, the poorest of the urban poor are often statistically elusive: they may be illegal immigrants, they may be transitory, and their health and vital statistics are often not recorded.

(3) Finally, urbanism can affect human health on a much larger spatial-temporal scale. Urban populations have increasingly large "ecological footprints" (Rees 1996). There are undoubted ecological benefits of urbanism, including various economies of scale, of proximity, and of shared use of resources, and the possibilities for reuse and recycling. Equally, though, there are great "externalities." Urban populations do not subsist on their own urban land. They depend on food grown elsewhere, on raw materials (timber, metals, fiber, etc.) and energy sources (especially fossil fuels) extracted from elsewhere, and on disposing of their voluminous wastes elsewhere.

Cities are thus sustained by economic infrastructures and supply lines that span vast areas of ecosystems. Resources are extracted from these natural and managed ecosystems, often at great distances from the cities. In Europe, we misguidedly celebrate the maximisation of "food-miles," as an index of trade liberalisation. Londoners drink wine from southern Australia, South Africa and Latin America. In Tokyo the disposable

Table 2.2. *Ecological footprints of Vancouver and the Lower Fraser Basin*

Geographic unit	Population	Land area (hectares)	Ecological footprint (hectares)	Overshoot factor
Vancouver city	472,000	11,400	2,360,600	207
Lower Fraser Basin	2,000,000	830,000	10,000,000	12

Source: Rees, 1996

chopsticks come from the Malaysian jungles. Our oil supplies come from distant sources. So too the year-round fresh fruits and vegetables that modern affluent western consumers now take for granted: out-of-season green beans from Kenya and strawberries from Brazil.

Urban populations thus depend on the natural resources of ecosystems that, in aggregate, are vastly larger in area than the city itself. The highly urbanised Netherlands consumes resources from a total surface area 15 times larger than itself. Folke and colleagues (1996) have studied the renewable resource appropriations by the cities of the Baltic Sea region. The estimated consumption of resources by 29 cities – wood, paper, fibers and food (including seafood) – depends upon a total area 200 times greater than the combined area of the 29 cities. That figure of 200 comprises 17 units of forest, 50 units of arable land and 133 units of marine ecosystems. Similarly, Rees (1996) has estimated that the almost half-million residents of Vancouver, Canada, occupying just 11,400 hectares, actually use the ecological output and services of 2.3 million hectares (Table 2.2). This ratio of 207 : 1 for the urban population is substantially greater than the ratio of 12 : 1 for the regional population of the Lower Fraser Basin as a whole. (Neither, points out Rees, would be sustainable if such levels of consumption applied to the world population.)

Viewed prospectively, the sustainability of urban populations and their health thus depends on the continued productivity of those distant ecosystems. Yet the scale of these externalities of urbanism is growing (Guidotti 1995). The externalities include massive urban contributions to the world's problems of greenhouse gas accumulation (especially via urban, car-dominated, combustion-engine transport, power generation for domestic and industrial use and cement production), stratospheric ozone depletion, land degradation and coastal zone destruction. The urbanised developed world, with one-fifth of the world population, currently contributes around three-quarters of all greenhouse gas emissions (IPCC 1996). Increasingly, these large-scale environmental changes are weakening the planet's life-support systems (McMichael 1993). Via this scaled-up process of externalised impacts, contemporary urbanism is thus jeopardising the health of current and future generations.

In the following sections, aspects of the second of these above mentioned three categories of impact – the direct health impact of urbanism – will be considered more fully.

The urban social environment

Large cities have long had their detractors. Last century, the American philosopher Henry David Thoreau said that cities are places where millions of people are lonely together; Emile Durkheim wrote of the anomie and alienation that afflicts many urban dwellers. More recently, Jacobs (1961) has emphasised the symbolic and emotional determinants of a sense of community within the urban environment: those cues and experiences that imbue the urban resident with a sense of belonging, a sense of social membership.

Much anecdotal and survey evidence indicates that many city dwellers do not feel engaged in a "community." Traditional rural or nomadic life, for all its hardships, has the social and emotional benefits of small-community living. So too does life in villages and other small settlements. However, to move into the city is to move from rural community-type society to urban association-type society. The former is conservative, hierarchical and stable; it entails social support from an extended kinship and close friends. Association-type life in the cities entails fluidity, more equality and more liberty but the sense of community is often greatly diminished.

This lack of social cohesion affects many health outcomes. Access to social networks influences an individual's prospects for health and survival (House *et al.* 1988). At the group level, various aspects of the community's social cohesion and economic equity, construed as "social capital" and as income relativity, respectively, influence the health prospects of the community at large (Wilkinson 1997). Also at the group level, Kaplan (1996) has concluded that neighborhoods with a high level of "strain" – that is, demands (crime rates, repetitive nature of work, etc.) relative to resources (income, self-perceived control, educational levels, etc.) – have consistently worse health indices.

Violence

Urban violence has become epidemic in many cities, especially in South and North America. Increasingly, it is becoming a problem in cities in Africa and Asia (Stephens 1996). The problem is concentrated among the urban poor, where particular foci of violence are the illegal drug trade, the desperate crimes of drug addicts, the violence due to alcoholism (and, relatedly, death on the roads), and racial and ethnic tensions (especially under contemporary conditions of structural unemployment). In New York city, death rates in 1990 were highest for black Americans in Harlem, where adult life expectancy in men was less than in Bangladesh (Stephens 1996).

Localised socioeconomic inequalities can affect rates of injury and violent death. For example, among various large neighborhoods in Chicago, of varying mean household income, the violence was greatest in neighborhoods with the most unequal internal distributions of income (Wilson and Daly 1997). The authors noted that studies of urban poor in the US "contain many articulate statements about the perceived risk of early death, the unpredictability of future resources, and the futility of long term planning."

Mental health

The overall balance of mental health between city and country remains a matter of contention. Indeed, misplaced generalisations about mental health disorders abound. In part they result from seeking a "hard" biological basis for mental disorders on the assumption that, irrespective of symptom patterns, the actual incidence of major psychiatric illnesses varies little between populations and cultures

There has also been a tendency to generalise across diagnostic categories about the risks to mental health posed by the urban environment (e.g. Barquero *et al.* 1982). A survey of over 9,000 British adults, during 1984–85, showed that the prevalence of psychiatric morbidity, as deduced from the 30-item General Health Questionnaire, was one-third higher in urban than rural dwellers (Lewis and Booth 1994). Among the urban residents in that survey, those with access to open space or gardens had a lower prevalence of morbidity. Webb (1984) drew attention to the generally higher rates of psychiatric disorders in urban versus rural populations in the Western world. In studies in developing countries, by contrast, it is not unusual to see higher reported rates of psychiatric morbidity in rural populations – for example, rates of depression and suicide appear to be higher in rural populations in both China (Desjarlais *et al.* 1995) and in Taiwan (Cheng 1989). From many studies in developed countries, it is clear that among city dwellers the poor, the socially marginalised and the isolated are particularly vulnerable to mental disorders.

The study of mental health disorders entails some particular difficulties. Are these differences within populations real, or are they due to differences in clinical presentation, diagnosis and recording? If real, might they be due to differential migration between city and country in relation to mental health status? If the rates are higher in lifelong urban dwellers, are there discernible variations between urban subpopulations? It is plausible that differential intra-urban migration could apply to severe psychotic disorders, but it seems less likely that minor neuroses would influence place of residence.

Studies have reported an increased prevalence of schizophrenia in urban populations relative to rural populations (Torrey and Bowler 1991), and particularly high rates in the centre of many modern industrialised cities (Thornicroft 1991). There have been suggestions that this inner-urban clustering may result from selective geographical drift. However, a study of almost 50,000 Swedish army recruits showed that the incidence of schizophrenia was 1.65 times higher among young men brought up in cities compared to those with a rural upbringing (Lewis *et al.* 1992), suggesting the importance of place of preadulthood residence – via, for example, a higher level of exposure to psychosocial stresses or environmental (infective or traumatic) factors in the urban setting.

Cardiovascular and other chronic diseases

Social experiences can be potent influences on other aspects of pathophysiology. Chronic unresolved stress can result from persistent uncertainties, the loss of collective morale, a basic sense of powerlessness and low reward : effort ratios in the workplace; the prevalence of such stress may be high in certain urban sub-populations (Marmot *et al.* 1995; Wilkinson 1996). This type of stress, and its corticosteroid consequences, may cause pathophysiological damage, and has been implicated in cardiovascular diseases. Studies in free-living primates corroborate both the higher cortisol levels in low-status, chronically anxious individuals, and the potentiation of coronary arterial damage that accompanies those raised cortisol levels (Sapolsky 1993). Meanwhile, other studies in humans show that strong social support networks attenuate mortality from cardiovascular and other degenerative diseases (House *et al.* 1988).

This type of interplay between life-stress and social support processes was evident in the study of trends in coronary heart disease (CHD) incidence in the town of Roseto in Pennsylvania, USA (Wolf and Bruhn 1993). This town formed *de novo* in the 1880s, when Italian immigrants settled there. For many decades they retained their low incidence of CHD, relative to the US population at large. This state of good health accompanied the retention of a traditional family-oriented social structure, within a cohesive and non-competitive community ethos. However, oncoming generations assimilated into the wider American culture, and embraced values of individualism and consumerist behavior. The incidence of CHD in Roseto then increased – indeed, it did so by substantially more than could be explained by changes in conventional risk behaviors (smoking, diet, blood pressure, sedentarism, etc.).

Urbanism, consumerism and behavior

Urban life has generally led the reshaping of society's value systems, consumer desires and personal behaviors. Health-related behaviors include: cigarette smoking, eating highly processed energy-dense diets, contraceptive practices and a generally sedentary lifestyle. The rise of modern mass "affluence" this century has resulted in increases in various degenerative diseases of later adulthood especially coronary heart disease, stroke, diabetes, gallstone disease, various cancers, chronic bowel disorders and arthritis (Hetzel and McMichael 1987). These diseases initially affected persons in upper socioeconomic strata. In recent decades, as knowledge of behavior-related health risks has emerged from epidemiological research and has become accessible to better-educated persons economically freer to exercise choices, the risks of these diseases have generally become greater in lower socioeconomic groups. Meanwhile, urbanisation in developed countries has also been associated with the decline of several major non-infectious diseases, including stomach cancer and peptic ulcer.

Obesity

Around the world the prevalence of obesity, particularly in urban populations, has increased markedly in recent decades (Table 2.3). Over the past two decades it has emerged as a huge contemporary public health problem (Seidell 1995). The metabolic disorders consequent upon obesity exacerbate the risks of hypertension, cardiovascular disease and diabetes.

The role of urbanism is clearly indicated in a recent study of seven populations of adult persons of African origin, living variously in Africa, the Carribbean and the USA (Cooper *et al.* 1997). Above age 34 years, the age-specific prevalence of hypertension was approximately twice as great in the US-based populations as in the other populations. Consistently, the main risk factors for high blood pressure in individuals were obesity and a raised blood sodium:potassium ratio. Earlier research in Kenya also demonstrated that, as rural migrants moved to Nairobi, so their blood pressure increased, again in association with micronutrient intake and weight

Table 2.3. *Increase in the prevalence of obesity in adults over past 2–3 decades*

England	Men (%)	Women (%)	USA (whites)	Men (%)	Women (%)
1980	6	8	1960–2	23	24
1986–7	7	12	1971–4	24	25
1993	13	16	1988–91	32	34

England, ages 16–64, body mass index (BMI) > 30, and USA (whites), ages 20–74, BMI > 27.8 (men), >27.3 (women).
Source: Seidell, 1995; Prentice & Jebb, 1996.

changes (Poulter *et al.* 1990).

In developed countries, the increase in obesity has been widespread, and particularly affects the less affluent – whereas in developing countries, it has thus far been most evident in the urban middle classes. In the UK, the prevalence of obesity has approximately doubled since the early 1980s (Prentice and Jebb 1996). Surveys indicate that poorer people tend to eat more energy-dense convenience-food diets, have less access to recreational facilities (and may feel safer indoors in violence-prone neighborhoods), and have a poorer understanding of the causes of obesity. The basic cause of obesity is a systematic imbalance in the energy budget of the urban lifestyle, entailing an increased uptake of dietary energy and a reduced output of exertional energy. Average daily caloric intake has increased by around 15% in US adults since 1980, but physical activity has not increased. Research in London has estimated that the average weekly distance walked by children has declined by around one-fifth over the past decade. Children also cycle much less and travel by car much more – the proportion of children walking to primary school in London has declined by nine-tenths since 1970 (Hillman 1993).

The urban physicochemical environment

Physicochemical exposures in the modern urban setting are diverse and widespread. Their health effects may be overt, as with road trauma or the increase in asthma hospitalisations during air pollution crises; more often they are non-acute and more insidious. The urban confluence of industrialisation, waste generation and concentrated transport systems has thus been a major force in altering the environment for human habitation (Schell 1991).

Environmental lead exposure, which blunts young children's intelligence, is a good example of an insidious urban exposure. Lead exposure has accrued over decades from industrial activity, dilapidated lead-based house-paint, and the use of leaded petrol for motor vehicles (Schell 1991). In today's cities, 80–90% of lead in air comes from leaded gasoline; meanwhile, the longstanding problem of drinking water contamination from old lead water pipes persists in some inner-city areas. The best estimate, based on cohort studies in several Western populations, is that children whose blood lead concentrations during early childhood differ by around 10 μg/decilitre will have a resultant difference of 2–3 IQ points, against an expected population mean of 100 (Tong and McMichael 1998). Such exposure differentials typically occur between the top and bottom quintiles of children within the urban environment.

The range of physicochemical hazards in the urban environment is protean. Other insidious exposures, that tend to be maximal in urban settings, and which often act cumulatively and in concert, may now be affecting prenatal and early-childhood growth (Schell 1991), immunological functioning, hormonal profiles and reproductive capacity (Sharpe and Skakkebaek 1993). Three contrasting examples of more overt environmental hazards are discussed below, to illustrate other aspects of this general problem category.

Air pollution

Urban air pollution has, in recent decades, become a worldwide public health problem (World Bank 1992). The earlier industrial/domestic air pollution from coal burning, which peaked in winter as combinations of black smoke and sulphur dioxide, has been replaced by pollutants from motorised transport which form photochemical smog in summer and a heavy haze of particulates and nitrogen oxides in winter.

Studies relating ambient air pollution levels to health risks were, until the 1970s, largely confined to examining particular extreme episodes of very high outdoors air pollution levels (e.g. the famous London smog of 1952). These episodes were associated with a marked increase in total mortality, especially cardiovascular and respiratory deaths, and with various respiratory disorders. Subsequent studies, based on daily mortality time-series, have elucidated the role of respirable particulates, ozone and nitrogen oxides in acute mortality (Schwartz 1996; Anderson *et al.* 1996) – although the quantitative interpretation of these studies remains problematic (McMichael *et al.* 1998). The more important question is whether longer-term exposures to elevated air pollution have a significant life-shortening effect. Long-term follow-up studies of populations exposed at different levels of air pollution indicate that the higher the background levels of particulates exposure the greater the mortality risk (Dockery *et al.* 1993; Pope *et al.* 1995).

Asthma, which has been increasing in Western countries for three decades, has an unresolved relationship to external air pollution. While some recent studies indicate a contributory role of air pollution as a trigger, if not an initiator, in asthma (Bates *et al.* 1990; Hoek *et al.* 1993), other studies are less conclusive. Meanwhile, the apparent increasing susceptibility of successive modern generations of children to asthma may derive from changes in human ecology that entail unusual configurations of early-life immunological experiences – such as allergenic domestic exposures (e.g. house-dust mites, cockroach dander, fungal spores), increased bronchial reactivity due to aspects of indoor air quality, reduced exposure to common childhood infections (due to smaller family sizes) or modern vaccinations (Holt and Sly 1997).

Urban transport

As the world's cities grow in size, urban transport systems are evolving continuously. In particular, car ownership and travel has increased spectacularly over the past half-century, creating new freedoms and new social and public health problems (McMichael 1996).

The reasons for the move from non-motorised and public motorised transport to heavy reliance on private cars are many. They include: gains in average personal wealth; an increase in leisure time; the seductions of advertising; the culture of road-building; deteriorating public transport facilities; greater mobility of women outside the home; and, in Western cities, the functional separation of place-of-residence, work, recreation and shopping. There are four broad categories of health hazard from car traffic:

(i)   exhaust emissions that cause local air pollution;
(ii) emissions that contribute to acid rain and to the global accumulation of carbon dioxide, each of which have consequences for health;
(iii) fatal and non-fatal injuries of car occupants, pedestrians and cyclists;
(iv) chronic traffic-related effects arise from fragmented neighborhoods, intrusive noise, and restrictions on physical exercise.

Inner-city populations in developed countries are increasingly aware of the domination and downside of the car (Fletcher and McMichael 1996). The extremity of the problem in Los Angeles has prompted a phased move away from combustion-engine cars to electric cars, and a major program of light-rail construction. Notoriously air-polluted downtown Athens went car-free in 1995 – and residents and retailers liked the results, despite initial misgivings. Government action in The Netherlands and Denmark, including providing extensive facilities for bicycles, has greatly boosted cycling.

The role and form of urban transport is undergoing intensified reappraisal. Not only have motor vehicles transformed the physical and social urban environment and the quality of its air, but they are also contributing increasingly to the declining sustainability of modern cities – including via pressures on the world's atmosphere and future climate.

Heatwaves, urban vulnerability and mortality

Severe heatwaves adversely affect health. This is particularly so in the center of large cities, where temperatures may be higher than in the suburbs and the surrounding countryside, and where the relief of nighttime cooling may be lessened. This "heat island" effect is due to the heat-retaining structures of most inner cities and the obstruction of cooling breezes.

For the world as a whole, 1995 was the warmest year on record. In July of that year, more than 460 extra deaths were certified as due to the effects of the extreme heatwave in Chicago in July, when temperatures reached 40°C (CDC 1995). Studies of such episodes have shown that those most vulnerable to heat-related illness and death are the elderly, the sick and the urban poor. In the Chicago heatwave, the rate of heat-related death was much greater in blacks than in whites, and in individuals who were bed-ridden or who were confined to poorly ventilated apartment-block housing (Semenza *et al.* 1996).

With its relatively stable maritime climate, temperature extremes are rarer in England and Wales than in the USA and other continental countries. Nevertheless, an estimated 768 extra deaths (an approximately 10% excess) were observed during the five-day heatwave in 1995 in England and Wales, relative to the equivalent period in 1993/94 (Rooney *et al.* 1998). Excess deaths were apparent in all age groups, being greatest for females and for deaths from respiratory and cerebrovascular disease. No more than half the excess mortality during the heatwave was attributable to the coexistent increases in air pollution. In Greater London, where daytime temperatures were higher (and where there was lesser cooling at night), mortality increased by around 20% during the heatwave. Preliminary analyses indicate that the excess mortality risks were generally greater in socioeconomically deprived groups (Megens *et al.* 1998).

Concluding comment: optimal size and form of cities

Animal behaviorists write about optimal population size in the wild, and how various physiological and behavioral mechanisms stabilise population size. The challenge for the majority of humankind in the twenty-first century will be to find the optimal size and form of human urban settlements as the foundation for good health, happiness and material well-being.

In the early 1970s, ideas emerged about the decentralisation of urban populations into small village-like settlements of about 500 people, aggregated into larger communities of around 50,000. In this way, "human-scale" communities would replace the social and ecological problem of the modern, everexpanding, megalopolis. They would have their own internal sense of identity, shared responsibility and cooperation. However, critics argued that such small and self-contained settlements would breed social pressures and moral coercion, would recreate the pettiness, rigidity and tedium of life in small villages. Others foresaw intellectual and cultural constraints, antithetical to the magnificent urban expression of human creativity, spirituality and striving.

The creation of small cohesive communities, nurturing a sense of belonging, must be balanced against considerations of efficiencies of scale for service provision. More generally, ecological considerations require greater population density, well-planned green areas, and better transit, biking and walking areas. Urban population density can be increased by redesigning cities as aggregates of semi-detached "urban villages," facilitating local self-sufficiency. The urban village concept, initially confined to European city planning, is now emerging in North America and Australia. The existing examples, while conservative, reveal an urban environment that is more integrated, less car-dependent and more oriented to the notion of "urban commons" than is low-density privatised suburbia.

Cities, as ever, are dynamic and exciting places in which to live. However, in various ways modern urbanism is becoming both ecologically and socially dysfunctional and therefore may jeopardise sustainable good health. Human biology, augmented by cultural ingenuity, is remarkably adaptable; but there are limits. Although we cannot reverse the mighty trend in urbanisation, the culmination of 10,000 years of history, we *can* modify our urbanism – the way we live in cities – to fit the needs of human biology and of the life-supporting biosphere.

References

Acheson, D. (1990). Edwin Chadwick and the world we live in. *Lancet* **336**, 1482–5.
Anderson, H. R., Ponce de Leon, A., Bland, J. M., Bower, J. S. and Strachan, D. P. (1996). Air pollution and daily mortality in London: 1987–1992. *British Medical Journal*, **312**, 665–69.
Allen, J. S. and Cheer, S. M. (1996). The non-thrifty genotype. *Current An-*

thropology, **37**, 831–42.
Ashton, J. (1992). The origins of healthy cities. In J. Ashton, ed., *Healthy Cities*, p.7. Milton Keynes: Open University Press.
Ashton, J. (1996). Evaluating the wider impact on health and quality of life. In T. Fletcher and A. J. McMichael, eds., *Health at the Crossroads: Transport Policy and Urban Health*, pp. 163–8. Chichester: Wiley.
Barquero, J. L. V., Munez, P. E. and Jaurequi, V. M. (1982). The influence of the process of urbanisation on the prevalence of neurosis. *Acta Psychiatrica Scandinavica*, **65**, 161–70.
Bates, D. V., Baker-Anderson, M. and Sizto, R. (1990). Asthma attack periodicity: a study of hospital emergency visits. *Environmental Research*, **51**, 51–70.
Burrows, R., Pleace, N. and Quilgars, D. (1997). *Homelessness and Social Policy*. London: Routledge.
CDC (Centers for Disease Control). (1995). Heat-related mortality. Chicago, July. *MMWR*, **44**, 577–9.
Cheng, T. A. (1989). Urbanisation and minor psychiatric morbidity: a community study in Taiwan. *Social Psychiatry and Psychiatric Epidemiology*, **24**, 309–16.
Cooper, R., Rotimi, C., Ataman, S., McGee, D., Osotimehin, B., Kadiri, S., Muna, W., Kingue, S., Fraser, H., Forrester, T., Bennett, F. and Wilks, R. (1997). The prevalence of hypertension in seven populations of west African origin. *American Journal of Public Health*, **87**, 160–8.
Desjarlais, R., Eisenberg, L., Good, B. and Kleinman, A. (1995). *World Mental Health. Problems and Priorities in Low-Income Countries*. New York: Oxford University Press.
Dockery, D. W., Pope, C. A., Xu, X, *et al.* (1993). An association between air pollution and mortality in six US cities. *New England Journal of Medicine*, **329**, 1753–9.
Faris, R. E. L. and Durham, H. W. (1939). *Mental Disorders in Urban Areas*. Chicago: University of Chicago Press.
Fletcher, T. and McMichael, A. J. (eds.) (1996). *Health at the Crossroads: Transport Policy and Urban Health*. Chichester: Wiley.
Floud, R., Wachter, K. and Gregory, A. (1990). *Height, Health and History. Nutritional Status in the United Kingdom, 1750–1980*. Cambridge: Cambridge University Press.
Folke, C., Larsson, J. and Sweitzer, J. (1996). Renewable resource appropriation. In R. Costanza and O. Segura, eds., *Getting Down to Earth*. Washington DC: Island Press.
Guha, S. (1994). The importance of social intervention in England's mortality decline: the evidence reviewed. *Social History of Medicine*, **7**, 89–113.
Guidotti, T. L. (1995). Perspective on the health of urban ecosystems. *Ecosystem Health*, **1**, 141–9.
Hancock, T. (1993). The evolution, impact and significance of the Healthy Cities/Communities movement. *Journal of Public Health Policy*, Spring, 5–18.
Hetzel, B. and McMichael, A. J. (1993). *The LS Factor. Lifestyle and Health*. Melbourne: Penguin.
Hillman, M. (1993). *Children, Transport and the Quality of Life*. London: Policy Studies Institute.
Hoek, G., Fischer, P., Brunekreef, B., Lebret, E., Hofschreuder, P. and Mennen, M. G. (1993). Acute effects of ambient ozone on pulmonary function of

children in the Netherlands. *American Review of Respiratory Diseases*, **147**, 111–17.

Holt, P. R. and Sly, P. D. (1997). Allergic respiratory disease: Strategic targets for primary prevention during childhood. *Thorax*, **52**, 1–4.

House, J. S., Landis, K. R. and Umberson, D. (1988). Social relationships and health. *Science*, **241**, 540–5.

Intergovernmental Panel on Climate Change. (1996). *Second Assessment Report, Volume I* (Working Goup I). Cambridge: Cambridge University Press.

Jacobs, J. (1961). *The Death and Life of Great American Cities*. New York: Modern Library.

Kalkstein, L. S. (1993). Health and climate change: direct impacts in cities. *Lancet*, **342**, 1397–9.

Karlen, A. (1993). *Plague's Progress. A Social History of Disease*. London: Gollancz.

Kaplan, G. A. (1996). People and places: contrasting perspectives on the association between social class and health. *International Journal of Health Services*, **26**, 507–19.

Kunitz, J. (1987). Making a long story short: a note on men?s height and mortality in England from the first through the nineteenth centuries. *Medical History*, **31**, 269–80.

Kunst, A. E., Looman, C. W. N. and Mackenbach, J. P. (1993). Outdoor air temperature and mortality in the Netherlands: A time-series analysis. *American Journal of Epidemiology*, **137**, 331–41.

Landers, J. (1993). *Death and the metropolis. Studies in the demographic history of London 1670–1830*. Cambridge: Cambridge University Press.

Landon, M. (1994). *Intra-urban health differentials in London*. London: London School of Hygiene and Tropical Medicine.

Lewis, G. and Booth, M. A. (1994). Are cities bad for your mental health? *Psychological Medicine*, **24**, 913–16.

Lewis, G., David, A., Andreasson, S. and Allebeck, P. (1992). Schizophrenia and city life. *Lancet*, **340**, 137–40.

MacFarlane, A. (1978). Daily mortality and environment in English conurbations. II Deaths during summer hot spells in Greater London. *Environmental Research*, **15**, 332–41.

Marmot, M. G., Bobak, M. and Davey Smith, G. (1995). Explanations for social inequalities in health. In B. C. Amick, S. Levine, A. R. Tarlov, and D. Chapman Walsh, eds., *Society and Health*, pp. 172–210. London: Oxford University Press.

McDowall, M. (1981). Long term trends in seasonal mortality. *Population Trends*, **34**, 16–19.

McKeown, T. (1976). *The Modern Rise of Population*. New York: Academic Press.

McKeown, T. (1979). *The Role of Medicine. Dream, Mirage or Nemesis?* Oxford: Basil Blackwell.

McKinlay, J. B. and McKinlay, S. M. (1977). The questionable contribution of medical measures to the decline of mortality in the United States in the twentieth century. *Milbank Memorial Fund Quarterly*, **55**, 405–28.

McMichael, A. J. (1993). *Planetary Overload. Global Environmental Change and the Health of the Human Species*. Cambridge: Cambridge University Press.

McMichael, A. J., Haines, A., Slooff, R. and Kovats, S. (eds.) (1996). *Climate*

Change and Human Health. Publication WHO/EHG/96.7. Geneva: WHO.

McMichael, A. J. (1996). Transport and health: assessing the risks. In T. Fletcher, A. J. McMichael, eds., *Health at the Crossroads: Transport Policy and Urban Health*, pp. 9–26. Chichester: Wiley.

McMichael, A. J., Anderson, R., Brunekreef, B. and Cohen, A. (1998). Inappropriate use of daily mortality analyses to estimate longer-term mortality effects of air pollution. *International Journal of Epidemiology*, 27, 450–63.

McNeill, W. H. (1976). *Plagues and Peoples*. New York: Doubleday.

Megens, T., Kovats, R. S. and McMichael, A. J. (1998). Mortality impact of heatwaves in the UK, 1976 and 1995, *The Globe*, 42, 4–5.

Office of Population Censuses and Surveys. (1994). *Mortality Statistics 1992: England and Wales*. Series DH1, No. 27. London: HMSO.

Parker, D. E., Horton, P. N., Cullum, P. N. and Folland, C. K. (1996). Global and regional climate in 1995. *Weather*, 51, 202–10.

Pope, C. A., Thun, M. J. and Namboodiri, M. M. (1995). Particulate air pollution as a predictor of mortality in a prospective study of US adults. *American Journal of Respiratory and Critical Care Medicine*, 151, 669–74.

Powles, J. (1973). On the limitations of modern medicine. *Science, Medicine and Man*, 1, 1–30.

Prentice, A. M. and Jebb, S. A. (1996). Obesity in Britain: Gluttony or sloth? *British Medical Journal*, 311, 437–9.

Rees, W. E. (1996). Revisiting carrying capacity: area-based indicators of sustainability. *Population and Environment*, 17, 195–215.

Romieu, I. (1992). Epidemiological studies of the health effects of air pollution due to motor vehicles. In *Motor Vehicle Air Pollution. Public Health Impact and Control Measures*, eds. D. T. Mage and O. Zali, pp. 13–62. Geneva: WHO.

Romieu ,I., Weitzenfeld, H. and Finkelman, J. (1990). Urban air pollution in Latin America and the Caribbean: health perspectives. *World Health Statistics Quarterly*, 43, 153–67.

Rooney, C., McMichael, A. J., Kovats, R. S. and Coleman, M. (in press). Excess mortality in England and Wales, and in Greater London, during the 1995 heatwave. *Journal of Epidemiology and Community Health*.

Poulter, N. R., Khaw, K. T., Hopwood, B. E. C., Mugambi, M. , Peart, W. S., Rose, G. and Sever, P. S. (1990). The Kenyan Luo migration study: observations on the initiation of a rise in blood pressure. *British Medical Journal*, 300, 967–72.

Sapolsky, R. M. (1993). Endocrinology alfresco: psychoendocrine studies of wild baboons. *Recent Progress in Hormone Research*, 48, 437–68.

Satterthwaite, D. (1996). *An Urbanizing World Global Report on Human Settlements*. Oxford: Oxford University Press.

Schell, L. M. (1988). Cities and human health. In G. Gmelch, W. Zenner, eds., *Urban Life*, pp. 18–35. Prospect Heights, IL: Waveland Press.

Schell, L. M. (1991). Effects of pollutants on human prenatal and postnatal growth: noise, lead, polychlorobiphenyl compounds and toxic wastes. *Yearbook of Physical Anthropology*, 34, 157–88.

Schwartz, J. (1996). Health effects of air pollution from traffic: ozone and particulate matter. In T. Fletcher, A. J. McMichael, eds., *Health at the Crossroads: Transport Policy and Urban Health*, pp. 61–82. Chichester: Wiley.

Second European Conference on Environment and Health. (1994). *The Declar-*

ation on Action for Environment and Health in Europe. Helsinki, Finland, June.

Seidell, J. C. (1995). Obesity in Europe: scaling an epidemic. *Journal of Obesity*, **19** (suppl 3), S1–S4.

Semenza, J. C., Rubin, C. H., Falter, K. H., *et al*. (1996). Heat-related deaths during the July 1995 heat wave in Chicago. *New England Journal of Medicine*, **335**, 84–90.

Sharpe, R. M. and Skakkebaek, N. (1993). Are oestrogens involved in falling sperm counts and disorders of the male reproductive tract? *Lancet*, **341**, 1392–5.

Shively, C. A., Laird, K. L. and Anton, R. F. (1997). The behaviour and physiology of social stress and depression in female cynomolgus monkeys. *Biological Psychiatry*, **41**, 437–68.

Stephens, C. (1996). Healthy cities or unhealthy islands? The health and social implications of urban inequality. *Environment and Urbanization*, **8**, 9–30.

Strachan, D. P., Taylor, E. M. and Carpenter, R. G. (1996). Family structure, neonatal infection, and hay fever in adolescence. *Arch Dis Childhood*, **74**, 422–6.

Szreter, S. (1988). The importance of social intervention in Britain's mortality decline c. 1850–1914: a re-interpretation of the role of public health. *Social History of Medicine*, **1**, 1–37.

Thornicroft, G. (1991). Social deprivation and rates of treated mental disorder. *British Journal of Psychiatry*, **158**, 475–84.

Tong, S. and McMichael, A. J. (in press). The magnitude, persistence and public health significance of cognitive effects of environmental lead exposure. *Journal of Environmental Medicine*.

Torrey, F. and Bowler, A. (1991). Geographical distribution of insanity in America: evidence for an urban factor. *Shizophrenia Bulletin*, **16**, 591–604.

Wallace, R. and Wallace, D. (1997). Community marginalisation and the diffusion of disease and disorder in the United States. *British Medical Journal*, **314**, 1341–5.

Webb, S. D. (1984). Rural–urban differences in mental health. In H. L. Freeman (ed.), *Mental Health and the Environment*, pp. 226–49. Edinburgh: Churchill Livingstone.

Weller, M. (1997). London's mental health problems. *Lancet*, **349**, 224.

Wilkinson, R. G. (1996). *Unhealthy Societies. The Afflictions of Inequality*. London: Routledge.

Wilkinson, R. G. (1997). Comment: income, inequality and social cohesion. *American Journal of Public Health*, **87**, 1504–6.

Wilson, M. and Daly, M. (1997). Life expectancy, economic inequality, homicide, and reproductive timing in Chicago neighbourhoods. *British Medical Journal*, **314**, 1271–4.

Wolf, S. and Bruhn, J. G. (1993). *The power of the clan: the influence of human relationships on heart disease*. New Brunswick, NJ: Transaction Publishers.

World Bank. (1992). *World Development Report. Development and the Environment*. Oxford: Oxford University Press.

World Commission on Environment and Development (1987). *Our Common Future*. Oxford: Oxford University Press.

World Health Organization (1990). *Diet, Nutrition, and the Prevention of Chronic Diseases*, Technical Report Series, No. 797. Geneva: WHO.

World Health Organization. (1995). *World Health Report.* Geneva: WHO.
World Health Organization. (1996). *Creating Healthy Cities in the 21st Century.*
  Background paper for UN Conference on Human Settlements. *Habitat II,*
  Istanbul, June.
Wrigley, E. A. and Schofield, R.S. (1989). *The Population History of England
  1541–1871: A Reconstruction.* Cambridge: Cambridge University Press.

3 Urban development and change: present patterns and future prospects

DAVID CLARK

Editors' introduction

There is extensive and remarkable diversity in urban forms and distributions around the globe; explaining rather than describing the diversity is difficult. Variations in urban forms do not correspond very closely to the economic development of the nations in which cities are located. Such terms as "developed" and "developing" are classificatory devices only, and do not explain variation in the distribution of urbanism any more than the antiquated classificatory terms Caucasoid or Negroid really explain variation in human biological characteristics. Variation in urban distributions is the result of history, culture and resources. Today, tremendous variation exists among nations in the growth of urban populations, the proportion of urban residents and the spread of urban lifestyles. With so much variety in urban change, it is exceedingly difficult to imagine a specific referent to the single category of urban. Thus, urban–rural comparisons are artificial contrasts that ignore the variation in urban places and distribution that Clark so ably describes. Future changes in urban population growth, urbanisation and urbanism will produce cities that are quite different from ones today, but, indisputably, the future is of *Homo sapiens* is an urban one.

One hundred years ago, in 1899, Adna Ferrin Weber published a thesis on *The Growth of Cities in the Nineteenth Century*. This masterpiece of data collection, analysis and explanation was the first comprehensive attempt to document and understand urban development at the global scale and its implications for health and social well-being. At the time, about 2% of the world's population lived in urban places, mostly in north-west Europe and the eastern seaboard of the United States. Today, there is significant urban development in most countries and the global level of urbanisation is estimated to be slightly above 50%.

Although the population of the globe is more urban than rural, levels of contemporary urban development vary widely (Brunn and Williams 1993).

Differences exist within and between regions and countries in the size and proportions of their populations that live in urban places, the number and size of cities and the ways of life of their residents. The urban pattern is changing in complex ways and at widely different rates as cities grow, house an increasing proportion of the population and extend their cultural influences deep into rural areas. This chapter focuses upon the evolution of the urban world. It seeks to document and account for present processes and patterns of urban development and change and to explore the implications for urban growth in the future.

Levels of urban development

The geography of the contemporary urban world is characterised by pronounced variations in the number and proportion of people who live in urban places. The map of urban population is dominated by China and India (Figure 3.1). China has by far the largest number of urban dwellers, and as such, merits particular attention in any balanced analysis of the contemporary urban world. One in five of the world's urban people lives in China and the population of Chinese towns and cities, at 499 million (in 1995), is similar to the urban population of Africa and South America combined. The urban population of India is some 279 million which is almost exactly the same as the urban population of the whole of Africa. Sizeable urban populations also occur in the USA, the former USSR, Mexico and Brazil, both because these countries are large and because a high proportion of their populations is urban. Japan and Indonesia are smaller countries with large urban populations. An urban population of some 378 million is spread across the 36 small nations which comprise the continent of Europe (excluding the former USSR). Comparatively few people however live in towns and cities in Africa. This is both because of sparse population overall and because the percentage of the population which lives in urban places is low.

Levels of urbanisation similarly are far from uniform. They are high across the Americas, most of Europe, parts of western Asia and Australia (Figure 3.2). South America is the most urban continent with the population in all but one of its countries (Guyana) being more urban than rural. More than 80% of the population live in towns and cities in Venezuela, Uruguay, Chile and Argentina. Levels of urban development are low throughout most of Africa, and south and east Asia. Fewer than one person in three in sub-Saharan Africa is an urban dweller. The figure is below 20% in Burundi, Rwanda, Ethiopia, Malawi, Uganda and Burkina Faso. Despite the presence of some large cities, levels of urban development throughout south and south-east Asia are low. An estimated 41 % of

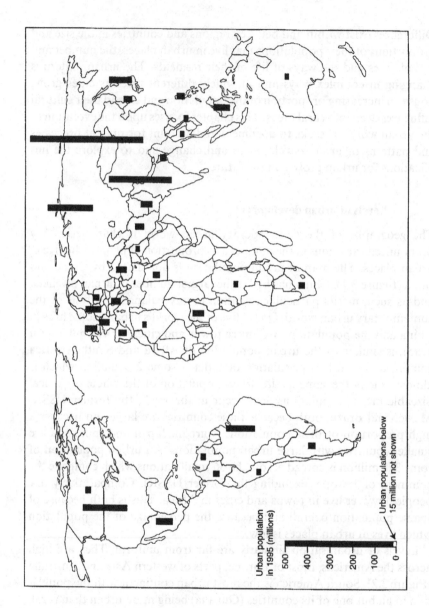

Urban population
in 1995 (millions)

500
400
300
200
100
0

← Urban populations below
15 million not shown

Figure 3.1 Urban population 1995

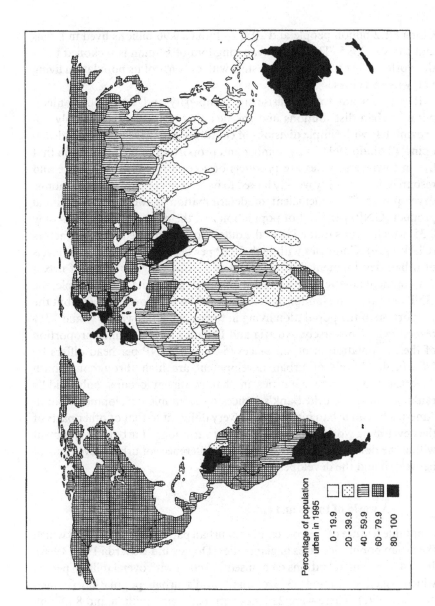

Figure 3.2   Percentage of population urban in 1995

China's 1.2 billion people and 29% of India's 0.96 billions lived in towns and cities in 1995. The Himalayan kingdom of Bhutan is reckoned to be the world's most rural sovereign state with only 6% of its population living in towns and cities.

It is important to emphasise the complexity and variety of contemporary urban distributions and the fact that this corresponds only approximately with simple divisions of countries into 'developed' and 'developing' (Torado 1994). The number and proportion of the population that live in towns and cities are products of a country's history, culture and resources and are only weakly linked to its level of contemporary economic development. The coefficient of determination between gross national product (GNP) per head of population and the percentage urban, is only 0.35, which means that 65% of the difference is explained by other factors (Clark 1996). Countries with high GNP per head tend to have high levels of urban development, but there is a very wide range of levels of urban development among countries with low GNP per capita. For example, the GNP per capita in Namibia and Peru in 1994 was below US$2,000 but the proportion of the population living in towns and cities was 70% and 27% respectively. Conversely, Austria and Ecuador have a similar proportion of their population in urban places (54%) but GNP per head differs by US$19,140. Levels of urban development are high throughout South America and yet the countries in that continent occupy only middle rankings on the World Bank's indices of economic development. South America has an urban history that is very different to that of other parts of the developing world. There are as many variations of urban development within the developing world as there are differences of urban development between it and the developed world.

A world of towns and cities

A notable feature of the contemporary urban pattern is the degree to which the urban population lives in giant cities (Dogan and Kasarda 1989, 1990). In 1990 the United Nations recognised 270 cities with over a million people which together housed 33% of the world's urban population (United Nations 1991). There were 22 cities with between 5 million and 8 million residents. Some 10% of the urban population lived in 20 mega-cities, or urban agglomerations, of size 8 million or more (Table 3.1). The best estimate is that Mexico City had some 20 million inhabitants in 1990, while Tokyo had 18 million, São Paulo, 17.4 million and New York, some 16 million. The way in which the population is distributed among cities of different size has important geographical implications. Not only is the urban population concentrated in a small number of countries, but, within

Table 3.1. *Urban agglomerations with 8 million or more persons,*
1950–2000

1950	1970	1990	2000
More developed regions			
New York	New York	Tokyo	Tokyo
London	London	New York	New York
	Tokyo	Los Angeles	Los Angeles
	Los Angeles	Moscow	Moscow
	Paris	Osaka	Osaka
		Paris	Paris
Less developed regions			
None	Shanghai	Mexico City	Mexico City
	Mexico City	São Paulo	São Paulo
	Buenos Aires	Shanghai	Shanghai
	Beijing	Calcutta	Calcutta
	São Paulo	Buenos Aires	Bombay
		Bombay	Beijing
		Seoul	Jakarta
		Beijing	Delhi
		Rio de Janeiro	Buenos Aires
		Tianjin	Lagos
		Jakarta	Tianjin
		Cairo	Seoul
		Delhi	Rio de Janeiro
		Manila	Dhaka
			Cairo
			Manila
			Karachi
			Bangkok
			Istanbul
			Teheran
			Bangalore
			Lima

Source: United Nations (1991: Table 11).

many of these countries, there is a disproportionate concentration in a
small number of cities.

Large cities occur in all parts of the world. Again, it is a pattern which is
independent of region, length of urban history and level of economic or
urban development. Metropolitan dominance is most pronounced in Latin
America and the Caribbean. The cities with over 1 million people in this
region house 45 % of the urban population. They include Mexico City, São
Paulo and Buenos Aires which are believed to be the largest, the 11th
largest and the 18th largest urban agglomerations in the world respectively. Large cities equally dominate the urban hierarchy in many of the

world's most rural regions and countries. Despite the low overall level of urban development, around 30% of the urban population of South Asia lives in cities with over 1 million people. They include Calcutta, Bombay and Delhi, cities which are among the 25 largest in the world. A similar percentage live in large cities in sub-Saharan Africa. Lagos is the biggest city in this region with a population of around 6 million.

Despite their enormous size, the world's mega-cities at present are viable and stable places which represent a significant social and economic achievement. They contribute disproportionately to national economic growth and social transformation by providing economies of scale and proximity that allow industry and commerce to flourish (Schteingrat 1990). They offer locations for services and facilities that require large population thresholds and large markets to operate efficiently. Mega-cities house many millions of people at extremely high densities and yet provide a range of opportunities and quality of life which is greater than that which is enjoyed in the surrounding rural area (Hardoy *et al.* 1992). Disturbing pictures of poverty, congestion and pollution in Calcutta, Bombay, Rio and Bangkok, and riots in Los Angeles and Beijing readily divert attention from what such places represent. The empirical evidence from around the world suggests that, rather than gigantic social mistakes, mega-cities are highly successful settlement forms (Dogan and Kasarda 1989, 1990).

Dimensions of urban change

The urban world is changing in three different and unconnected ways; through urban growth, urbanisation and the spread of urbanism. Urban growth occurs when the population of towns and cities rises. Urbanisation refers to the increase in the proportion of the population that lives in towns and cities. Urbanism is the name which is most commonly used to describe the social and behavioral characteristics of urban living which are being extended across society as a whole as people adopt urban values, expectations and lifestyles. Each development has profound implications for health affecting the nature and spread of disease and the provision and perception of health care.

Urban growth

By far the most important characteristic of contemporary urban change is the sheer scale of urban population growth. Each year between 1985 and 1995 the world's urban population rose by a staggering 73 million. Urban growth correlates strongly with overall population growth and so it is not surprising to find that the greatest gains occurred in highly populated countries where large numbers are added to the national population each year. The urban population of China alone rose by 226 million over the decade. This vast increase in only 10 years is almost as large as the present population of all the towns and cities of South America. Major increases also occurred in India (87 million), Brazil (32 million), Indonesia (26 million) and Nigeria (22 million). Little or no urban growth took place in Europe where national population levels are virtually static. For example, the urban population of Germany rose by around 1 million between 1985 and 1995.

Urban populations are growing most rapidly in Africa and southern Asia (Gugler 1988). The rates are highest in parts of sub-Saharan Africa and the Middle East (Figure 3.3). Afghanistan leads the way with an annual average rate of growth of urban population, between 1990 and 1995, of 8.5 %, and high rates of growth in the same region also occurred in Oman and Yemen. The countries of sub-Saharan Africa are predominantly rural and have few sizeable cities but they are also gaining urban populations rapidly. Botswana, Swaziland, Mozambique, Rwanda and Tanzania all had rates of urban growth in excess of 7 % over the period.

Urban growth in most countries is a product of both high rates of natural increase of the urban population and net in-migration. Despite the attention which is commonly paid to migration, there is little difference in their relative importance and the two processes compound and reinforce each other. The relative contribution of the components of urban growth is shown in an analysis of urban demographic change in 29 countries reported by Preston (1988). He found that in 24 cases the rate of natural increase of the population in cities exceeded that of net in-migration. The mean percentage of urban growth which was attributable to natural increase was 60.7 %.

Natural increase is an important component in urban growth despite the fact that both birth rates and death rates in cities are normally lower than in rural areas. One reason for the lower birth rate is the substantial number of young male migrants in the city who are unmarried or who have left their wives at home in their village. Another is that artificial birth control methods are more widely available in urban areas. Death rates are lower despite the poverty and squalor of many cities because urban populations tend to be young. Urban places in the developing world are generally more healthy than their rural hinterlands. They benefit from investment in drinking water and sanitation and are places where medical and scientific techniques, expert personnel and funds from the developed nations are first imported and where most people can be reached at least cost (Berry

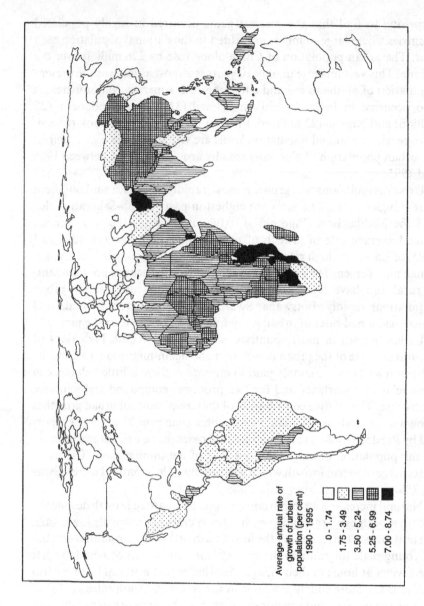

Figure 3.3   Annual average rate of growth of urban population 1990–1995 (%)

1973). A specific factor is that more births tend to take place in hospital and so rates of perinatal and infant mortality are lower (Gilbert and Gugler 1992). Diseases such as malaria and cholera can more easily be controlled or treated in cities than in the countryside.

Little or no growth is occurring in urban populations throughout most of the developed world. Growth rates of less than 0.5% per annum were recorded over the 1990–95 period in the United Kingdom, Belgium, Germany, Italy, Denmark, Sweden and Finland. Such low overall rates of growth are a product of differential size-related trends. Generally, metropolitan centres are losing population while towns and small cities are gaining. A pronounced fall in the population of major cities has occurred over the past 20 years (Clark 1989); 19 major cities in the US manufacturing belt lost population during the eighties as did 22 of the largest 24 cities in the UK (Clark 1985). Sizeable losses were similarly incurred by the metropolitan centres of north-western Europe (Fielding 1982). A slight overall increase in the urban population of these countries took place, however, because of a compensatory rise in the numbers of people living in small and medium-size towns and cities.

Urbanisation

The proportion as well as the total number of people who live in towns and cities is also increasing rapidly at the global scale. Urbanisation involves a shift in the distribution of population from rural to urban locations. Each year some 312 million more people are added to the world's towns and cities than to its rural areas. Although at 0.54% per annum the global rate of urbanisation seems comparatively modest, it has profound long-term implications. After several millennia of urban development only 25% of the population lived in towns and cities in 1950. That figure has doubled in the last 50 years to the present level of around 51%.

Urbanisation in the last 200 years appears to be a cyclical process through which nations pass as they evolve from agrarian to industrial societies. For Davis (1969) the typical course of urbanisation in the modern world can be represented by a logistic curve. The first bend in the curve is associated with very high rates of urbanisation as a large shift takes place from the country to towns and cities in response to the creation of an urban economy. It is followed by a long period of consistent moderate urbanisation. As the proportion climbs above about 60% the curve begins to flatten out, reaching a ceiling of around 75%. This is the level at which rural and urban populations appear to achieve a functional balance. At any one time, individual countries are at different stages in the cycle so it is necessary, in making sense of present rates of urbanisation

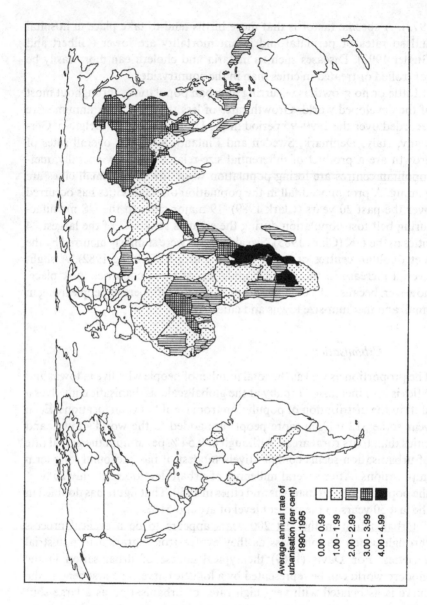

Average annual rate of
urbanisation (per cent)
1990–1995

☐ 0.00 - 0.99
▨ 1.00 - 1.99
▤ 2.00 - 2.99
▦ 3.00 - 3.99
■ 4.00 - 4.99

Figure 3.4   Annual average rate of urbanisation 1990–1995

(Figure 3.4), to take account also of the overall level of urbanisation (Figure 3.2). Some countries, such as Botswana, Mozambique, Tanzania, Rwanda and Nepal are presently experiencing high rates of urbanisation as their populations switch rapidly from rural to urban locations. In others the rate is modest, either because the cycle of urbanisation is complete, as in the United Kingdom and Saudia Arabia, or because, as in Sudan and Mali, it is only just beginning. Further substantial shifts in the distribution of population can be expected in many countries with presently low levels of urbanisation as the proportion of the population that lives in towns and cities rises to ceiling levels.

The present rapid urbanisation of the developing world is being driven by external processes. It is an internal locational adjustment to the progressive absorption of countries within an integrated global economy (Timberlake 1984, 1987; Clark 1998). The switch in the location of population from rural to urban is presently quickest in many of the countries of Africa and Asia. It is most marked in east Africa where the annual average rate over the period 1990 to 1995 was in excess of 4.0% (Figure 3.4). Several of the countries in this region are presently going through the high rate phase of the logistic cycle and the social and economic tensions associated with urbanisation are, in consequence, severe (Gugler 1997: Hardoy *et al.* 1992). Elsewhere in Africa and Asia the rate is lower because the cycle of urbanisation has only recently started. Countries with annual average rates of urbanisation below 2.0% and levels of urbanisation below 20% include Cambodia, Afghanistan and Sri Lanka.

Little change is taking place in the urban and rural balance in the developed world, because, in most countries, the cycle of urbanisation has run its course (Berg *et al.* 1982). Present levels of urbanisation in the UK, USA and much of Western Europe have changed little since 1950 (Clark 1996). More detailed analysis in fact shows that in many developed countries, the processes responsible for urbanisation have turned around (Champion 1989: Cross 1990). After many decades of relative decline, the rural component of the population is now increasing relative to the urban component. The net effect in geographical terms is that there is a shift of population at the national level from a state of more concentration to one of less concentration. For Berry (1976) this amounts to counter-urbanisation in so far as the traditional processes which favour towns and cities at the expense of rural areas are now working in reverse.

The spread of urbanism

The lifestyles of increasing numbers of people are being affected and transformed by urban growth and urbanisation. The concept of urbanism is that of a set of social and economic relationships and patterns of behavior which arise in cities and follow from their impact on society (Wirth 1938). It is expressed and reflected through activities, interactions, tastes, fads, fashions, aspirations and achievements. The spread of urbanism is the most difficult of the principal processes of global urban change to document since few countries hold censuses of the lifestyles of their citizens. Highly subjective evidence on behavior and consumption, however, suggests that lifestyles are increasingly independent of place. Many people who live in remote locations have lifestyles which are similar to those in cities. They participate fully in a telemediated urban culture and their attitudes and values have much in common with those of residents of the world's principal cities (Curran and Gurevitch 1992). Conversely, there are a large number of people, especially in the million and mega-cities of the developing world, who retain patterns of association and behaviour that are more akin to those in rural areas. Their attitudes and values are permeated and constrained by traditional influences of religion, the family and geographical parochialism. The lifestyles of many first and second-generation in-migrants have not yet been affected by incorporation within urban society.

The spread of urbanism, like urbanisation, is a finite process. It begins with a wholly rural society in which urban influences are absent and ends when everyone everywhere lives an urban way of life. Most insights into urbanism are the product of in-depth sociological studies of which there is a long tradition in the United States and Europe but a deficiency in developing countries. Such work as does exist needs to be evaluated critically since it is difficult if not impossible for researchers to shed their own cultural values and so study alien societies, and indeed their own, objectively and dispassionately (Tracey 1993; Lull 1995). What it suggests, however, is that society in Western Europe, much of North America and parts of Australasia, the Middle East and South America is deeply permeated by urban values and dominated by urban institutions, but elsewhere the incidence of urbanism is limited. Vast tracts of Africa and Asia, and extensive areas in other parts of the developing world are largely unaffected by urban influences. Rural ways of life predominate over large parts of the contemporary urban world.

The urban future

The scale and direction of present urban development point to the emergence by 2025 of an urban world that will bear little resemblance to the urban present. A recent estimate by the United Nations (1991) is that by the year 2025 there will be some 5.5 billion people out of a world population of 8.5 billion living in urban places. This projected future urban population is roughly the same as the total population of the world today. Some 4.4 billion will be living in towns and cities in what are presently classified as developing countries. The population of China's urban places will be close to 1 billion and in India is expected to be some 740 million.

Urban growth will be accompanied by increased urbanisation. Some 65% of the world's population is expected to be urban by the year 2025. Most of this increase will occur because of the urbanisation of the population across large parts of Africa and Asia. These regions will be most radically affected by urban development in the next quarter century.

The most striking feature of the predicted urban geography of the year 2025 is the uniformly high level of urban development in the Americas (Figure 3.5). The population of all of the principal countries of North, Central and South America is expected to be over 60% urban and in most it will be in excess of 80%. The Americas are presently highly urbanised, so this change represents a consolidation of existing patterns. Similarly high levels of urban development are anticipated in Australia, Japan, parts of the Middle East, north Africa and most of Europe. Urbanisation levels in excess of 60% are expected across the whole of Asia north of the Himalayas.

The United Nations forecast suggests that levels of urban development in Africa and southern Asia will be very much higher than today, but will vary considerably from country to country. Although the population in most countries will be more urban than rural, the proportion living in towns and cities in Burundi, Malawi, Rwanda, Ethiopia, Uganda, Burkina Faso, Afghanistan, Nepal, Bhutan, Cambodia, East Timor and Papua New Guinea is expected to be less than 40%. Such countries have yet to go through the phase of rapid urbanisation that is a characteristic feature of the cycle of urban development. They will be the world's last remaining rural territories. At 51%, India is expected to be only marginally more urban than rural in 2025.

The number of mega-cities is expected to rise to 28 in 2000 and to be more than 60 in 2025 (United Nations 1991). Strong patterns of urban dominance already exist in many African countries and these are likely to be reinforced, leading to a further polarisation of the settlement hierarchy. The problem will not so much be the overconcentration of the urban population in one city as the wide dispersal of people in settlements too small to support even basic services and non-agricultural economic activities. Even by 2025, in the least urbanised countries in Africa there are expected to be too few centres which are large enough to provide adequate levels of local servicing. A widening of the gap between town and country

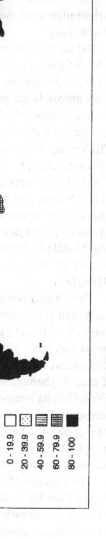

Predicted
percentage of population
urban in 2025

	0 - 19.9
	20 - 39.9
	40 - 59.9
	60 - 79.9
	80 - 100

Figure 3.5   Predicted percentage of population urban 2025

is expected to be one of the major consequences of the urban transition in Africa over the next 30 years.

The urban geography of the developed world is likely to be very different. Here, urban populations are presently high and the principal shifts will take place among cities. Rather than a concentration in a small number of large cities, which is the current pattern, the population is expected to be more evenly spread across many smaller centres. Cities of about 200,000 are likely to be the most attractive as they are large enough to sustain an acceptable range of services without the congestion and pollution that are associated with life in the mega-city. The spatial structure of large cities is likely to be transformed with a fall in central densities as people and businesses move to the suburbs and beyond. Many of the areas which are vacated will become parkland and open space so that cities will take on a "doughnut" form. In the longer term it is possible that central area populations will rise as cities begin to reurbanise (Berg et al, 1982). Decentralisation of population at the local scale, and deconcentration at the national level will significantly reduce urban/rural differences, so producing a "rurban" arrangement in which urban and rural patterns of settlement, and lifestyles, become fused (Clark 1989). The expected trends in different parts of the world will lead to a progressive inversion of the contemporary urban pattern at the global scale. A small-city and rural orientation will increasingly characterise the landscape in developed economies but strongly urban-centred mega-city societies will emerge and predominate in the developing world.

A sustainable urban future?

The preceding forecasts paint a disturbing picture of major population increase and further massive urban development in the first quarter of the twenty-first century. Such is the scale of urban development which is predicted that it raises questions as to whether urban development of this magnitude can be sustained, especially as the highest rates of growth are occurring in the poorest nations which are least equipped to cope with its consequences. It is difficult, given the current pressures on resources and the environment, to see how urban populations in Africa and Asia can double over the next 25 years without some form of social, economic or ecological breakdown (Johnson 1993; Haughton and Hunter 1994).

Some of the principal sustainability issues in the developing world concern service provision. The lack of a piped water supply is especially significant as many health problems are linked to water (Cairncross 1991). The problems are serious in almost all developing countries as studies overviewed by Hardoy, Mitlin and Satterthwaite (1992) emphasise. They

Table 3.2. *Estimates of levels of servicing in selected Third World cities*

	Percent of households estimated to be served by		
	piped water	central sewage system	garbage collection
Accra (Ghana)	100	30	10
Bankok (Thailand)	66	2	80
Dakar (Senegal)	28	n.a.	0
Dar Es Salaam (Tanzania)	53	13	33
Jakarta (Indonesia)	33	0	60
Kampala (Uganda)	18	19	10
Karachi (Pakistan)	38	n.a.	33
Khartoum (Sudan)	n.a.	5	n.a.
Manila (Philippines)	n.a.	15	50
São Paulo (Brazil)	n.a.	n.a.	33
Kinshasa (Democratic Republic of Congo)	50	0	n.a.

n.a. = not available
Source: Abstracted from Hardoy, Mitlin and Satterthwaite (1992)

paint a depressing picture of levels of provision which, in the opinion of the authors, is more accurate than that portrayed in official government statistics. For example, most of the population in Accra has access to piped water but the system is often not operational (Table 3.2). In Bangkok, Dar es Salaam and Kinshasa, over half of all households are connected and the occupants of the remainder must either use standpipes or buy water from vendors. The situation in Dakar is worse as only 28 % of households have private water connections while 68 % rely on public standpipes and 4 % buy from water carriers.

A similar pattern of low but variable provision characterises sewage disposal and garbage collection. Hardoy, Mitlin and Satterthwaite (1992) estimate that around two-thirds of the urban population in the developing world have no hygienic means of disposing of sewage and an even greater number lack an adequate means of disposing of waste water. Most cities in Africa and many in Asia have no sewers at all and human waste and waste water end up untreated in canals, rivers and ditches. Where sewage systems exist they rarely serve more than the population that lives in the richer residential areas. Some 70 % of the population of Mexico City live in housing served by sewers but this leaves some 3 million people who do not. In Buenos Aires it is estimated that 6 million of the 11.3 million inhabitants are not connected to the sewer system. The fact that levels of public service provision are low today raises far-reaching questions as to how they can cope with the vast increase in numbers which are expected in the future.

Despite the largely pessimistic tone of the sustainability literature, there are however some grounds for believing that the arguments have been overstated (Orrskog and Snickars 1992). The debate on sustainability is in its infancy and there is a lack of detailed empirical research (Cadman and Payne 1990; Blowers 1993). Sceptics point out that the present concerns are merely the most recent in a string of doomsday predictions from Malthus in the late eighteenth century to Meadows *et al.* (1972) in their highly influential work on the *Limits to Growth*, but famine and social breakdown have yet to occur. A crisis of food supply is likely at some stage in the future but it seems some way off.

The success of past attempts to deal with urban environmental problems provides grounds for some optimism about the future. Most cities in the developed world are cleaner today than they were twenty years ago and their residents enjoy higher levels of health and amenity as a result (McMichael 1993). Air and water quality are two areas in which improvements have been dramatic. Cleaner cities are the products of major investment in public services, especially water supply and sanitation, and the introduction and enforcement by governments of regulations covering environmental discharges. They are helped by the switch from coal to oil and gas as energy sources, although this introduces new and as yet unresolved problems of exhaust emissions. The experience in developed countries suggests that with appropriate intervention and direction, a sustainable future for cities is possible. Whether this can be achieved in the developed world, where the urban problems are presently more acute, and where there is neither economic wealth nor strong government, is however debatable.

Conclusion

This chapter has analysed present and expected future processes and patterns of urban development and change. It has drawn attention to the massive concentrations of population in the towns and cities of China and India, the recent rapid urbanisation of the developing world, and the scale of urban changes which are presently affecting parts of Africa and Asia. It has highlighted the urban characteristics of the developed world where change is occurring through the spread of urbanism rather than through population redistribution. Recent forecasts suggest that urban growth and urbanisation will continue, even accelerate, at the global scale so transforming the population geography of those parts of the world which are presently predominantly rural. The changes which are anticipated raise

64   *David Clark*

fundamental questions concerning urban viability. Some query the ability of cities, especially in the poorest countries, to cope with explosive expansion. Others, reflecting on the lessons of history and the experiences of cities in the developed world, view the urban future with optimism. The challenge for analysts is to develop a comprehensive understanding of urban development and change so as to enable governments to act to secure a sustainable urban future.

Note

1. This chapter is based on estimates of urban populations abstracted from the United Nations 1991 report on *World Urbanisation Prospects.* The highly variable quality and reliability of world urban data are emphasised in United Nations (1994) and in the World Bank (1993). For a general discussion see Goldstein (1994). The data and the maps refer to the countries which existed on January 1 1989.

References

Berg, L. van den, Drewett, R., Klassen, L. H., Rossi, A. and Vijverberg, C. H. T. (1982). *A Study of Growth and Decline.* London: Pergamon.
Berry, B. J. L. (1973). *The Human Consequences of Urbanization.* New York: St Martin's Press.
Berry, B. J. L. (1976). *Urbanization and Counterurbanization.* London: Sage.
Blowers, A. (1993). *Planning for a Sustainable Environment.* London: Earthscan.
Brunn, S. D. and Williams, J. F. (1993). *Cities of the World: World Regional Urban Development.* New York: Harper Collins.
Cadman, D. and Payne, G. (1990). *The Living City: Towards a Sustainable Future.* London: Routledge.
Cairncross, F. (1991). *Costing the Earth: the Challenge for Governments, the Opportunities for Business.* Boston: Harvard Business School Press.
Champion, A. G. (1989). *Counterurbanisation.* London: Arnold.
Clark, D. (1982). *Urban Geography: An Introductory Guide.* Beckenham: Croom Helm.
Clark, D. (1985). *Post-Industrial America.* London: Routledge.
Clark, D. (1989). *Urban Decline.* London: Routledge.
Clark, D. (1996). *Urban World/Global City.* London: Routledge.
Clark, D. (1998). Interdependent urbanisation in an urban world: an historical overview. *The Geographical Journal,* **164,** 85–95.
Cross, D. F. W. (1990). *Counterurbanisation in England and Wales.* London: Avebury.
Curran, J. and Gurevitch, M. (1992). *Mass Media and Society.* London: Arnold.
Davis, K. (1969). *World Urbanization.* Los Angeles: University of California.
Dogan, M. and Kasarda, J. D. (1989). *The Metropolis Era Vol 1: A World of Great Cities.* London: Sage.
Dogan, M. and Kasarda, J. D. (1990). *The Metropolis Era Vol 2: Mega-Cities.*

London: Sage.
Fielding, A. J. (1982). Counterurbanization in Western Europe. *Progress in Planning*, **17**, 1–52.
Gilbert, A. and Gugler, J. (1992). *Cities, Poverty and Development*. Oxford: Oxford University Press.
Goldstein, S. (1994). Demographic issues and data needs for mega-city research. In *Mega City Growth and the Future*, eds. R. J. Fuchs, E. Brennan, J. Chamie, F. Lo and J. I. Uitto, pp. 32–61. New York: United Nations University Press.
Gugler, J. (1988). *The Urbanisation of the Third World*. Oxford: Oxford University Press.
Gugler, J. (1997). *Cities in the Developing World: Issues, Theory and Policy*. Oxford: Oxford University Press.
Hardoy, J. E. and Satterthwaite, D. (1990). Urban change in the Third World: are recent trends a useful pointer to the urban future? In *The Living City*, eds. D. Cadman and G. Payne, G. pp. 75–110. London: Routledge.
Hardoy, J. E., Mitlin, D. and Satterthwaite, D. (1992). *Environmental Problems in Third World Cities*. London: Earthscan.
Haughton, G. and Hunter, C. (1994). *Sustainable Cities*. London: Regional Studies Association.
Johnson, S. P. (1993). *The Earth Summit: The United Nations Conference on Environment and Development*. London: Graham and Trotman.
Lull, J. (1995). *Media, Communication and Culture: A Global Approach*. Cambridge: Polity Press.
McMichael, M. (1993). *Planetry Overload:Global Environmental Change and the Health of the Human Species*. Cambridge: Cambridge University Press
Meadows, D. H., Meadows, D. L., Randers, J. and Behrens, W. W. (1972). *The Limits to Growth*. New York: University Books.
Orrskog, L. and Snickars, F. (1992). On the sustainability of urban and regional structures. In *Sustainable Development and Urban Form*, ed. M. Breheny. London: Pion.
Preston, S. H. (1988) Urban growth in developing countries: a demographic reappraisal. In *The Urbanization of the Third World*, ed. J. Gugler, pp. 11–32. Oxford: Oxford University Press.
Schteingrat, M. (1990). Mexico City. In *The Metropolis Era Vol 2: Mega-Cities*, eds. M. Dogan and J. D. Kasarda, pp. 268–329. London: Sage.
Timberlake, M. (1987). World-system theory and the study of comparative urbanization. In *The Capitalist City*, eds. M. P. Smith and J. R. Feagin, pp. 37–64. Oxford: Blackwell.
Timberlake, M. (ed) (1984). *Urbanization in the World-Economy*. London: Academic Press.
Torado, M. P. (1994). *Economic Development*. London: Longman.
Tracey, M. (1993). A taste of money: Popular culture and the economics of global television. In *The International Market in Film and Television Programs*, eds. E. M. Noam and J. C. Millonzi, pp. 163–98. Norwood New Jersey: Ablex.
United Nations. (1991). *World Urbanization Prospects*. New York: United Nations.
United Nations. (1994). *Demographic Yearbook*. New York: United Nations.

Weber, A. F. (1899). *The Growth of Cities in the Nineteenth Century*. New York: Macmillan; reprint 1962, Ithica, NY: Cornell University Press.
Wirth, L. (1938). Urbanism as a way of life *American Journal of Sociology*, **44**, 1–24.
World Bank. (1993). *World Development Report*. New York: World Bank.

4   *Transitional environments: health and the perception of permanence in urban micro-environments*

R. HUSS-ASHMORE AND C. BEHRMAN

Editors' introduction

This chapter, and the one preceding it, by Clark, address the issue of diversity and character of "urban" places. Trailer parks, which Huss-Ashmore and Behrman term "permanently transitional communities," defy classification in common categories of urban places, demonstrating Clark's point about the uselessness of simple categories for either description or analysis. The macrolevel perspective provided by Clark is complemented by the microlevel approach taken here. The analysis works on several levels. Like Clark, Huss-Ashmore and Behrman address the issue of what an urban place is, by examining the culture, environment and biology of residents of trailer parks. The approach is remarkably similar to that used in adaptability studies which were usually conducted with remote and isolated communities and sought to understand how the biological and cultural systems worked adaptively. A key theoretic and methodological problem for such studies has been defining the boundaries of the study population. Today more then ever before, communities are linked to larger networks through economic relations. In an urban world where boundaries are often indistinct, the authors are able to define a community for intense, ethnographic and human biologic study. Ironically, the community defined in this chapter is characterised by an ideology of transition, although residence in trailer parks may be long-term and anything but transient. The interrelationships among ideology, the physical environment of trailer parks and human health and biology are made clear and serve as a model for investigating urban places using an adaptability framework.

Environment has played a central role in human biological research over the past 35 years. Indeed, most of our theoretical models include something called "the environment," sometimes as a vague set of background conditions, sometimes as precisely measured parameters that can alter human biological functioning (Huss-Ashmore 1997). However, despite the primacy of environment in our theories, human biologists have not agreed

67

on how to operationalise the concept. What do we measure, and how? Following the lead of biological ecologists and environmental physiologists, we have often simplified the environment to single climatic or geophysical stressors such as heat, cold or hypoxia. Alternatively, we have seen it as a spectrum of resources, or, more often, simply as available energy. This is not to say that we have ignored the presence of people. A thing called the "social environment" is often part of our models, but it has rarely been measured, at least not by human biologists (Thomas 1997).

With a shift in venue to research in urban areas, the problem of defining the effective environment becomes more acute. Heat and cold remain, of course, as physical stressors, and energy capture is still important. But the spectrum of stressors and resources changes, as do the potential biological impacts. One of the obvious changes is in the importance of *human* aspects of the environment. Urban areas are species-specific, built environments, conceived of and constructed by humans for humans. Understanding the biological importance of urban environments, and the microenvironments within them, requires attention to the social and psychological landscapes that they present.

Urban microenvironments can be defined on the basis of a number of variables, including location, class, ethnicity, religion, sexual orientation or housing type. In addition, the permanence or transitoriness of a given population may be another variable with biological significance in the urban environment. For example, research on environmental hazards has shown that the perception of constancy of the environment greatly affects the strategies that humans use for dealing with their surroundings (Burton *et al.* 1978). Similarly, psychologists have argued that "rootedness" and "uprootedness" may be related to certain types of mental illness (Godkin 1980).

In this chapter we explore the idea of transitoriness or impermanence as a factor affecting biological well-being in populations of the urban poor. While the idea can be applied to populations as diverse as the homeless and the elderly in retirement communities, our focus here is on residents of trailer parks. Our purpose is to define trailer parks as distinctive environments for human biological research. We talk first about trailer parks as urban environments with the potential to affect health. We then use pilot data from four urban mobile home communities in the western US as an example.

Why trailer parks?

We came to trailer parks via our research on periurban areas in eastern and southern Africa. We found these to be areas on the outskirts of towns and cities where migrants from both rural and inner-city backgrounds converged. Periurban residents were not always poor, but shared a number of other traits with people in squatter settlements or shantytowns in other parts of the world. In general, quality of housing was poor, with little space and little incentive to invest in improvements. As with squatter settlements, the communities could be described as "makeshift," "temporary" or "transitional" (Harpham *et al.* 1988). A high percentage of residents lived alone or in single-generation households. There was a relative lack of services, such as water, sanitation, electricity and adequate health care. In their attempt to achieve economic success, periurban migrants often had multiple income sources, some of which were semi-legal or blatantly illegal. This occupational multiplicity gave them greater access to cash than much of the rural population, but (as in the case of illegal beer brewing and prostitution) came with its own health risks.

In Swaziland, we found that periurban women and children differed biologically from their urban and rural counterparts (Behrman and Huss-Ashmore 1994). When matched for age, periurban women had higher Body Mass Index than either rural or urban women. They had higher energy intakes during the preharvest season, a period often described as the "hungry season" in rural Swaziland (Huss-Ashmore and Goodman 1989). In addition, they ate more sweets, fats and purchased foods. Periurban children were also taller and heavier for their age than rural children. This suggests that periurban households use cash to even out the seasonal effects of food supply, with accompanying changes in body size.

One of the most distinctive features of these periurban areas was the perception by residents that they were only there temporarily, despite what was often long-term residence. People came seeking a foothold in the urban economy, or hoping to make their fortune and move back to the rural area. They were always going to "make it" and move on. Through choice or the inability to secure permanent employment, they often stayed, becoming petty entrepreneurs or shanty landlords, building substandard housing and renting single rooms. Although they were sometimes politically important members of the community, this wasn't home. Because of these characteristics, we defined these periurban areas as permanent transitional zones, with distinct social features and implications for health.

In looking for a First World analogue, we clearly could not use geographic criteria. Suburbs, while geographically peripheral to urban areas, share few other features with African periurban zones. Our interest in the transitional aspect of periurban residence led us to consider trailer parks, a type of urban community with similar environmental and social characteristics. Geographically, trailer parks need not be peripheral to cities (there are several in the middle of Anchorage, Alaska), but they often are. They may be outside the city limits, where urban services such as water, sewage disposal, road maintenance or trash collection are not provided. They are frequently invisible to municipal authorities, located on private land, unzoned and thus unregulated (Knox 1993). They have become, in American folk sociology, a haven for the temporarily rootless and financially unstable: single mothers, truck drivers and the elderly, among others. On the basis of public perception and our own experience, we hypothesised that trailer parks might be First World urban environments characterised by temporariness or a state of permanent transition, and with special health risks.

Health and the mobile home

Trailer demographics

Trailers have been a part of the American landscape since the end of World War I. Many of the early forms were merely tents on a cart, but these gave way rapidly to small, home-made wheeled houses and eventually to factory-manufactured trailers. A magazine story published in 1937 proudly announced "We have lived in trailers since 1919," although the trailer in question was a converted circus wagon (Slagle 1937). Prior to the 1950s, the majority of trailers were used by well-to-do families for vacation travel, or by salesmen, journalists, construction workers or others whose work required them to travel. The housing shortage experienced during World War II led to a sudden increase in the number of trailers used as full-time dwellings and gave many American families their first experience of trailer living. Retirees took to trailer life from the beginning, and by 1940, almost 20% of trailer owners were retired (Cowgill 1941).

While the initial charm of trailers was their mobility, they could not spend all their time on the road, and so the trailer park was born. These were initially campgrounds, a piece of land on which the itinerant could park between journeys (Thornburg 1991). By the 1950s, owners of trailer parks provided sewage hookups, water and electricity, and more families lived in trailer communities full time. In the 1970s these became mobile home parks, or mobile home courts. Today the preferred term is manufactured housing, and planned communities of manufactured housing are appearing in middle-class suburbs. These communities regulate the height of grass in the yard, the places and times where laundry may be dried, and the number of pets (one) to be owned (Sack 1997). These are not trailer parks, in the classic sense. We define a trailer park as a community of five

Figure 4.1   High mobility: a recreational vehicle or motor home

or more mobile homes *originally designed to be moveable*, located on land which the residents do not own.

The changes in terminology applied to these dwellings reflect a belated recognition that most trailers no longer trail anything, let alone the family car, and that mobile homes are increasingly immobile. There is of course still a range of mobility in manufactured housing, from motor homes (Figure 4.1) to genuine trailers to factory-built double-wides with cement-block foundations. Some units start out mobile, but grow increasingly immobile with time, as foundations or site-built additions are added (Figure 4.2).

As trailers have altered their mobility, they have also changed their location in social space. In 1940, a study by Donald Cowgill found that most American trailerites had a secure economic status, with very few owners in the low-income range (Cowgill 1941). One-third of the trailerites had attended college, and 20% had bachelor's degrees or higher. By contrast, the median income of mobile-home households in 1990 was $20,026, compared to $30,531 for other households (O'Hare and O'Hare 1993). Two-thirds of mobile-home households made less than $25,000 per year. Mobile-home residents also have less education than people who live in other housing types; only 20% have any college education, compared to 40% of other American householders. One aspect of trailer demographics

Figure 4.2   Low mobility: a trailer near Palmer, Alaska, over which a permanent structure is being built. Porches, sheds and extra rooms are common trailer additions that increase the impression of permanence

that may not have changed is ethnicity. Minorities make up 10% of mobile home households now, and Cowgill found 1 African-American in a sample of 93 households.

The downward social mobility of mobile homes undoubtedly has to do with their low cost relative to other housing. In 1990, the average new manufactured house cost $27,800, compared to $149,000 for the average new site-built house (O'Hare and O'Hare 1993). A used double-wide mobile home in Florida can be purchased for $12,000, and older and smaller models carry correspondingly lower prices. This has made them so attractive that there were 7.4 million occupied units in 1990; 1 in 16 Americans lives in a mobile home (about 15 million people). Half of them are in urban areas.

Whether urban or rural, the majority of mobile homes are in the "sunbelt" states of the South and the West. Florida, California, Texas, North Carolina and Georgia have the largest number (O'Hare and O'Hare 1993). However, 25 urban areas across the country each have more than 20,000 occupied mobile home units, and 8 have more than 50,000 each (Table 4.1); 4% of these are in central cities.

Table 4.1. *Cities with more than 50 000 occupied mobile homes each*

Occupied mobile homes: Metropolitan area	N
Los Angeles–Anaheim–Riverside, CA	190 154
Tampa–St Petersburg–Clearwater, FL	108 796
San Francisco–Oakland–San Jose, CA	64 417
Houston–Galveston–Brazoria, TX	58 588
Phoenix, AZ	56 437
Detroit–Ann Arbor, MI	54 928
Seattle–Tacoma, WA	54 603
Dallas–Ft Worth, TX	53 152

Source: O'Hare and O'Hare, 1993.

Factors affecting health

Geographic location is of course a factor that might affect health or biological status of mobile home dwellers, through the effects of climate or natural hazard. In addition, there are three other sets of factors that may have health effects. These are demographic factors, physical aspects of housing, and social and psychological factors related to the trailer park environment.

The demographic composition of any population is related to its pattern of health risks. Taken as a whole, the trailer-dwelling population of the US has a similar age profile to that of the general population. While trailer parks are often thought of as a refuge for the elderly, and some are indeed limited to that group, less than 23% of total mobile home residents are aged 65 and older (Knox 1993). In fact, 45% of residents are between 25 and 44 years of age, and the median age is 44. A third of mobile home households are headed by someone between the ages of 18 and 34, and 37% of the total have children under the age of 18 (O'Hare and O'Hare 1993). Note that these census figures are for the entire population of mobile home dwellers. We have no information on the age profile of urban trailer residents, nor on the portion of that population living in trailer parks. If these subpopulations differ in age or sex composition, we would expect different patterns of health risk.

Probably the most widely recognised health problems for mobile home dwellers are related to physical aspects of their housing. The main hazards are air pollution, fire and wind damage. There is a considerable body of epidemiological research on these factors, and legislation has attempted to address the problem, but the relative health risk for trailer residents remains high.

Air pollution in mobile homes is primarily a problem of formeldehyde exposure, but nitrogen dioxide is also a concern. Formaldehyde (HCHO) is an irritant chemical that also causes hypersensitivity reactions, with human subjects reporting eye irritation, respiratory problems, headache, dizziness, nausea and dermatitis on exposure to varying concentrations (Liu *et al.* 1991). The main source of HCHO in residential environments is plywood and particle board, both widely used in the construction of mobile homes. Because of their air-tight structure and relatively small volume, mobile homes are likely to have higher concentrations of pollutants than other housing types. In a large study of mobile homes in California, indoor formaldehyde levels were related to an array of physical symptoms, including burning eyes and skin, fatigue, sleeping problems, dizzyness, chest pain and sore throat (Liu *et al.* 1991). The effect held even when controlling for age, sex, smoking status and chronic respiratory or allergy problems. Persons with chronic respiratory problems had an elevated response to formaldehyde exposure, even at low concentrations. Formaldehyde may have multiple effects on health, as it has also been shown to lower T-lymphocyte concentrations and blastogenesis (Thrasher *et al.* 1987). Nitrogen dioxide levels in mobile homes have also been found to be high enough to cause respiratory problems, with mobile homes using gas cooking fuels having significantly higher levels of air pollution (Petreas *et al.* 1988). While fuel type was the most significant explanatory factor in nitrogen dioxide levels, differences in indoor levels were also associated with total dwelling volume and location in the Los Angeles basin.

Mobile homes have long had a reputation for elevated fire risk, having twice the rate of fire deaths of all other types of homes combined (Zuckman 1990). Not only do mobile homes have a high proportion of flammable materials, but the structure is often smaller and the fire builds faster. A recent study in North Carolina found that fires in mobile homes carried 1.7 times the risk of fatality, and that risk was higher for children under 5 years of age and for persons with disabilities (Runyan *et al.* 1992). While smoke detectors usually decrease risk of death, in the mobile home population they had little protective effect. Similar results have been found in other studies; in New Mexico, children in mobile homes had triple the fire mortality of those in houses or apartments, despite the fact that smoke detectors have been required equipment in new homes for at least a decade (Parker *et al.* 1993). Age of the home was also correlated with probability of death, indicating that older trailer homes are more dangerous. Smoking and alcohol consumption increase the risk of fire-related death in all types of housing, including mobile homes (Mobley *et al.* 1994).

Wind damage is frequent in mobile homes, with tornadoes and hurricanes providing the most dramatic examples. In 1992, Hurricane Andrew destroyed or damaged 11,500 mobile homes in Florida alone (*Morbidity and Mortality Weekly Report* 1993; Smith 1993). Epidemiological studies of tornadoes have indicated that mobile home residents are 22–27 times more likely to be killed or seriously injured by such an event than are residents in other types of housing (Carter *et al.* 1989; Eidson *et al.* 1990). The main cause of death is head injury sustained when the person or the dwelling becomes airborne and strikes another object. Improper anchorage and the loss of floors and walls are associated factors. New standards for wind-resistant construction and increased anchorage have been put in place in parts of the country where tornadoes or hurricanes are frequent, but the standards for anchorage have been difficult to enforce, and often do not apply to older units. Some communities in the tornado belt of the southern US have provided occupants of trailer parks with tornado shelters, but this is far from universal (Carter *et al.* 1989).

While the structure of the housing itself poses risks for mobile home dwellers, there are other factors which may affect health. We suggest that factors in the social environment of trailer parks may be stressful in themselves, and should be examined as part of the overall environmental characterisation. There are three issues here that we wish to explore: zoning, land ownership and rootedness.

The popular image of trailer parks as havens for the unemployed, with weedy yards and rusting junk cars, affects the zoning for mobile homes (Knox 1993; Wallis 1991). Many communities ban mobile homes outright, even on single lots. Others have limits for the length of time that a mobile home may be parked in a given place. In urban areas, mobile home communities are generally zoned "commercial-industrial," so that they end up flanked by gravel pits and railroad tracks. As one writer described the reasoning: "mobile home parks are not very attractive and these are not very nice people. So you approve low development standards and the whole thing is circular" (Knox 1993 : 64).

Land ownership presents related problems for mobile home dwellers. While many rural mobile home residents live on their own lots, urban ones frequently do not. More than half of all mobile home owners live on someone else's land, and few states regulate the relationship between resident and landlord (Knox 1993; Wallis 1991). Absentee landlords are common, so that there is little incentive to invest in maintenance of the property. Conversely, there are few regulations that restrain landlords from evicting their trailer tenants. Twenty-two states have not defined what constitutes "good cause" for eviction from a trailer park. This means that there is a large grey area in terms of acceptable conduct, but also that tenants have no security. Being evicted from a trailer park means moving your house (assuming that you own it), a process that can cost between $2,500 and $10,000 (Knox 1993; Wallis 1991).

Insecurity of tenure is one of the barriers to a sense of permanence in any type of housing. For mobile home residents, this impermanence may be reinforced by the negative social image of the places where they live. Social mobility is one of the basic myths of American culture, and having the option of improved future housing has been shown to be important to mobile home residents (Paulus *et al*. 1991). But transience or impermanence may be important in other ways. Wenkart (1961) has shown that attachment to place is important in forming self-identity, and Godkin (1980) has argued that "uprootedness" is a critical factor in the psychiatric histories of alcoholic patients. Buttimer (1980) further argues that a sense of "centeredness," of having a home, is essential for well-being. She starts from the premise that all humans need a home base and an environment of reach beyond that, by which they gain access to other resources. She expands this to say that cultures vary in their emphasis on either "home" or "reach," and that this emphasis can be read from the landscape. Modern American cities, she argues, are landscapes devoted to "reach," dominated by roads, rail lines, highways, airports, factories and communications technology. Using this argument, trailer parks epitomise the culture of "reach," with even home designed to be mobile. The psychological impact of this cultural homelessness has received little attention (Hager 1954; Wallis 1991).

Trailer parks in Walla Walla

We are interested in trailer parks as social spaces and as environments with potential health impact. We have only just begun to explore these issues, by interviewing residents of four mobile home communities in the vicinity of Walla Walla, Washington. Walla Walla is a city of about 35,000, connected to surrounding towns by commercial strip developments. The urban population of the region is about 50,000, with an economy based on agribusiness and the state prison. There are a number of mobile home communities within the environs of Walla Walla, and the four we have picked show the range of physical and socioeconomic conditions. Here we present data on the physical environment of the communities, and the demography, employment, health and housing experience of the residents. The data we present are preliminary and descriptive, but show some distinctive features of this environment.

Two of our study communities are located within Walla Walla, both in areas of commercial strip development. Trailer Harbor is a small, privately owned trailer park wedged behind the owner-landlord's old frame house on the east side of Walla Walla. The property fronts on a busy street which is filled with fast-food restaurants, discount stores, small businesses and parking lots. Twenty-six trailers form a *U* around the perimeter of the rectangular lot. There is a paved drive and a small grassy area in the middle, with a common bathhouse and laundry building. A grocery store and other conveniences are within easy walking distance. The dwellings range from fairly new recreational vehicles and well-kept older mobile homes to derelict structures that would collapse if moved.

Royal Trailer Court is a large property on the same side of the city, located between a shopping center and a state highway and railroad right-of-way. This property is owned by a real estate company and managed by a hired person living on the site. There are about 90 trailers currently occupied. The access road is unpaved and poorly maintained, and the residents hold it up as proof that the city discriminates against the poor. The park is laid out like a suburban development, but there is a social division, with more Anglo-American residents and trailer owners toward the front and more Hispanics and renters toward the back. This is a source of local tension, with young Hispanics blamed for noise and petty crime. There is no real common space, and the bath and laundry building is a graffiti-covered hull at the side of the park. The quality of the trailers themselves varies widely, and many have porches or other extensions built on.

Cottonwood Trailer Park is a small, privately owned park outside the city, set just off a main road lined with junk shops, abandoned businesses, produce stands and mobile home dealers. The owner of the park lives in a trailer and oversees an adjacent horse farm. A muddy, pitted road runs down the center, literally lined with cars. There are no trees, and the backs of the trailers can be seen from the main road. There are about 40 trailers, some of which have small gardens.

Pelican Place is a cooperative trailer park in a satellite town, where 75% of the residents have purchased both their trailers and the lots on which they sit. The area is mixed commercial and residential, a type of urban sprawl with no identifiable center or "downtown" area. There are 150 lots in the park, with wide streets, trees and large lots. Trailers look well cared for and there is a common area with a playground, picnic tables and a bath and laundry facility.

In this exploratory phase of our research we have interviewed 50 households, 31 from the two communities in Walla Walla and 19 from the parks outside. While this is not a random sample, we have purposely included both owners and renters, as well as trailers of different sizes, ages and states of repair, in order to capture the variability in the environment. In one of the locations, every trailer was visited and all residents at home during three successive visits were interviewed. In a second location, we proceeded house-by-house from the front of the park, visiting as many of the households as could be located and interviewed in a single visit. In the other two

Table 4.2. *Employment status of trailer park residents sampled*[a]

Employment	Self	Second earner
FT/Service–Industrial	17	13
Unemployed	11	4
PT/Service–Industrial	9	3
Retired	7	1
Multiple	5	1
FT/Professional	1	1
PT/Professional	0	1
Informal	0	1
Pick-up work	0	3
Total	50	28

[a] PT = part-time employment, FT = full-time employment.

parks, there were maps at the entrance that listed the lots by number. We divided the lots among our three interviewers (C.B. and two students), so that each had a section of the park. Each interviewer started from a randomly chosen dwelling and talked to as many of the residents in her section as she could locate during a single visit.

There are minor differences among the three locations in terms of number of residents per household, education and median income, with higher levels of college educated residents and slightly higher incomes in the cooperatively owned community. On average, our sample households have 2.7 residents each, with a range from 1 to 11. There are 1.16 wage earners per household, and median household income is between $15,000 and $20,000; 50% have no residents with any college education; 82% of the sample are Anglo-American and 14% are Hispanic, with 2% each of Asian Americans and Native Americans.

Of our respondents, 36% were not currently working, either unemployed or retired (Table 4.2). The majority of the sample worked in the service or industrial sectors, either full time or part time. Only six of the wage earners reported multiple occupations at the time of the interview, but work histories show a striking pattern of occupational instability. One woman reported that she was currently a housewife, but had previously worked selling Avon products (cosmetics and the like) door-to-door, was a live-in housekeeper and did childcare, packed cherries, labeled plums, trained hops (tied growing hops vines to trellises), worked at restaurants and bars, had done filing on a local military base, and had worked for a microfilm company and at the county auditor's office. Another man reported that he was currently unemployed, but had previously worked as a maintenance man for a real estate firm, and at different times in his life had done photo finishing, managed a restaurant, worked for the city, driven a truck and investigated accidents for the state. This occupational multiplicity is even more striking at the household level. One typical household reported that the wife works part time in the evenings tending bar, but also takes in sewing and does odd jobs. Previously she worked in food preparation, at a bakery and in a restaurant. Her husband, a former mechanic, manages a photo processing shop. In addition, twenty-six respondents commented on their future job plans. When we compared these to their work histories, 35% planned to stay in the same line of work, while 65% planned to take up yet another type of job in the future.

Occupational instability in this group is undoubtedly a factor in the low-income levels seen. In addition, the types of jobs taken and the shifting from job to job mean that few of the sample households can count on social benefits such as retirement pensions or health insurance. Health insurance was a particular concern for this population, with only 28% reporting full insurance coverage for either themselves or their families, and 22% reliant on state or federal health care subsidies.

We asked our respondents to recall any illnesses experienced by them or their families in the past month, and any major medical events last year.[1] Twenty-one households reported illness during the last month, with respiratory infections and accidents or injuries being the most common complaints (Table 4.3). Pregnancy and chronic disease were the chief concerns during the last year. Use of health care facilities was relatively high: 48% of our respondents had had some type of medical check-up in the past six months, and a total of 80% had done so in the past two years. Other research has shown that the urban poor rely disproportionately on hospital emergency rooms for routine health care (Schappert 1997). However, in this sample, only 18% routinely sought emergency room care (Table 4.4). The largest single source of care was the Veterans' Administration, with private doctors second; these two sources were used habitually by 58% of households.

Because of the associated health risks, we asked our sample about the prevalence of smoking in these households. Of households interviewed, 66% contained one or more persons who smoked. Those households contained a total of 107 people, thus exposing 73% of our sample, either actively or passively, to tobacco smoke; 30 of these (28%) were children aged 10 or less, and another 18 (17%) were between 11 and 20 years of age. We do not have comparable data for the population as a whole, however, 27% of Americans reported themselves as smokers in 1991, down from 44% in 1964 (American Heart Association 1997). Given the air-tight nature of mobile homes and the interaction of tobacco smoke with other air pollutants, the level of smoking reported here could present a significant health hazard, especially for children. Smoking also increases the

Table 4.3. *Illnesses experienced by sample households (*N = 50*) during the past month and last year**

Reported illness	Past month		Last year	
	N	(%)	N	(%)
Respiratory	7	33	2	10
Other infectious	1	5	1	5
Accident/injury	4	19	2	10
Heart/circulation	2	10	2	10
Cancer	2	10	2	10
Other chronic	3	14	4	19
Pregnancy/childbirth	2	10	6	28
No. of illnesses reported	21		19	

* See endnote.

Table 4.4. *Routine or habitual sources of medical care for sample households*

Usual source of care	N	(%)
Private doctor	18	36
Veteran's administration	11	22
Hospital emergency room	9	18
Family Medical Center	7	14
Public Clinic	2	4
Not sure	3	6

already elevated risk of fire for mobile home dwellers.

Since we were interested in the question of rootedness or permanence, we asked residents about their housing history, perceptions and goals: 20% had lived in their current residence for 1 year or less, and median length of stay was 2 years; 62% were unsure how long they would continue to live in their current place, and 26% felt they would only stay for another 2 years or less. Comments on quality of life in their current residence were both positive and negative. The best features of the trailer communities were thought to be friendliness and convenience, while crime, noise and drug trafficking were the negative aspects. Although 32% had lived in a trailer as their previous residence, only 20% expected to live in one as their next residence. The desire to own a home in the future was clearly expressed, as 54% expected to live in either their own trailer, house or condominium in the future.

The desire for ownership and for some more permanent type of housing came across even more clearly when we asked respondents to describe their ideal home. A simple content analysis of their responses shows the words used most frequently to describe their ideal living situation (Table 4.5).

Table 4.5. *Words used to describe respondent's ideal home: all words used by more than 1 respondent*

Description of residents' ideal home

Words used	No. times
House	24
Big	17
Bedrooms	15
Yard	11
Country/rural	7
Safe	5
Mountains	4
Farm/ranch	4
Cabin	3
Kitchen	3
Basement	2
Warm	2
Clean	2

"House" was used 24 times, "big" 17 times, "bedrooms" (usually 2 or 3) 15 times and "yard" 11 times. Other major concerns were that it be safe, warm or clean. One of the most striking features of their answers was the desire for rural residence. Country, rural, farm, ranch and land were mentioned a total of 12 times. It seems clear that this group of mobile home dwellers aspire to a very different situation from the one in which they currently live.

Directions for future research

The data presented here suggest a number of distinctive features that characterise trailer parks as urban microenvironments. Probably the most notable feature is the nature of the dwellings themselves. The history of trailers as mobile units, only recently completely immobilised, differentiates them from other more permanent types of housing. Their insubstantial construction, necessary to a light-weight mobile dwelling, means that their life-span is limited. It also makes them subject to fire and wind damage, and contributes to indoor air pollution. But the small size and cheap materials also make them inexpensive, and thus attractive to people with few economic resources. The limited economic options of mobile home residents contribute, in turn, to the social environment of trailer parks.

Poverty is a distinctive feature of trailer park residents as a group. But there are important factors that separate them from the other urban poor. One obvious factor is location. The zoning of trailer parks means that the majority of urban trailers end up on the periphery of cities, in commercial strip developments or areas of interurban sprawl. Another factor is ethnicity; trailer residents are overwhelmingly white, while the inner-city poor are predominantly African-American. Just these two factors have a profound effect on the options available for transportation, health care, employment and education. A third distinctive feature is the rural orientation of mobile home dwellers, in many cases despite their own urban origins.

A yearning for rural life can be seen as part of an ideology of impermanence, similar to our experience in periurban areas of Africa. For trailer park residents, the focus on an imagined rural future may be a coping strategy, a denial that the present is permanent. Two other aspects of our data speak to this point: the overwhelming emphasis on having one's own house, and the instability of employment in our sample. In both housing and jobs, there is a clear mythology that we are only here for now, until we "make it" and move on.

As researchers, our next step is to look at three aspects of mobile home life. First, we need to characterise in greater detail the stressors and resources, both experienced and perceived, for trailer park residents in a variety of geographic regions. Second, we need to examine the coping strategies used by residents to gain access to resources and to avoid or mitigate hazards. An ideology of impermanence may select for flexibility, and the occupational multiplicity that we see may be part of that pattern. Finally, we need to look at health outcomes, in terms of nutritional status, child growth, blood pressure, immune function and other markers of biological status.

Trailer parks may seem initially to be an exotic locus for human biological research. We would argue, however, that this work fits comfortably into the tradition of human adaptability studies. The tripartite research agenda given here mirrors the strategy set out by the International Biological Programme for its classic work in human adaptability (Weiner 1966). While human biology and human adaptability researchers are now shifting their focus, including more behavioral data and novel environments, we are still concerned with explaining human variation. Whether in the Andes or the Arctic or the periurban slums of the developing world, we are still concerned with how human populations respond to environmental challenges. We continue to see human biology as reflecting the varying capacity of different populations to cope with environmental hazards and opportunities. The linking of transitional environments, such as trailer parks, to human health outcomes can help us to understand the biological signficance of the urban experience in both Third and First World populations.

Note

It is difficult to measure illness frequency retrospectively, due to problems of recall, self-perception, definitions of illness, etc. Therefore, while we asked about all illnesses experienced by the household last month, we restricted our data on last *year* to major illnesses or medical events. Thus figures for the two time periods (Table 4.3) reflect different types of illness experience.

References

American Heart Association (1997). Web Site (http://www.amhrt.org/hs96/cig-stats.html).

Behrman, C. and Huss-Ashmore, R. (1994). Peri-urban nutrition: diet and energy stores in Swazi women. *American Journal of Physical Anthropology*, Supplement **18**, 54–55.

Burton, I., Kates, R. W. and White, G. F. (1978). *The Environment as Hazard.* New York: Oxford Unitersity Press.

Buttimer, A. (1980). Home, reach, and the sense of place. In *The Human Experience of Space and Place*, eds. A. Buttimer and D. Seamon, pp. 166–87. London: Croom Helm.

Carter, A. O., Millson, M. E. and Allen, D. E. (1989). Epidemiologic study of deaths and injuries due to tornadoes. *American Journal of Epidemiology*, **130**(6), 1209–17.

Cowgill, D. O. (1941). *Mobile Homes.* Philadelphia: University of Pennsylvania Press.

Eidson, M., Lybarger, J. A., Parsons, J. E., MacCormack, J. N. and Freeman, J.I. (1990). Risk factors for tornado injuries. *International Journal of Epidemiology*, **19**(4), 1051–6.

Godkin, M. A. (1980). Identity and place: clinical applications based on notions of rootedness and uprootedness. In *The Human Experience of Space and Place*, eds. A. Buttimer and D. Seamon, pp. 73–85. London: Croom Helm.

Hager, D. (1954). Trailer towns and community conflict in Lower Bucks County. *Social Problems*, **1**(1), 34–41.

Harpham, T., Lusty, T. and Vaughan, P. (1988). *In the Shadow of the City.* Oxford: Oxford University Press.

Huss-Ashmore, R. (1997). Human adaptability research among African agriculturalists. In *Human Adaptability: Past, Present and Future*, eds. S.J. Ulijaszek and R. Huss-Ashmore, pp. 61–81. Oxford: Oxford University Press.

Huss-Ashmore, R. and Goodman, J. L. (1989). Seasonality of work, weight, and body composition for women in highland Lesotho. In *Coping with Seasonal Constraints*, ed. R. Huss-Ashmore with J .J. Curry and R. K. Hitchcock, pp. 28–44. MASCA Research Papers in Science and Archaeology, vol. 5. MASCA, University Museum, University of Pennsylvania, Philadelphia.

Knox, M. L. (1993). Home sweet mobile home. *Mother Jones*, Jan. Feb, 62–5.

Liu, K-S., Huang, F-Y., Hayward, S. B., Wesolowski, J. and Sexton, K. (1991). Irritant effects of formaldehyde exposure in mobile homes. *Environmental Health Perspectives*, **94**, 91–4.

Mobley, C., Sugarman, J. R., Deam, C. and Giles, L. (1994). Prevalence of risk factors for residential fire and burn injuries in an American Indian community. *Public Health Reports*, **109**(5), 702–5.

MMWR (1993). Comprehensive assessment of health needs 2 months after Hurricane Andrew–Dade County, Florida 1992. *Morbidity and Mortality Weekly Report*, **42**(22), 435–7.

O'Hare, W. and O'Hare, B. C. (1993). Upward mobility. *American Demographics*, January, 26–33.

Parker, D. J., Sklar, D. P., Tandberg, D., Hauswald, M. and Zumwalt, R. E. (1993). Fire fatalities among New Mexico children. *Annals of Emergency Medicine*, **22**, 517–22.

Paulus, P. B., Nagar, D. and Camacho, L. M. (1991). Environmental and psychological factors in reactions to apartments and mobile homes. *Journal of Environmental Psychology*, **11**, 143–61.

Petreas, M., Liu, K-S., Chang, B-H., Hayward, S. B. and Sexton, K. (1988). A survey of nitrogen dioxide levels measured inside mobile homes. *Journal of the Air Pollution Control Association*, **38**, 647–51.

Runyan, C. W., Bangdiwala, S. I., Linzer, M. A., Sacks, J. J. and Butts, J. (1992). Risk factors for fatal residential fires. *New England Journal of Medicine*, **327**(12), 859–88.

Sack, K. (1997). Mobile homes go upscale: one pet per plot, please. *New York Times*, February 23, E3–E4.

Schappert, S. M. (1997). Ambulatory care visits to physicians offices, hospital outpatient departments, and emergency departments: United States 1995. *Vital and Health Statistics*, series 13, no. 129. Hyattsville, MD: National Center for Health Statistics.

Slagle, E. M. (1937). We have lived in trailers since 1919. *Trailer Travel*, II (4), 20.

Smith, A. K. (1993). Hardly a trailer park. *US News & World Report* April 5, 77–79.

Thomas, R. B. (1997). Wandering toward the edge of adaptability: adjustments of Andean people to change. In *Human Adapability: Past, Present, and Future*, eds. S.J. Ulijaszek and R. Huss-Ashmore, pp. 183–232. Oxford: Oxford University Press.

Thornburg, D. A. (1991). *Galloping Bungalows: The Rise and Demise of the American House Trailer*. Hamden, CT: Archon Books.

Thrasher, J. D., Wojdani, A., Cheung, G. and Heuser, G. (1987). Evidence for formaldehyde antibodies and altered cellular immunity in subjects exposed to formaldehyde in mobile homes. *Archives of Environmental Health*, **42**(6), 347–50.

Wallis, A. D. (1991). *Wheel Estate*. New York: Oxford University Press.

Weiner, J. S. (1966). Major problems in human population biology. In *The Biology of Human Adaptability*, eds. P. T. Baker and J. S. Weiner, pp. 1–24. Oxford: Clarendon Press.

Wenkart, A. (1961). Regaining identity through relatedness. *American Journal of Psychoanalysis*, **21**, 227–33.

Zuckman, J. (1990). New fight likely over change in mobile home standards. *Congressional Quarterly*, June 2, 1732.

Part II
Epidemiology

5 Emerging infectious diseases: biology and behavior in the inner city

GEORGE DIFERDINANDO

Editors' introduction

In McMichael's account of urban change and modern development (chapter 2) infectious disease is seen to recede in importance as modern public health engineering and regulation reduced exposure to infectious agents in water and food, and the food supply itself became more plentiful and healthy. The recent rise of infectious diseases in urban settings is addressed by DiFerdinando who lucidly demonstrates changes in the frequency and distribution of the two most prominent infections, tuberculosis (TB) and human immunodeficiency virus (HIV). Changes in the frequency of these diseases are related to characteristics of urban populations, particularly urban social relations. Both diseases have been associated throughout their histories with urban populations, but during the emergence of HIV and the re-emergence of TB, each was originally associated with different social segments of urban populations. An important change occurred when interaction between the two segments increased. One of the themes expressed in several chapters of this volume is that urban populations inhabit overlapping microenvironments. In urban populations this overlap has changed the risk factors for the diseases, and changed the notion of their causes. One hundred years ago many urban residents were exposed to TB, and the presence of disease depended on variation in susceptibility rather than exposure. People defended themselves against infection, and the disease was caused by susceptibility. Now, in HIV affected populations, TB is 'caused' by exposure to the infective agent since HIV increases susceptibility so greatly.

The causative agent of tuberculosis disease (TB), Mycobacterium tuberculosis (M.tb), has undergone changes in its relationship with genetic and environmental factors over the past 50 years in urban centers around the world (DeCock 1996) and in the United States in particular (CDC 1996; Frieden et al. 1993; DeCock 1996; Snider 1997). Understanding the nature of these relationships in New York can inform the efforts in TB control elsewhere, and offer insights into changing disease ecologies in urban environments. The changing nature of M.tb is due to a number of factors, including: (1) changes in the control of TB in society at large (Brudney and Dobkin 1991); (2) social and behavioral changes in the urban environment that have created a slow and sustained expansion of populations at risk of TB (Spense *et al.* 1993); and (3) the exposure of at-risk populations to new cofactors for both M.tb exposure and TB disease (Markowitz *et al.* 1997), which are linked to behaviorally transmitted infectious diseases, especially HIV infection (Selwyn *et al.* 1992). These factors have turned the slow expansion of the population at risk of TB in urban environments in New York into one of dramatic increase (Edlin *et al.* 1992). In this chapter, change in TB occurrence and character are examined in the context of urban phenomena, including HIV infection.

Tuberculosis history within urban environments

The risk factors for infectious disease in general and for M.tb in particular are subject to change. Current risk factors for M.tb include urban crowding, as well as different urban physical environments where there is variation in both degree of ventilation, and likelihood of exposure (Ochs 1962; Houk *et al.* 1968). Human behavior strongly influences the nature and degree of M.tb transmission. Furthermore, racial and ethnic differences are associated with variation in individual ability to become infected if exposed, and of developing active TB once infected (Stead *et al.* 1990), although the causal basis of the associations are unproven. Variation in socioeconomic status is another factor that modifies the risk for many infectious diseases (Terris 1948), including M.tb. Many risk factors, including differences between urban and rural settings, between different urban physical environments, and between different racial, ethnic and behavioral groups, manifest themselves differentially between different socioeconomic groups. In general, members of higher socioeconomic groups live in larger physical space in either apartments or houses; this decreases contact with large numbers of other people on an intimate basis. In addition, they have greater access to various types of health care and prevention, and better nutrition and improved immunity to infection as a consequence. These factors can mediate either possible exposure to M.tb, or the development of active TB, if exposed and infected.

Different occupational groups and subgroups have different levels of risk of exposure associated with them. In the past, one way in which TB was thought to be associated with occupational exposure was the level of stress the job induced; thus the difficulty or strenuousness of the job, or the degree of intellectual or emotional stress experienced, was believed to be related to the development of active TB (Hawkins *et al.* 1957). Although this idea has never disappeared, over the years the nature of occupational exposure in relation to TB risk has moved away from stress and towards exposure (Chilress 1951). Thus, disease controllers examine how occupational categories are related to likelihood of M.tb exposure, rather than the likelihood of developing TB disease if exposed and infected. The earlier, stress-based idea presumed that all of the population was infected with M.tb; if this were so, then the variables which were susceptible to public health control were those related to disease development, and not exposure. As the incidence of TB declined, exposure to M.tb became more important as a controllable public health measure, since stress was then thought to act out within certain occupational domains, and not across all of them.

The distribution of TB cases in New York State (Figure 5.1) varies markedly across counties, with some having no cases of reported illness, many having very few cases, and the larger urban centers (Nassau, Suffolk, the five boroughs of New York City (NYC), Westchester, Erie, Monroe County, Onondaga) having larger numbers of cases (New York State Department of Health 1996). There is substantial variation in the number of TB cases according to NYC metropolitan area, and suburban and rural areas of New York. The rate and the incidence of TB disease reached its lowest point in 1978 after decades of traditional control activity (Table 5.1), after a dramatic decrease in the number of TB disease cases in both NYC and New York State (NYS) between 1955 and 1978. The decrease in the number of deaths due to TB was even more dramatic, because TB treatment was more effective than ever before. However, after this time, TB rates increased again in NYC and the rest of NYS (Figure 5.2).

One of the striking differences between TB in the United States between the late 1940s and the early 1980s was the spread of the disease into new population groups that either spread sexually transmitted diseases, or were indirectly affected by the sexual transmission of diseases, especially newborns and children under the age of 5 years (Table 5.2).

Racial and ethnic differences

The incidence of TB among different racial groups varies dramatically, regardless of location in NYS (Figures 5.3, 5.4). Rates are higher in non-white populations and in Hispanic white groups than among whites, while among black non-Hispanics, the rate in NYC is almost double that of the rest of NYS. In NYC, there is a 6-fold difference in TB rates between white-non-Hispanics and black-non-Hispanics, while in upstate New

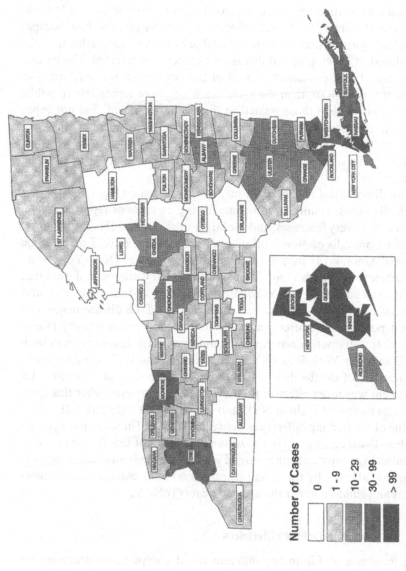

Figure 5.1 Distribution of tuberculosis cases, New York, 1996.
Source: New York State Department of Health, Bureau of Tuberculosis Control

Number of Cases

0
1 - 9
10 - 29
30 - 99
> 99

Table 5.1. *Tuberculosis cases 1940–1996*

Year	New York State (exclusive of New York City)		New York City		New York State (Total)	
	N	Rate per 100,000	*N*	Rate per 100,000	*N*	Rate per 100,000
1940	5,158	85.6	9,005	120.8	14,163	105.1
1945	4,768	74.4	7,062	91.9	11,830	83.9
1950	4,776	68.8	7,717	97.8	12,493	84.2
1955	3,502	43.6	6,214	79.2	9,716	61.2
1960	2,376	26.4	4,699	60.4	7,075	42.2
1961	2,052	22.3	4,360	56.3	6,412	37.8
1962	2,005	21.4	4,437	56.7	6,442	37.5
1963	1,865	19.6	4,891	61.7	6,756	38.7
1964	1,715	17.8	4,207	52.7	5,922	33.6
1965	1,627	16.6	4,242	53.0	5,869	33.0
1966	1,633	16.5	3,663	45.7	5.296	29.5
1967	1,527	15.2	3,542	44.4	5,069	28.1
1968	1,475	14.5	3,224	40.5	4,699	25.9
1969	1,384	13.5	2,951	37.4	4,335	23.9
1970	1,275	12.3	2,590	32.8	3,865	21.2
1971	1,180	11.3	2,572	32.5	3,752	20.4
1972	1,176	11.2	2,275	29.0	3,451	18.8
1973	1,009	9.6	2,101	27.4	3,110	17.1
1974	844	8.1	2,022	26.6	2,866	15.9
1975	1,041	9.9	2,893	38.6	3,934	21.8
1976	916	8.7	2,156	29.0	3,072	17.1
1977	829	7.9	1,605	22.0	2,434	13.6
1978	753	7.1	1,307	18.2	2,060	11.6
1979	699	6.6	1,530	21.5	2,229	12.6
1980	780	7.4	1,514	21.4	2,294	13.1
1981	641	6.1	1,582	22.4	2,223	12.7
1982	674	6.4	1,594	22.5	2,268	12.9
1983	658	6.2	1,651	23.1	2,309	13.1
1984	616	5.8	1,630	22.6	2,246	12.7
1985	638	6.0	1,843	25.5	2,481	13.9
1986	615	5.8	2,223	30.6	2,838	15.9
1987	615	5.8	2,197	30.1	2,812	15.7
1988	688	6.5	2,317	31.8	3,005	16.8
1989	657	6.2	2,545	34.8	3,202	17.8
1990	656	6.1	3,520	48.1	4,176	23.2
1991	748	7.0	3,673	50.2	4,421	24.6
1992	763	7.2	3,811	52.0	4,574	25.4
1993	717	6.7	3,235	44.2	3,952	22.0
1994	641	6.0	2,995	40.9	3,636	20.2
1995	621	5.8	2,445	33.4	3,066	17.0
1996	535	5.0	2,053	28.0	2,588	14.4

Figures after 1974 include reactivated cases.
Source: New York State Department of Health Bureau of Tuberculosis Control.

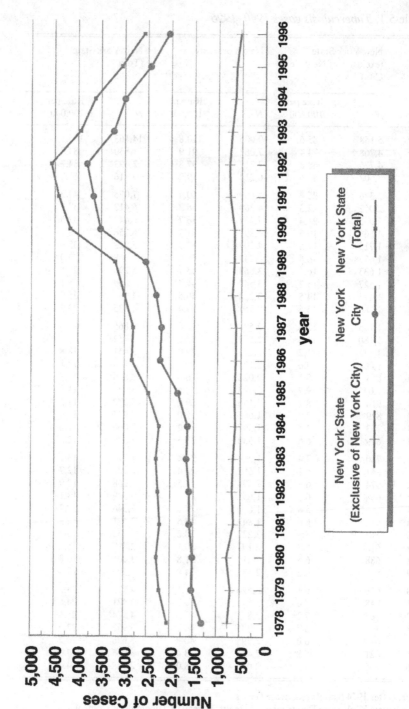

Figure 5.2   Incidence of tuberculosis in New York City, New York State (exclusive of New York City) and New York State, 1978–1966.
Source: New York State Department of Health, Bureau of Tuberculosis Control

Table 5.2. *Tuberculosis cases and rates by gender, age, and race/ethnicity, Yew York State, 1996*

Year	New York State (exclusive of New York City)		New York City		New York State (Total)	
	N	Rate per 100,000	N	Rate per 100,000	N	Rate per 100,000
Gender						
Males	324	6.3	1,269	36.9	1,593	18.5
Females	211	3.9	784	20.2	995	10.6
Age						
Under 5 years	9	1.2	55	10.4	64	5.0
5–9	9	1.3	30	6.6	39	3.3
10–14	8	1.2	14	3.1	22	1.9
15–19	20	2.6	58	12.1	78	6.3
20–24	36	4.3	136	23.1	172	12.1
25–34	91	5.2	417	30.3	508	16.2
35–44	110	6.9	527	47.8	637	23.5
45–54	68	6.0	349	45.8	417	22.0
55–64	46	4.6	195	32.3	241	14.8
65+	138	9.8	272	29.0	410	17.5
Race/ethicity						
White, non-Hispanic	190	2.0	220	6.9	410	3.3
Black, non-Hispanic	162	22.4	894	48.2	1,056	40.9
Hispanic	97	22.5	603	33.8	700	31.6
American Indian	2	5.6	0	0.0	2	3.0
Asian	84	46.0	336	63.8	420	59.2
Total	535	5.0	2,053	28.0	2,588	14.4

Source: New York State Department of Health Bureau of Tuberculosis control.

York, there is a 15-fold difference between black non-Hispanics and whites. In non-urban NYS, TB appears to have been eliminated.

HIV

While most persons in NYS with TB do not have a reported HIV test, the absolute impact of HIV coinfection is well documented. For example, of the 2,000 TB cases in NYC, two-thirds have known HIV status; half of them are HIV positive. This has an impact on the ability to control the disease, since rendering people non-infectious is more difficult if they suffer impaired immunity as a consequence of HIV infection (Figure 5.5).

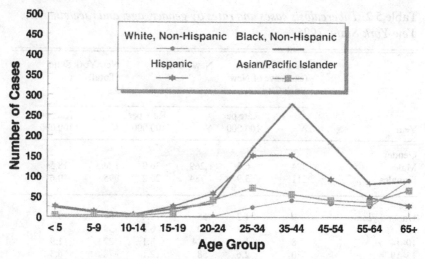

Figure 5.3   Tuberculosis cases (incidence) by age and race ethnicity, New York City, 1996.
Source: New York State Department of Health, Bureau of Tuberculosis Control

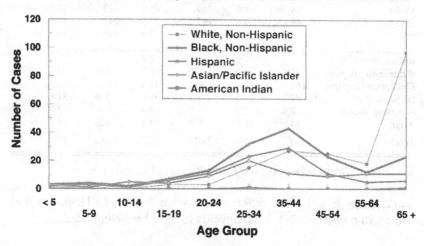

Figure 5.4   Tuberculosis cases (incidence) by age and race ethnicity, New York State (exclusive of New York City), 1996.
Source: New York State Department of Health, Bureau of Tuberculosis Control

The historical development of TB control systems in New York

The patterns of TB occurrence in urban and rural NYS and existing models of its causality have determined systems of control both past and present. Distinct phases can be discerned and described thus: (1) the sanatorium period prior to TB medications; (2) sanatorium plus anti-tuberculous medications period; (3) a period of outpatient treatment of

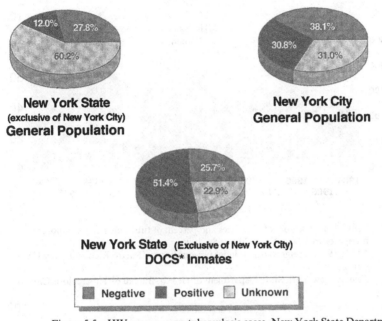

New York State
(exclusive of New York City)
General Population

New York City
General Population

New York State (Exclusive of New York City)
DOCS* Inmates

■ Negative ■ Positive ▨ Unknown

Figure 5.5   HIV cases among tuberculosis cases. New York State Department
of Correctional Services.
Source: New York State Department of Health, Bureau of Tuberculosis Control

drug sensitive disease; and (4) the present system of directly observed therapy (DOT) for multiple drug registrant (MDR) disease. The current system is the one used in the rest of the world, and will continue in the United States for the foreseeable future.

In the sanatorium period from about 1910 onward, the ubiquity of M.tb exposure led to the vast majority of the population being infected. Since exposure was essentially controlled for, other factors were deemed relevant in the risk of developing disease, if infected. While current knowledge suggests that the "inheritance" of TB disease was then largely due to intense exposure, at that time there was a belief in a vaguely delineated genetic predisposition to disease. A second risk factor considered important then was occupational stress, typified by high stress occupations versus low stress occupations. Other factors considered important were personality type, and life stress.

Prior to and during the sanatorium period, models of TB causation emphasized the influence of overcrowding, and immigration, which was thought to include the stress of changing social context and increased likelihood of disease if infected. In this view, living in a poor, stressful, unhygienic environment lowered resistance to disease. Even today, a high proportion of people who immigrate to the United States are infected with

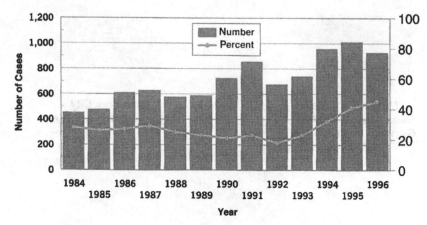

Figure 5.6   Number of new cases and percent of tuberculosis cases among foreign-born,* New York City, 1984–1996
*After 1990, foreign-born excludes persons born in Puerto Rico and other US territories.
Source: New York State Department of Health, Bureau of Tuberculosis Control

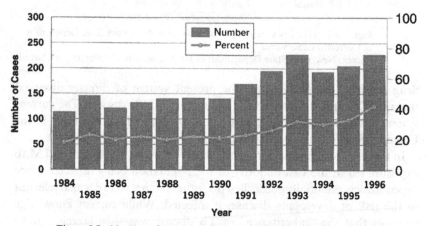

Figure 5.7   Number of new cases and percent of tuberculosis cases among foreign-born,* New York State (exclusive of New York City), 1984–1996
*foreign-born excludes persons born in Puerto Rico and other US territories.
Source: New York State Department of Health, Bureau of Tuberculosis Control

M.tb (CDC 1990), and many immigrants who develop active TB after immigration to the United States develop it within the first two to five years of their moving. This has been attributed by some to faulty screening of active cases at the time of immigration; however, it is seen by others as a manifestation of the stress of movement and dislocation. Most immigrants settle in urban areas initially; even in upstate New York, immigrants are more likely to live in urban rather than more rural areas (Figures 5.6, 5.7). During the sanatorium period when the models of disease determination emphasised stress rather than exposure, TB control involved the isolation of individuals from their stressful environment, placing them in a situation where they would be well fed, exposed to plenty of sunlight, and allowed calmness and relaxation (Caldwell 1988). Removal from the household had the beneficial effect of decreasing exposure to those still healthy and removing the stress of care giving.

Occupational risk was not believed to be related to exposure, but to the stress of the job. With a very high proportion of the general population infected, occupational exposure was not considered to add to risk. It had to be demonstrated, much to the surprise of some, that health care was an additive risk factor for disease (Badger and Ayvazian 1949). The response to this model of exposure and disease made occupational retraining important, especially during the Depression, for immediate and long-term stress relief. The idea was to retrain people to do less stressful jobs when they returned to the workforce. These behaviors motivated many social reformers to consider how occupational retraining might be done, usually in urban settings.

In the late 1940s and early 1950s, the first anti-TB drugs became available in New York and the rest of the United States (McDermott *et al.* 1947). These medications were added to the treatment environment of long-term hospitalisation, isolation, stress reduction and nutrition. Anti-TB drugs were added singly at first, and, when it was found that single drug treatment led to drug resistance (British Medical Research Council 1948), anti-TB cocktails of two and then three drugs were developed (Medical Research Council 1950; Johnston and Wildrick 1974). As this was a time of institutionalisation of patients in sanatoria, even after treatment, patients were kept in hospital for months after apparent "cure" (Tyrell 1956). These cures rapidly decreased the death rate, lowered the infectiousness of survivors, and slowed the spread of TB in the population. Spread of TB within institutions was minimal, since the rapid sterilisation of patients' sputa occurred when they were placed on effective medications. Programs established to retrain people and families to decrease stress were maintained during the 1950s and early 1960s, although they were increasingly redirected to post-hospitalisation medical care.

By the mid-1960s and early 1970s, the success of this treatment model of extended hospitalisation and drug treatment changed the response of society to TB disease (Arden House Conference on Tuberculosis Recommendations 1960). Economic forces drove the initially hesitant, then rapid, "deinstitutionalisation" of TB patients, with hospitalisation durations dropping from several months to just a few weeks (Yeager and Medinger

1986). These changes eventually produced a situation in which Multiple Drug Resistant TB (MDR-TB) could develop and spread (Reichman 1991; Dandoy and Elman 1972).

With deinstitutionalisation, the decrease in disease incidence was maintained through the 1960s and into the early 1970s. Some social workers were appointed as outpatient caseworkers, to assist patients with the rapid adjustment to a world that still often considered them dangerous and marked (Curry 1968; Sbarbaro 1970). However, even as the number of cases decreased, various types of fiscal crises occurred in New York City and State that undermined the new treatment paradigm. The greatest cost of the new control system was associated with the personnel involved with social follow-up of the outpatients. Thus, in the early and mid-1970s, outpatient follow-up of patients was cut back to marginal or non-existent levels (Brudney 1993).

People were now often left to their own devices in their treatment of a formerly deadly disease (Sumartojo 1993). Psychologically, the memory of the former deadliness of TB likely worked, at first, to motivate patients to maintain their medication. The system continued to run, with TB control being maintained by the early period of use of medication either in the hospital or in the few weeks after hospitalisation. Hospitalisation continued to be of a 6 to 8 week period, in which about 80% of cases were cured; the post-hospital therapy which continued for another 4 to 10 months was to cure the remaining cases (Iseman and Sbarbaro 1991).

Urban microenvironments, TB risk and sexually transmitted diseases

As TB was being deinstitutionalised and individuals were being made responsible for the maintenance of their own treatment, different urban disease microenvironments merged. This was to radically change the dynamics of TB disease treatment, and would, in turn, encourage drug resistance to develop, M.tb infection to spread, and the persons to whom it was spread to be uniquely susceptible to infection and other disease (Bloom and Murray 1992; Selwyn 1993; Hansel 1993). The overlap of people with TB and people at high risk of sexually transmitted diseases (STD) increased during this period, and this led to the change of the genotype of M.tb and how it interacted with urban human life.

As TB disease had been better treated and controlled within society between the 1940s and 1960s, rates had deceased, but had become concentrated in specific communities. These included groups characterised by individuals with less access to health care, less healthful living styles, as well as being more likely to have migrated from other countries, and more likely to be members of marginalised racial and ethnic groups. The once universal risk of infection became concentrated, based on the rate of disease within subpopulations, and likelihood of an individual sharing a confined space with a diseased person (Reider *et al.* 1989). Racial differences in TB rate had always existed, and in earlier times these were explained by malnutrition, and by ill-defined notions of racial predisposition. However, both racial and socioeconomic differences accelerated with the increased TB rates of recent times (New York City Department of Health, Bureau of Tuberculosis 1972, 1993). Disease spread increased as poverty became more concentrated, and some behavioral consequences of that poverty further facilitated disease transmission. With this, the selective pressure on M.tb mutations for resistance and decreased response to treatment increased; and this increased, in turn, the likelihood of continued spread of infection.

In many urban areas, TB took on the characteristics of STD in terms of risk and populations at risk. The "typical" TB disease patient could now be described as a person of poverty, often of a racial, ethnic or immigrant subgroup, with lack of continuous exposure to health care (New York City Department of Health, Bureau of Tuberculosis 1987). There was also an increased likelihood (although far from typical) that TB patients might have substance abuse problems (such as use of intravenous drugs (IV), crack cocaine, or alcohol), might be living in marginal or no housing, with a history of frequent mental or criminal justice institutionalisation. These are all characteristics of STD core groups (Rothenburg 1983; Brunham and Plummer 1990) within populations. As TB decreased in the rest of the population, it became concentrated in the same groups that had hyperendemicity of STD infections (Braun *et al.* 1993). Thus, within STD core groups there was now an incubator of TB, with the possibility that some STDs might have a biological effect on M.tb infection susceptibility and TB occurrence (Nunn *et al.* 1990).

Compliance with medications under the new paradigm

While the term "compliance" was not commonly used to describe medication-taking behavior in the 1950s and 1960s, patients had many motivations and opportunities to be "compliant" in this way. Early in this period, five-month hospitalisation left little opportunity for non-compliance. Later, continued fear of disease by patients and outpatient follow-up supported medication taking after discharge (Fox 1961). With a reduction of hospitalisation to two months, there was still a significant period of enforced medication compliance, which cured up to 80% of patients. In the 1970s, hospitalisation decreased to periods of weeks and not months, and the uncured rate from hospitalisation was higher. Those patients still diseased upon discharge were then left on their own to take their medication for longer periods of time than at any time previously since medication became available (American Thoracic Society 1970). Without support, medication compliance fell (Ormerod and Prescott 1991), leaving largely uncured TB ex-patients in their urban environments, in contact with large numbers of uninfected individuals, and more concerned with economic survival (Wardman *et al.* 1988) than with controlling their disease.

Follow-up, when attempted, is always harder in an urban environment than in a rural one. In urban settings, if patients do not take their medications and they have a reoccurrence of illness, they may preferentially take one medication above another and are more likely to develop resistance. They are also less likely to be noticed by authorities if they become ill, as it is easy to be anonymous in an urban environment. If patients remain infectious, or revert to infectiousness, they also have great potential for infecting large numbers of other people in the urban environment. Patients are also much more likely to be in environments that encourage spread of disease, such as group housing, tenements, housing projects, jails, holding cells or prison.

Links between TB and HIV in New York

Sexually transmitted disease endemicity has always been associated with urbanism, and specifically with a large number of geographically close individuals who may act as hyperendemic core groups (Rothenburg 1983; Brunham and Plummer 1990). The human immunodeficiency virus was the specific disease agent that connected the urban STD core groups and the groups in which M.tb had become concentrated. The gay white male population, most visible and likely in greater concentration in urban centers, was the epicenter of HIV infection in the late 1970s and early 1980 in NYC. A hallmark of HIV infection, and subsequent development of AIDS, was the development of opportunistic infectious diseases. Spread of syphilis and gonorrhea were also common in this core group. During this period, TB disease was not among the common opportunistic infections, as the middle-class white male population had a low prevalence of M.tb (Murray *et al.* 1984). Rather, Mycobacterium avium intercelluare was the mycobacterium of concern in the early HIV core group spread (Horsburgh and Selik 1989). Without common occurrence of TB in the early phase of the AIDS epidemic, there was no perceived need to mobilise control in relation to this illness. The two infections, TB and HIV, seemed to be running in different circles. Thus, while STD control was highlighted by the HIV epidemic, the TB control infrastructure continued to deteriorate during the late 1970s and early 1980s, while the development of drug resistance within minorities and the poor continued apace (Given *et al.* 1994).

There were two circumstances of commonality for these two epidemics. One was proximity within the overcrowded NYC hospital system (Edlin *et al.* 1992). As AIDS patients were hospitalised with Pneumocystitis carinii pneumonia, they found themselves in the same respiratory or medical intensive care units as the patients with difficult-to-treat drug resistant TB disease, who were often still infectious (Sepkowitz 1995). Spread occurred as hospital workers did not know that the physical isolation of TB patients was necessary, having been lulled into a false sense of safety through decades of effective TB drug therapy (Beck-Sague *et al.* 1992). Thus, one pathway of M.tb movement from the traditional TB groups to the gay male population was by health care proximity in hospitals.

More important than TB spread in hospitals was the effect of the "cross over" of HIV from gay white males to males who also used IV drugs (New York State Department of Health, Bureau of AIDS Epidemiology 1997). An overlap existed between IV users and the various populations of color in NYC, HIV being spread rapidly within the black and Latino populations of men, and finally, to female populations of NYC. The crossover of HIV took place from a population that had a very low prevalence of TB infection (the gay white male population) to populations that had never really had good control of TB, the black male population and various drug-using groups. As cross-over took place, TB disease in HIV-infected patients became recognised by health professionals (Murray *et al.* 1987).

Populations that had significant TB rates now experienced spread of STDs on a person to person basis, while groups with significant HIV infection experienced the spread of TB from individuals to populations, as a consequence of crowding. Crowded institutions offered many opportunities for the "sharing of air" in group settings of at-risk persons for both HIV and M.tb. Given the overstretched nature of public hospitals in New York, people often lay in emergency rooms for days at a time, in close proximity to others. Homeless shelters, increasing in number and size as the economy worsened, led to new applicants being warehoused in administrative offices of various social service agencies (Lerner 1993; Bricknew *et al.* 1993). Prisons became much more crowded, with a back-up of people in local and precinct jails, and in holding cells (Braun *et al.* 1989). Added to these new levels of crowding among at-risk groups, a wave of immigration swept the United States, NYC absorbing many of these migrants (Bloch *et al.* 1989). New migrants came predominantly from countries where there was a very high incidence of TB infection. This lead to an increased population of persons at-risk for TB disease in the lower socioeconomic strata of New York City and urban New York State.

HIV and M. tuberculosis mutation in urban environments

As HIV destroys the immune system, conjoint TB and HIV infection decreases the time between TB exposure and infection to the development of active TB (Fischl *et al.* 1992; Hopewell 1992). The more rapid development of active disease after exposure decreases the cycle time from exposure, infection, development of disease and placement on TB medications. The cycle of drug resistance can be illustrated at the level of the individual patient in the following way. Taking incomplete or inappropriate medication, they develop resistance to some or all of those medications, and then, if still infectious, expose and infect others with a now resistant strain (Neville *et al.* 1994; Sumartojo 1993). Newly infected individuals take time to develop disease, upon which the sequence of events can start anew in those individuals. By shortening the time between infection and disease (CDC 1989), HIV coinfection acts to speed up what was previously a relatively slow process of development of single drug resistance (months), and multiple drug resistance (years) (Llibre *et al.* 1992).

At the time that HIV/TB interaction began, the infrastructure for out-patient control in NYC had been eroded, and there were large increases in both immigration, and physical overcrowding (Wehane *et al.* 1989). This created greater opportunity for the spread of drug resistance. Until the early 1990s, there was true endemicity of M.tb spread, often with multiple drug resistant strains, within certain populations in urban environments, including clinic populations and workers therein, inpatients and workers exposed to certain patients in hospitals, prisoners and workers in prisons, as well as individuals in holding cells, homeless shelters, low-cost housing, and crack houses. These were all characterised as being microenvironments of high exposure to TB infection and multiple drug resistant TB (Paul *et al.* 1993), with lower immunity to TB infection among those with HIV infection (Orme *et al.* 1992) and/or poorly nourished. Thus, TB and HIV infection were linked, with overcrowding as the major cofactor in this association. In the climate of coincident TB and HIV infection, the development of TB in new cases is hastened. For example, in tenement housing and other crowded urban environments in San Francisco, 25% of new cases occurred as a consequence of exposure within the previous 12 months (Small *et al.* 1994). It was possible to track this time period by contact tracing in conjunction with molecular fingerprinting of specific

Table 5.3. *Number and percent of tuberculosis cases placed on directly observed therapy (DOT)*

	New York State (exclusive of New York City)			New York City			New York State (Total)		
	Alive at diagnosis	DOT *N*	DOT %	Alive at diagnosis	DOT *N*	DOT %	Alive at diagnosis	DOT *N*	DOT %
1991	706	297	42	n.a.	n.a.	n.a.	—	—	—
1992	731	335	46	3,651	555	15	4,382	890	20
1993	687	344	50	3,085	1,227	40	3,772	1,571	42
1994	612	415	68	2,900	1,289	44	3,512	1,704	49
1995	604	391	65	2,378	1,225	52	2,982	1,616	54

strains of TB, all of which developed since the time that TB–HIV interaction began (Alland *et al.* 1994).

Response to crisis

This new situation and its associated crisis of infection led to the development of interventions targeted towards location, population and behavior change. The aims were to identify and treat new cases (Freiden *et al.* 1995), make changes in the environments of specific locations associated with TB transmission (Nardell *et al.* 1996) including the management of air flow and human traffic patterns through the different environments associated with TB transmission, to decrease exposure time and to segregate people who had a higher likelihood of having TB. The major intervention was neither environmental nor medical, but one of human behavior modification using a technique called "directly observed therapy" (DOT) (Iseman *et al.* 1993). This involved the direct observation of outpatients, rather than the implicit directly observed therapy that occurred during hospitalisation and medication use. This method has been shown to reduce exposure time, and with it, the development of disease, initiation of therapy, the development of resistance and exposure to others. The use of DOT (Table 5.3, Figure 5.8) became the central method for the recontrol of TB in New York in the 1990s (Figure 5.9). DOT continues to be used both to treat individuals and to control the spread of immediate disease.

Summary

In this chapter, the effect of urbanism on TB disease among urban populations has been described in terms of disease control. The urban model of

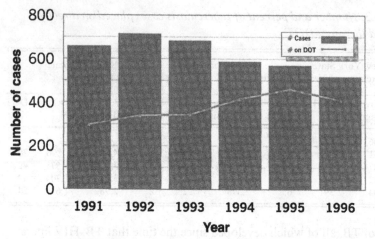

Figure 5.8   Number of new cases of tuberculosis and number receiving any
directly observed therapy, New York State (exclusive of New York city),
1991–1996.
Source: New York State Department of Health, Bureau of Tuberculosis Control

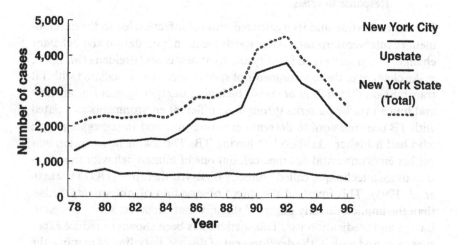

Figure 5.9   Tuberculosis: rise and fall, New York 1978–1996

sexually transmitted disease (HIV) transmission (the core group model)
has modified the airborne disease model of "traditional" M.tb spread. The
interaction of HIV and TB created a new disease pattern for TB and
subsequently increased the rate of change of the biology of M.tb. From the
first effective use of antibiotics against it, M.tb was able to mutate to
resistance at a slow and predictable rate, related to the use of efficacious
drug regimens, the persistence of individuals in taking their medications,
and the effectiveness public health workers in monitoring patients while on

therapy. The interaction of the more traditional TB population with the STD core group has lead to faster change of M.tb and spread of TB between people, as well as more rapid infection and development of disease, less robust response of the individual's immune system to M.tb, and less "assistance" to the medications taken. These changes have created a more rapid development of single and multiple drug resistance.

The current pattern of tuberculosis infection creates areas of danger in the urban environment that are not synonymous with other ideas of danger from infectious diseases. The new areas of urban risk are not general, and include prison cells and inmate holding areas, medical and respiratory intensive care units, emergency rooms and outpatient clinics, homeless shelters, parole offices, and schools and school buses. These and other crowded areas within urban environments become increasingly dangerous as M.tb mutates rapidly and spreads. This change in risk redirects other types of resources in urban environments towards the control of TB. Currently, M.tb is seen as an environmental danger to dispossessed people and poor people with generally poor health, as well as to providers of human services to those populations. It is also seen as a danger to anyone with HIV infection, whether that person is part of a "traditional" risk group or not. The mental model of the control of infectious disease has changed as the disease has changed. TB, a disease that was once considered to be one that poor people contracted, now forces social, clinical, environmental and legal change, since it can now impact on many more people, living in overlapping urban environments. STD endemicity has always been associated with urbanism, in that urbanism can provide large numbers of geographically close individuals who can act as hyperendemic core groups. In contrast, prior to the HIV epidemic, TB endemicity was associated with location or general health rather than behavior. With the spread of HIV as an STD, the overlaps in behavior and population due to sexual orientation, sexual behavior and drug use led to the spread of M.tb and TB disease in a way that was closer to the STD core hyperendemicity model. With this, TB has taken on some of the characteristics of STD in terms of risk and populations at risk. The likelihood of TB if infected, and the possibility of infection if exposed, is associated as much with HIV and immune status as with some of the more traditional risk factors. This is especially true in the predominantly urban populations with high rates of HIV infection. There is now the confounding of two previously distinct risk pools, the crowding-risk pool of TB, and the behavior-risk pool of STD, in varying temporal and cultural sequences. This situation might allow some cross-urban and cross-national comparisons, because HIV is spreading among populations in many parts of the world at different rates, and affecting TB rates differently. However, given that the TB–HIV link is well established in populations in Southeast Asia (Narian and Pattanayak 1994 and Africa (Nunn *et al.* 1992; DeCock *et al.* 1991), and is developing rapidly in India (Lalit 1993), cross-cultural comparisons of the interaction of TB and HIV in the urban context will need to be initiated immediately. Regardless of whether such studies are carried out or not, the link between transmission of TB and HIV has altered the human–M.tb interaction, one that had been established since prehistory (Bates and Stead 1993), forever.

References

Alland, D., Kalkut, G. E., Moss, A. R., McAdam, R. A., Hahn, J. A., Bosworth, W., Drucker, E. and Bloom, B. R. (1994). Transmission of tuberculosis in New York City: an analysis by DNA fingerprinting and conventional epidemiologic methods. *New England Journal of Medicine*, **330**, 1710–16.

American Thoracic Society (1970). Bacteriologic standards for discharge of patients. *American Review of Respiratory Disease*, **102**, 470–3.

Arden House Conference on Tuberculosis Recommendations (1960). American Review of Respiratory Disease, **81**, 481–4.

Badger, T. L. and Ayvazian, L. F. (1949). Tuberculosis in nurses: clinical observations on its pathogenesis as seen in a fifteen-year follow-up of 745 nurses. *American Review of Tuberculosis*, **60**, 305–27.

Bates J. H. and Stead, W. W. (1993). The history of tuberculosis as a global epidemic. *Medical Clinics of North America*, **77**, 1205–17.

Beck-Sague, C., Dooley, S.W., Hutton, M.D., *et al.* (1992). Hospital outbreak of multidrug-resistant Mycobacterium tuberculosis infections: factors in transmission to staff and HIV-infected patients. *Journal of the American Medical Association*, **268**, 1280–86.

Bloch, A. B., Reider, H. L., Kelly, G. D., Cauthen, G. M., Hayden, C. H. and Snider, D. E. (1989). The epidemiology of tuberculosis in the United States. Implications for diagnosis and treatment. *Clinics of Chest Medicine*, **10**, 297–313.

Bloom, B. R. and Murray, C. J. L. (1992). Tuberculosis: commentary on a reemergent killer. *Science*, **257**, 1055–64.

Bricknew, P. W., Scharer, L .L. and McAdam, J. M. (1993). Tuberculosis in homeless populations. *Tuberculosis: A Comprehensive International Approach*. eds. L. B. Reichman and E. S. Hershfield, pp. 433–54. New York: Marcel Dekker.

Braun, M. M., Cote T.R. and Rabkin, C. S. (1993). Trends in death with tuberculosis during the AIDS era. *Journal of the American Medical Association*, **269**, 2865–8.

Braun, M. M., Truman, B. I., Maguire, B., DiFerdinando, G. T., Jr., Wormser, G., Broaddus, R. and Morse, D. L. (1989). Increasing incidence of tuberculosis in a prison inmate population. Association with HIV infection. *Journal of the American Medical Association*, **261**, 393–7.

British Medical Research Council. (1948). Streptomycin treatment of pulmonary tuberculosis. A Medical Research Council investigation. *British Medical Journal*, **2**, 769–82.

Brudney, K. (1993). Homelessness and TB: a study in failure. *Journal of Law and Medical Ethics*, 21, 360–7.

Brudney, K. and Dobkin J. (1991). Resurgent tuberculosis in New York City: HIV, homelessness, and the decline of TB control programs. *American Review of Respiratory Disease*, 125,745–9.

Brunham, R. C. and Plummer, F. A. (1990). A general model of sexually transmitted disease epidemiology and its implications for control. *Medical Clinics of North America*, 74, 1339–52.

Caldwell, M. (1988). *The Last Crusade. The War on Consumption 1862–1954*. New York: Atheneum Macmillan.

Centers for Disease Control (1990). Tuberculosis among foreign born persons entering the United States. *Morbidity and Mortality Weekly Report*, 39, RR–18, 1–21.

Centers for Disease Control (1989). Tuberculosis and human immunodeficiency virus infection. Recommendations of the Advisory Committee for Elimination of Tuberculosis. *Morbidity and Mortality Weekly Report*, 38, 236–50.

Chilress, W. G. (1951). Occupational tuberculosis in hospitals and sanatorium personnel. *Journal of the American Medical Association*, 66, 16–27.

Curry, F. J. (1968). Neighborhood clinics for more effective outpatient treatment of tuberculosis. *New England Journal of Medicine*, 279, 1262–7.

Dandoy, S. and Elman, S. B. (1972). Current status of general hospital use for tuberculous patients in the United States. *American Review of Respiratory Disease*, 106, 580–6.

DeCock, K. (1996). Editorial: Tuberculosis control in resource-poor settings with high rates of HIV infection. *American Journal of Public Health*, 86, 1071–3.

De Cock. K. M., Gnaore, E., Adjorlolo, G. *et al.* (1991). Risk of tuberculosis in patients with HIV-1 and HIV-2 infections in Abidjan, Ivory Coast. *British Medical Journal*, 302, 496–9.

Edlin, B. R., Tokars, J. I., Grieco, M. *et al.* (1992). An outbreak of multiple-drug resistant tuberculosis among hospitalized patients with acquired immunodeficiency syndrome. *New England Journal of Medicine*, 326, 514–21.

Fischl, M. A., Daikos, G. L., Uttamchandani, R. B., Poblete, R. B., Moreno, J. N., Reyes, J. N., Boota, A. M., Thompson, L. M., Cleary, T. J., Oldham, S. A. *et al.* (1992). Clinical presentation and outcome of patients with HIV infection and tuberculosis causes by multiple-drug resistant bacilli. *Annals of Internal Medicine*, 117, 184–90.

Fox, W. (1961). The problem of self-administration of medicaments: a review of published work and a study of the problems. *Bulletin of the International Union Against Tuberculosis*, 32, 307–31.

Frieden, T. R., Fujiwara, P. I., Washko, R. M. *et al.* (1995). Tuberculosis in New York City–turning the tide. *New England Journal of Medicine*, 333, 229–33.

Frieden, T. R., Sterling, T., Pablos-Mendez, A., Kilburn, J. O. *et al.* (1993). The emergence of drug-resistant tuberculosis in New York City. *New England Journal of Medicine*, 328, 521–6.

Given, M. J., Khan, M. A. and Reichman, L. B. (1994). Tuberculosis among patients with AIDS and a control group in an inner-city community. *Archives of Internal Medicine*, 154, 640–8.

Hansel, D. A. (1993). The TB and HIV epidemics: history learned and unlearned. *Journal of Law and Medical Ethics*, 21, 360–7.

Hawkins, N. G., Davies, R. and Homes, T.H. (1957). Evidence of psychosocial factors in the development of pulmonary tuberculosis. *American Review of Tuberculosis and Pulmonary Diseases*, **75**, 768–80.

Hopewell, P. C. (1992). Impact of human immunodeficiency virus infection on the epidemiology, clinical features, management and control of tuberculosis. *Clinics of Infectious Disease*, **15**, 540–7.

Horsburgh, C. R. and Selik, R. M. (1989). The epidemiology of disseminated nontuberculous mycobacterial infection in the acquired immunodeficiency syndrome (AIDS). *American Review of Respiratory Disease*, **139**, 4–7.

Houk, V. N., Baker, J. H., Sorenson, K. and Kent, D. C. (1968). The epidemiology of tuberculosis infection in a closed environment. *Archives of Environmental Health*, **16**, 26–35.

Iseman, M. D., Cohn, D. L. and Sbarbaro, J.A. (1993). Directly observed treatment of tuberculosis. We can't afford not to try it. *New England Journal of Medicine*, **328**, 576–8.

Iseman, M. and Sbarbaro, J. (1991). Short-course chemotherapy of tuberculosis. Hail Britannia (and Friends)! *American Review of Respiratory Disease*, **143**, 697–98.

Johnston, R. F. and Wildrick, K. H. (1974). State of the art review. The impact of chemotherapy on the care of tuberculosis. *American Review of Respiratory Disease*, **109**, 636–64.

Lalit, K. (1993). Upsurge in tuberculosis: HIV effect. *Indian Journal of Tuberculosis*, **40**, 43–56.

Lerner, B. H. (1993). New York City's tuberculosis control efforts : the historical limitations of the "war on consumption." *American Journal of Public Health*, **83**, 758–65.

Llibre, J. M., Tor, J., Manterola, Carbonelli, C. and Roset, J. (1992). Risk stratification for dissemination of tuberculosis in HIV-infected patients. *Quarterly Journal of Medicine*, **298**, 149–57.

McDermott, W. *et al.* (1947). Streptomycin in the treatment of tuberculosis in humans. *Annals of Internal Medicine*, **27**, 769–822.

Markowitz, N., Hansen, N. I., Hopewell, P. C., Glassroth, J., Kvale, P. A., Mangura, B. T. *et al.* (1997). Incidence of tuberculosis in the United States among HIV-infected persons. *Annals of Internal Medicine*, **126**, 123–32.

Medical Research Council. (1950). Treatment of pulmonary tuberculosis with streptomycin and para-aminosalicylic acid. *British Medical Journal*, **2**, 1073–85.

Murray, J. F., Garay, S. M., Hopewell, P. C., Mills, J., Snider, G. L. and Stover, D. E. (1987). Pulmonary complications of the acquired immunodeficiency syndrome: an update. *American Review of Respiratory Disease*, **135**, 504–9.

Murray, J. F., Felton, C. P., Garay, S. M., Gottlieb, M. S., Hopewell, P. C., Stover, D. E. and Teirstein, A. S. (1984). Pulmonary complications of the acquired immunodeficiency syndrome. Report of a National Heart, Lung, and Blood Institute Workshop. *New England Journal of Medicine*, **310**, 1682–8.

Nardell, D. A., Barnhart, S. and Permutt, S. (1996). Control of tuberculosis in health care facilities: the rational application of patient isolation, building ventilation, air filtration, ultraviolet air disinfection, and personal respirators. In *Tuberculosis*, eds. W.N. Rom and S.M. Garay, pp. 873–91. New York: Little Brown.

Narian J. P. and Pattanayak, S. (1994). AIDS and tuberculosis in Asia: a public health priority for the region. In *Proceedings of the Tenth International Conference on AIDS*. Yokohama, Japan, August 7–12, **10**, 337 (abstract no. PC0284).

Neville, K., Bromberg, A., Bromberg, R., Bonk, S., Hanna, B. and Rom, W. N. (1994). The third epidemic-multidrug resistant tuberculosis. *Chest*, **105**, 45–8.

New York State Department of Health, Bureau of AIDS Epidemiology. (1997). *AIDS in New York, 1996*. Albany, New York: New York State Department of Health.

New York State Department of Health (1997). *Tuberculosis in New York: Data Summary, 1996*. Albany, New York: New York State Department of Health.

New York City Department of Health, Bureau of Tuberculosis (1993). *Tuberculosis in New York City, 1992: Information Summary*. New York City: New York City Department of Health.

New York City Department of Health, Bureau of Tuberculosis (1987). *Tuberculosis in New York City, 1984–85*. New York City: New York City Department of Health.

New York City Department of Health, Bureau of Tuberculosis (1972). *Tuberculosis in New York City, 1971*. New York City: New York City Department of Health.

Nunn, P., Gicheha, C., Hayes, R., Gathua, S., Brindel, R., Kibuga, D., Mutie, T., Kamunyi, R., Omwega, M., Were, J. et al. (1992). Cross-sectional survey of HIV infection among patients with tuberculosis in Nairobi, Kenya. *Tubercle Lung Disease*, **73**, 45–51.

Nunn, P., Kibuga, D., Elliott, A. and Gathua, S. (1990). Impact of human immunodeficiency virus on transmission and severity of tuberculosis. *Transactions of the Royal Society of Tropical Medicine and Hygiene*, **84**(Suppl 1), 9–13.

Ochs, C. W. (1962). The epidemiology of tuberculosis. *Journal of the American Medical Association*, **179**, 247–52.

Orme, I. M., Miller, E. S., Roberts, A. D., Furney, S. K., Griffen, J. P., Dobos, K. M., Chi, D., Rivoire, B. and Brennan, P. J. (1992). T lymphocytes mediating protection and cellular cytolysis during the course of mycobacterium tuberculosis infection. *Journal of Immunology*, **148**, 186–9.

Ormerod, L. P. and Prescott, R. J. (1991). Inter-relations between relapses, drug regimens and compliance with treatment in tuberculosis. *Respiratory Medicine*, **85**, 239–42.

Paul, E. A., Lebowitz, S. M., Moore, R. E., Hoven, C. W., Bennett, B. A. and Chen, A. (1993). Nemesis revisited: tuberculosis infection in a New York City men's shelter. *American Journal of Public Health*, **83**, 1743–5.

Reichman, L. B. (1991). The U-shaped curve of concern. *American Review of Respiratory Disease*, **144**, 741.

Reider, H. L., Cauthen, G. M., Comstock, G. W. and Snider, D. E. (1989). Epidemiology of tuberculosis in the United States. *Epidemiology Review*, **11**, 79.

Rothenburg, R. B. (1983). The geography of gonorrhea: empirical demonstration of core group transmission. *American Journal of Epidemiology*, **117**, 688–94.

Sbarbaro, J. A. (1970). The public health tuberculosis clinic: its place in comprehensive health care. *American Review of Respiratory Disease*, **101**, 463–5.

Selwyn, P. A. (1993). Tuberculosis and AIDS: epidemiologic, clinical and social dimensions. *Journal of Law and Medical Ethics*, **21**, 279–88.

Selwyn, P. A., Schell, B., Alcabes, P., Friedland, G. N., Klein, R. S. and Schoenbaum, E. C. (1992). High risk of active tuberculosis in HIV-infected drug users with cutaneous anergy. *Journal of the American Medical Association*, **268**, 504–9.

Sepkowitz, K. A. (1995). AIDS, tuberculosis, and the health care worker. *Clinic of Infectious Disease*, **20**, 232–42.

Small, P. M., Hopewell, P. C., Singh, S. P., Paz, A., Parsonnet, J., Ruston, D. C., Scheter, G. F., Daley, C. L. and Schoolnik, G. K. (1994). The epidemiology of tuberculosis in San Francisco. A population-based study using conventional and molecular methods. *New England Journal of Medicine*, **330**, 1703–9.

Snider, G. I. (1997). Tuberculosis then and now: a personal perspective on the last 50 years. *Annals of Internal Medicine*, **126**, 237–43.

Spence, D. P. S., Hotchkiss, J. and Davies, P. D. O. (1993). Tuberculosis and poverty. *British Medical Journal*, **307**, 759–61.

Stead, W. W., Senner, J. W., Reddick, W. T. and Lofgren, J. P.(1990). Racial differences in susceptibility to infection with mM. tuberculosis. *New England Journal of Medicine*, **322**, 422.

Sumartojo, E. (1993). When tuberculosis treatment fails. A social behavioral account of patient adherence. *American Review of Respiratory Disease*, **147**, 1311–20.

Terris, M. (1948). Relation of economic status to tuberculosis mortality by age and sex. *American Journal of Public Health*, **38**, 1061–70.

Tuberculosis morbidity–United States. (1996). *Morbity and Mortality Weekly Report*, **45**, 366–70.

Tyrell, W. F. (1956). Bed rest in the treatment of pulmonary tuberculosis. *Lancet*, **1**, 821.

Wardman, A. G., Knox, A. J., Muers, M. F. and Page, R. L. (1988). Profiles of non-compliance with antituberculosis therapy. *British Journal of Diseases of the Chest*, **82**, 285–9.

Wehane, M. J., Snukst-Torbeck, G. and Schraufnagel, D. E. (1989). The tuberculosis clinic. *Chest*, **96**, 815–18.

Yeager, H. and Medinger, A. I. (1986). Tuberculosis long-term care beds: have we thrown out the baby with the bathwater. *Chest*, **90**, 752–3.

6 Fecundity and ovarian function in urban environments

P. T. ELLISON

Editors' introduction

The future of the human species is dependent on reproductive ecology, so it is logical for us to want to know how reproduction is affected by features of urban life, both old and new. Ellison makes it clear that past ecological transitions, from foraging to agriculture, and from agriculture to industrialisation, altered reproductive ecology dramatically. Both transitions involved changes in residential pattern that are part of urbanisation. The post-industrial city also can have an impact, and Ellison describes several mechanisms and pathways, termed "proximate determinants," by which urban factors can influence parameters of reproductive function, specifically ovarian function. Population density *per se* is not the salient factor, but urbanism includes behaviors that can affect ovarian function. Among these are urban patterns of maturation and aging, anxiety, dietary composition, pollutants, as well as energy expenditure and energy balance, a topic that Ulijaszek also addresses in chapter 13. Although most of these factors are not unique to cities, they may be more varied there and so make a stronger contribution to variation in ovarian function. By recognising that cities are not a unitary phenomenon but show great variation, the focus turns towards identifying the specific determinants of ovarian function. The ability to define these determinants depends on their accurate measurement. Once measured and their effects defined, it becomes possible to generalise beyond any particular urban population. Lessons that Ellison and others learn with urban populations can have broad application to other, different populations, whether or not they can be classified as being urban, suburban, periurban, or rural in nature.

Introduction

Human reproductive ecology has been defined as "the study of reproduction as an aspect of human biology that is responsive to ecological context" (Ellison 1995). Recent efforts in this area have primarily focused on rural populations in the developing world rather than on urban environ-

111

ments or populations in more developed countries. This focus derives to some extent from an interest in conditions that more closely approximate the formative environments of the human past, or particular features of those environments. In addition, an emphasis on the reproductive physiology of rural populations has been useful in expanding our understanding of the range of natural variation in human reproductive function beyond the predominantly clinical perspective that is heavily based on observations of urban populations in developed countries (Ellison *et al.* 1993). By pushing the venue of research in human reproductive physiology out of the laboratory and clinic and into the field, human reproductive ecologists have begun to make observations that not only affect our understanding of the ecology of rural populations, but that have important significance for contemporary health issues in urban and developed populations as well (Ellison 1998; Eaton *et al.* 1994; Wasser 1990). By escaping from the dominant paradigms of clinical medicine which view reproductive physiology primarily as a "closed," homeostatically buffered system and replacing them with a paradigm of reproductive physiology as an "open," ecologically sensitive system, human reproductive ecologists are now in a position to return to the study of urban populations with new insight.

It is extremely important for the paradigm of reproductive ecology to be applied to urban populations and developed societies as well as to rural populations. While rural environments may provide important opportunities for extrapolating into our past and for broadening our understanding of human variability, urban environments constitute the ecological conditions under which increasing numbers of human beings live today in both the developed and developing world. Our ability to elucidate our present condition and to anticipate the future of our species depends crucially on incorporating an understanding of urban environments and their consequences for human biology. In particular, there is every reason to expect that urban environments have significant consequences for human reproductive ecology. This becomes manifest if we consider the two great transformations of human ecology that have occurred in the last 15,000 years. Both have been significantly associated with the growth of urbanism as a residential pattern, and both appear (in one case we suspect it and in the other we know it) to have involved dramatic changes in human reproduction.

The first transformation occurred with the shift from mobile foraging and mixed foraging/horticulture to settled agriculture as a subsistence regime. In the wake of this progressive shift in subsistence came a host of biological and social changes: changes in diet and nutrition (Cohen 1977); changes in disease patterns and mortality (Cohen 1989); changes in patterns of growth and development (Cohen and Armelagos 1984); changes in social structure, social stratification, and gender roles (Bentley 1996; Cohen *et al.* 1980). It also seems increasingly likely that this transformation affected human reproductive patterns as well. Changes in disease patterns, particularly the increase in directly transmitted infectious diseases and water-born pathogens, may have increased rates of infant and early childhood mortality leading to increased fertility. Such an increase might have been mediated both socially, through an increased parental incentive to replace lost children, and physiologically, through a removal of the suppressive effect of lactation on the mother's reproductive system. An increased availability of cereal-based weaning foods may also have reduced the dependence of growing children on maternal milk, leading to a more rapid postpartum recovery of ovarian function (Buikstra *et al.* 1986). Changing gender roles in subsistence may have lessened female workloads, leading to increases in female fecundity, while the changing nature of subsistence and economic activities and changes in social support systems may have added to the incentive for larger families (Bentley 1985; Peacock 1990). All of these changes may have combined to increase fertility rates and contribute to more rapid population growth (Bentley *et al.* 1993).

The second great transformation of human ecology has been equally as profound and far-reaching, transforming our biology and socioecology at virtually every level. This is the transformation that flows from the shift from agriculture to industrialisation as a dominant subsistence mode. Like the first great transformation, the second is not accomplished suddenly, nor is there a single pathway of development that all populations follow. To identify this shift as in some sense a unified phenomenon is not to claim that it is simple in either its causes or its effects. But the effects it exerts on and through human biology are profound. The secular trend in growth and development is one effect, itself related to changes in diet, nutrition, childhood workloads, and disease burdens (Eveleth and Tanner 1990). Major changes in disease patterns and causes of mortality constitute an "epidemiological transition" at least as dramatic as that associated with the shift to agriculture, with diseases of aging replacing infectious diseases as the dominant causes of mortality (Omran 1971). Dramatic changes in reproductive biology are reflected not only in earlier reproductive maturation, but also in the decreasing effectiveness of lactation in spacing births, and in increasing levels of fecundity as a consequence of changes in nutrition and activity patterns (Bentley *et al.* 1993; Short 1987; Ellison *et al.* 1993). The motivation for family building is reduced by the combination of changes in child mortality and the socioeconomic costs and benefits of children (Caldwell 1982). Conscious use of contraception spreads, and with it the age-pattern of female fertility begins to diverge markedly from the age-pattern of female fecundity (Henry 1961). With lower levels of both mortality and fertility the entire age-structure of the population is transformed. Once again we find consequent changes in social structure, social stratification, and gender roles. The sequelae of this modern transformation of human ecology continue to have ramifications in every aspect of public and domestic life.

Both of these transformations have been associated with shifts in patterns of residence that we identify as "urbanisation," increasingly dense patterns of human habitation that are directly related to the changes in subsistence base. These shifts in residential patterns themselves can contribute to some of the biological consequences of the ecological transformations of which they are a part, such as changing the opportunities for disease transmission. But many of the significant reproductive consequences of urbanisation as an ecological transformation (e.g. early maturation, increasing fecundity, increasing use of contraception, delays in age at first birth) are not simple or direct consequences of residential density. "Urban environment" is, to this extent, a rather loose concept. Part of the mission of this volume is to identify more explicitly those features of "urban environments" with greatest importance for human biology. In approaching this problem, reproductive ecologists employ a strategy that may be different from that adopted by other human biologists concerned with urban environments, but it is one that may usefully be generalised to other domains.

The "proximate determinants" paradigm

Rather than work "forward," trying to identify and define aspects of urban environments that seem noteworthy or unique or potent and tracing their effects toward reproductive endpoints, reproductive ecologists tend to work "backward," identifying "proximate determinants" of reproductive endpoints, then studying how those determinants are themselves affected by slightly more distal environmental and constitutional variables, those in turn by more distal social and ecological determinants, and so on until at last some determinants can be directly related to urban environments and inferences about the entire chain of causation can be drawn (Davis and Blake 1956; Bongaarts and Potter 1983; Campbell and Wood 1988). There are several advantages to this basically Cartesian approach. Complex causal pathways are more easily understood by reducing them to simpler links of more direct causation. Working backward along those chains of causation from dependent toward independent variables also helps to reduce futile research efforts that start with features of urban environments which may ultimately turn out to have no effect on reproduction whatsoever.

For the purposes of this chapter I will focus on ovarian function as an important determinant of female fecundity, or the physiological capacity for reproduction. Ovarian function is not the sole determinant of female fecundity, but it is one of the most important. It integrates the effects of both central and systemic regulation of the reproductive axis and is demonstrably sensitive to environmental, constitutional, and behavioral variables (Ellison 1990, 1995). In addition, it is an important determinant of other aspects of female health and well-being, including the risks of heart disease, osteoporosis, reproductive cancers, and Alzheimer's disease (Ellison 1998). Finally, sensitive assessments of ovarian function are possible through non-invasive techniques under a wide variety of conditions (Ellison 1988, 1993; Campbell 1993; Lasley *et al.* 1994; Worthman *et al.* 1991). These techniques, such as the measurement of steroid hormones in saliva, urine, or small volumes of blood, allow for the accumulation of directly comparable data necessary from different populations.

Other reproductive endpoints could be chosen which might show different sensitivities to environmental factors. However, many of these are unsatisfactory for one reason or another. Other indices of female reproductive function, such as menstrual patterns, that are self-reportable, are not particularly sensitive or reliable indices of female fecundity. Reproductive outcome variables, such as fetal loss rates or birthweight, are themselves contingent on female fecundity and on contraceptive practice. Indices of male fecundity, such as semen analyses, are difficult to obtain on representative populations, nor is their relationship to male fecundity well worked out (Campbell and Leslie 1995). Difficulties inherent in obtaining and interpreting these data, however, do not preclude their usefulness in elucidating environmental effects on human reproductive physiology. However, in what follows, I will rely primarily on the comparative evidence derived from measures of female ovarian function.

Variation in ovarian function and its relationship to variation in fecundity

The first point that can be established is that ovarian function does indeed vary to a considerable degree within and between healthy women in urban environments. This may seem an obvious point to human biologists used to studying human variability, but directly conflicts with a dominant clinical view that conceives of ovarian function as tightly buffered by homeostatic mechanisms. According to the latter view, disregulation of ovarian function may occur under highly stressful ecological circumstances. While such circumstances may be prevalent in the developing world, they are often imagined to be unimportant in the privileged environments of urban centers in the developed world. However, we have documented substantial variation in ovarian steroid profiles among middle-class women in Boston who are all of comparable age, normal height and weight, and who are neither exercising nor gaining or losing weight (Ellison 1995). Approximately two-thirds of the variation in our samples occurs between women, indicating that some individuals have characteristically high profiles of estradiol or progesterone while others have characteristically low ones. One-third of the total variance occurs within women, however, indicating that steroid profiles can vary substantially month-to-month within the same women (Sukalich et al. 1994).

Variation in ovarian function of the magnitude observed in these studies can be associated with significant variation in female fecundity. Conception cycles among urban women have been found to have higher levels of progesterone than non-conception cycles (Lenton et al. 1988; Stewart et al. 1993). However, because progesterone levels may themselves be affected by chorionic gonadotropin secreted by the trophoblast of an implanting embryo, it is difficult to be sure that elevations in progesterone are antecedants of conception rather than consequences. Nevertheless, the effectiveness of exogenous progesterone supplementation in increasing conception rates, especially among older women, suggests that quantitative variation in progesterone profiles is significant (Meldrum 1993). The impact of variation in follicular estradiol levels on conception probabilities is less ambiguous. Clinical studies have established that in follicular estradiol levels are correlated with endometrial thickness, itself correlated with implantation success, with the cytoplasmic and nuclear maturation of the oöcyte, and hence with the fertilisability of the oöcyte and the viability of the resulting zygote (Yoshimura and Wallach 1987; Liu et al. 1988; Dickey 1993; Dubey et al. 1995; Arnot et al. 1995). In a study of Boston women attempting to conceive, we found that both absolute and relative variation in midfollicular estradiol levels had a significant effect on the probability of conception among cycles exposed to intercourse during the fertile period (Figure 6.1, Lipson and Ellison 1996). It is particularly noteworthy that the per cycle probability of conception varies from some 90% when estradiol levels were 10–15 pmol/L higher than usual for a given woman to virtually zero when her estradiol levels were as little as 5 pmol/L lower than usual. In other words, not only is the relationship between ovarian function and fecundity statistically significant, it is biologically significant as well. It is also striking that there is such a range of positive variation in conception probabilities above average. Reproductive ecologists often think in terms of conditions that suppress fecundity, whereas these results suggest that conditions which temporarily enhance fecundity may be more important.

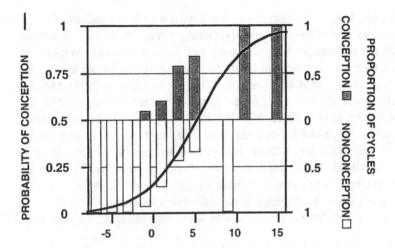

RELATIVE MIDFOLLICULAR OESTRADIOL (pmol/l)

Figure 6.1 Logistic regression of the probability of conception against relative mid-follicular oestradiol levels (level for given month minus average level for each woman) in 50 cycles from 10 women attempting to conceive. All cycles included intercourse at least once within two days of the mid-cycle oestradiol drop. See also Figure 6.2. For details, see Lipson and Ellison (1996)

Determinants of variation in ovarian function

Energetics

Having established, then, the significance of ovarian function as a determinant of female fecundity in urban environments, we can begin to work our way back toward more distal determinants. In this we can be guided to a certain extent by results that have already been established for women in rural environments. For example, the sensitivity of ovarian function to energy balance and energy expenditure has now been amply demonstrated in rural populations. In Zaïre and Nepal, when women lose even moderate amounts of weight over a period of several months, hormonal indices of ovarian function decline and conception rates drop (Ellison *et al.* 1989; Bailey *et al.* 1992; Panter-Brick and Ellison 1993, 1994). Among rural Polish farm women, heavy workloads during the agricultural season are correlated with suppressed ovarian function, even though energy balance is maintained through increased energy intake (Jasienska and Ellison 1998).

Energy balance and energy expenditure are also important determinants of ovarian function in urban environments. Moderate voluntary weight loss in Boston is associated with reductions in progesterone profiles comparable to those observed in rural populations (Lager and Ellison 1990). German researchers have also found that "restrained eating," i.e. eating significantly below appetite, is also associated with lower indices of ovarian function, even if weight is stable (Schweiger *et al.* 1992). Even month-to-month fluctuations in weight among our conception study subjects in Boston, women who were not trying to lose or gain weight, are associated with changes in midfollicular estradiol (Lipson and Ellison 1996). Moderate differences in static weight or fatness between women within the normal range, as contrasted with dynamic changes in weight and fatness, have not been linked to differences in ovarian function in any study. However, in Washington State, nulliparous women below 85% of ideal weight standards were observed to be nearly five times as likely to be unable to conceive in a year as normal weight controls (Green *et al.* 1988). In this case, since the dataset is limited to status variables, it is difficult to know whether differences in energy balance are confounded in the contrasts.

Energy expenditure has been linked to suppressed ovarian function among athletes in many studies, but there is substantial evidence that even moderate levels of physical exercise can suppress ovarian function among women in urban environments. Recreational joggers, for example, have been found to have levels of progesterone and estradiol well below those of otherwise similar non-exercising women (Figure 6.2, Ellison and Lager 1986; Ellison 1998). This observation has been linked to the epidemiological evidence for reduced cancer risk associated with exercise, and provides one of the most hopeful prospects for the development of practical intervention strategies to reduce breast cancer prevalence (Bernstein *et al.* 1994; Ellison 1998).

Lactation

Lactation has a clear suppressive effect on ovarian function. While this effect was originally thought to be mediated exclusively by the temporal distribution of suckling events, more recent evidence suggests that the energetic stress of lactation may be a more important aspect of nursing "intensity." The energetic stress of lactation varies both with the absolute level of milk production and with the energetic status of the mother. For example, Worthman *et al.* (1993) found that well-nourished Amele women in lowland New Guinea resume ovarian function much earlier than would be predicted from their pattern of frequent and sustained breastfeeding. Ford *et al.* (1989) reported that postpartum maternal weight and weight gain are significantly associated with the duration of lactational amenorrhea in rural Bangladesh while other studies in the same population failed to find any relationship between the duration of lactational amenorrhea and nursing frequency (Huffman *et al.* 1987.) Tay *et al.* in Edinburgh

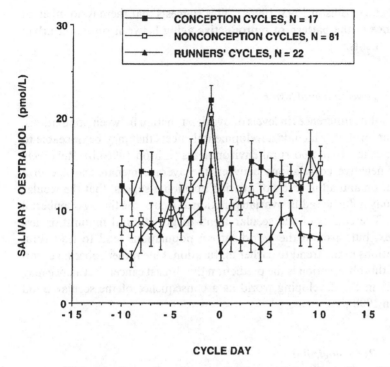

CYCLE DAY

Figure 6.2 Comparison of average (±SE) oestradiol profiles in 22 recreational runners compared with profiles in conception and non-conception cycles from non-exercising women of comparable age, weight, and height

(1996) have recently found that while frequent suckling is associated with elevated prolactin among urban Scottish mothers at all stages of lactation, there is no relationship between prolactin levels and the return of menses. Resumption of ovarian activity is instead highly correlated with the introduction of supplemental foods.

Age

Ovarian function varies significantly with age in all populations, contributing significantly to the age-patterns of both female fecundity and natural fertility (Lipson and Ellison 1992; Ellison *et al.* 1993; Ellison 1996a; O'Rourke and Ellison 1993; O'Rourke *et al.* 1996). Where directly comparable data exists, it appears that the age pattern of ovarian function is basically the same in rural and urban populations, despite differences in the absolute levels of ovarian function. For example, average luteal progesterone levels in Boston, rural Zaïre, and rural Nepal display parallel trajectories by age that are superimposable when expressed as standardised values (Ellison 1993, 1996a). This suggests that there is no inherent difference in the pattern of reproductive aging between rural and urban environments.

Developmental history

The absolute differences in levels of ovarian function between populations, however, probably include developmental effects that may be traceable to difference in urban and rural environments (Ellison 1996a,b). Relatively strong negative correlations between the average menarcheal age in a population and adult levels of ovarian hormones suggest that the secular trend may influence adult ovarian function as well as the timing of pubertal events. The causes of the secular trend in growth and maturation are complex, but, around the world, urban populations tend to lead rural populations in the trend to earlier maturation. One rather sobering corollary of this observation is the prediction that breast cancer prevalence may increase in the developing world as a consequence of the secular trend (Ellison 1998).

Psychological stress

The effects of energetics, lactation, and age on ovarian function are well established. The effects of several other factors which may be relevant to urban environments are somewhat less well characterised and for that reason represent important areas of current research. One such area concerns the potential effects of psychosocial stress on ovarian function. Folk wisdom concerning the effects of psychological stress on fertility is widespread. Anthropologists have sometimes extrapolated these speculations to suggest that anxiety over the coming harvest might affect the fecundity of agricultural populations (Malina and Himes 1977; Mosher 1979). Rigorous research in this area that clearly discriminates psychosocial stress from potentially confounding variables is less abundant. There is compelling evidence that dramatic acute stress, such as trauma or surgery, can interfere with reproductive axis activity (Wasser and Isenberg 1986; Wasser 1990; Schenker et al. 1992). But evidence of an effect of more moderate and/or chronic stresses is less compelling. Several studies have found that women in treatment for infertility score higher on psychological stress inventories than their peers who are not in treatment, but the direction of causation in these studies is highly ambiguous (Eisner 1963; Mai et al. 1972; Harrison et al. 1981; Nijs et al. 1984; Demyttenaere et al. 1988; Wright et al. 1991; Downey and McKinley 1992; Schover et al. 1992, 1994;

Wasser *et al.* 1993, 1994). Domar *et al.* (1990) report on the treatment of infertile couples with a generalised regimen of group therapy and lifestyle modification focusing on stress reduction, noting that some 30% of the subjects conceive within the following year. The absence of control data, however, makes it impossible to be sure whether this represents a statistically significant effect of the treatment.

The techniques of reproductive ecology can be directly applied to this issue, by using direct measures of ovarian function rather than conception as a dependent variable. Greater attention can also be paid to the independent variable with designs that can avoid samples biased by their fertility status. In particular, it is important to determine whether moderate psychological stress, such as urban dwellers may routinely face on a broad scale in their daily lives, is capable of affecting reproductive axis function, and if so, whether the effect has a dose-response relationship to measures of psychological stress. If psychological stress is only capable of disrupting reproductive function when it exceeds some critical level it may still be of interest, but would fall more appropriately under the category of pathology.

In our laboratory we are currently conducting a series of studies designed to assess the effects of so-called "state," or acute, and "trait," or chronic, anxiety on ovarian function. State anxiety can be studied by using recognised acute stressors that are experienced by relatively homogeneous groups of women. We have been using standardised medical school admissions tests (MCAT) as one such stressor. Levels of ovarian function can be assessed in women taking these tests and in otherwise matched women who are not, both in the month in which the exam occurs and in a non-test month several months later. Concurrent measurements of cortisol levels and the administration of standardised psychological inventories allow independent assessment of the degree of physiological stress evoked by the test. The effect of the test can be examined both within and between women, with or without control for elevated cortisol levels or psychological test scores.

These studies are currently ongoing, but initial data suggests that even acute sources of anxiety, like MCAT tests, do not necessarily affect ovarian function the way moderate energetic stresses do (Figure 6.3). A group of 24 women who studied for and took the MCAT exam in the spring of 1997 had cortisol levels that were slightly, but not significantly, higher than 19 women matched for age who were not taking the MCAT (7892 ± 604 vs. 6962 ± 586 pmol/L, mean ± SE). The slight difference in mean cortisol levels was primarily a consequence of elevated levels among the test takers on the day of the test. Average mid-luteal progesterone levels measured in the luteal phase immediately before and immediately after the test in both

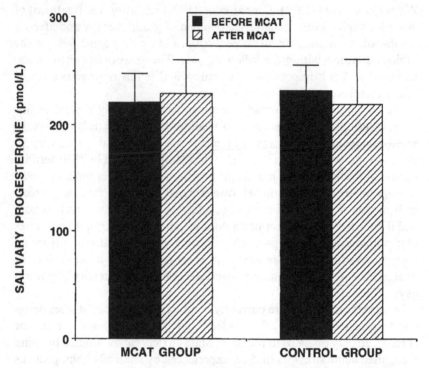

Figure 6.3   Comparison of average mid-luteal progesterone levels from luteal phases immediately before and immediately after the MCAT test in 24 MCAT takers and 19 controls matched for age. There were no significant differences either between groups within time periods, or between time periods within groups

test takers and controls showed no significant differences either within or between groups (MCAT group before: 221 ± 27 pmol/L; MCAT group after: 229 ± 31; control group before: 232 ± 24; control group after: 219 ± 42).

Chronic, or trait, anxiety, however, needs to be assessed separately. Differences in trait anxiety, as the term implies, probably represent differences in individual temperament more than differences in environmental factors. The ambiguity deriving from ascertainment bias inherent in clinical studies can be avoided in this case by recruiting a random sample of women blind to their own levels of ovarian function, assessing their levels of trait anxiety through standardised inventories, and measuring their ovarian function. The important contrast in this case is between women rather than within women, since trait anxiety is assumed to be stable over time. We are currently conducting a large study following this design. Preliminary data, however, once again suggest that ovarian fun-

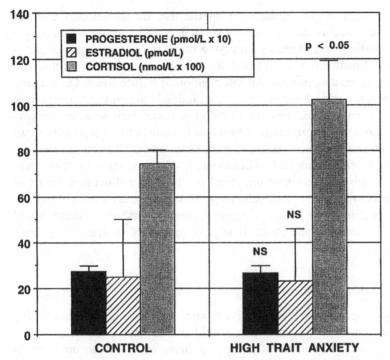

Figure 6.4 Comparison of average (±SE) indices of luteal progesterone, follicular oestradiol, and morning cortisol assayed in saliva samples from 6 women scoring ⩾ one standard deviation above the population standardised average for trait anxiety on the State-Trait Anxiety Inventory compared with similar indices from 12 women with scores within one standard deviation of the standardised average. Cortisol indices are significantly different between the two groups (p 0.05); progesterone and oestradiol indices are not significantly different

ction may be insensitive to variation in chronic anxiety (Figure 6.4). If we compare indices of ovarian function among a sample of 6 women who score at least 1 standard deviation above the average in trait anxiety with similar indices from 12 women with average or low scores on this scale, we find no significant differences. The groups do, however, differ in their average cortisol levels, suggesting that a high score for trait anxiety is associated with this physiological indicator of stress.

The demonstration of causation, as opposed to mere association, is always difficult, and usually involves experiments where the independent variable is manipulated directly and subsequent changes in the dependent variable assessed. In most research in human biology this ideal can only be approximated. In the first research design described above relative to anxiety, the administration of the medical school admissions test is used to approximate the application of a situational stress. It is only an approximation, because the subjects taking the test are self-selected and not randomly assigned to that condition. In a second approximation we are collaborating with Domar in studying the effect of a program of group therapy directly on ovarian function. Therapy as a treatment in this case approximates an experimental intervention to reduce stress. Once again, however, the subject population is not randomly drawn and may overrepresent women whose ovarian function is particularly sensitive to psychosocial stress. In an attempt to partially control for this possibility we will also study subjects undergoing the same therapy for conditions unrelated to reproduction, such as headaches and hypertension. Further extensions of this research direction could involve longitudinal monitoring of ovarian function, cortisol levels, and self-assessed stress arising from daily life. The cumulative weight of evidence from studies such as these should help to illuminate the effect, if any, of psychosocial stress on ovarian function.

Dietary composition

Dietary composition is implicated in some studies as a variable affecting ovarian function. For example, several studies have documented lower levels of ovarian function among vegetarian compared to omnivorous women (Pirke *et al.* 1986). Other studies have suggested a link between specific nutrients, such as fat, fiber, or soy products, and ovarian steroid levels (Lustig *et al.* 1989, 1990; Bentley 1987). In some cases the effects of these variations in dietary composition are confounded with differences in energy intake, energy expenditure, or restrained eating habits, so that the results are somewhat ambiguous. This is particularly true of studies that document differences in steroid levels among different ethnic groups. Native Chinese populations have been shown to have lower estradiol levels, for example, than populations of European descent, while immigrant Chinese populations to North America have levels that approach those of sympatric European populations as they become assimilated (MacMahon *et al.* 1971, 1974; Dickson *et al.* 1974; Trichopoulos *et al.* 1984; Goldin *et al.* 1986). The cause of the initial differences and their subsequent reduction is often attributed to diet. However dietary composition changes cannot easily be disentangled from a host of concurrent lifestyle changes, many of which are also known to affect ovarian function. The fact that the same populations show rapid convergence of growth and development patterns to those of sympatric European populations suggests broad-based changes in ecology and behavior and not isolated dietary differences.

The effects of dietary composition on ovarian function can also be somewhat paradoxical and counter-intuitive. For example, high dietary fiber intake has been associated with reduced circulating estradiol levels (Goldin 1982). In this case, the effect is likely the result of increased gut absorption of free estradiol and reduced enterohepatic recirculation (Goldin 1982, 1986). A reduction in circulating estradiol that is a consequence of increased clearance rates, however, is very different from reduced ovarian function. In fact, a predictable effect of increased clearance of estradiol is reduced levels of negative feedback at the level of the hypothalamus and pituitary, leading to increased gonadotropin release and increased levels of ovarian function. This principle is used clinically when estradiol antagonists, like clomiphene, are used to stimulate follicular development in anovulatory subjects. A recent study in my laboratory of women with differing levels of dietary fiber intake but comparable levels of calories, protein, and fat, revealed that high fiber consumers actually have shorter follicular phase lengths than controls, consistent with the hypothesised effect of increased peripheral clearance of estradiol (Figure 6.5, Halperin 1997). There were no observed differences between the groups in ovulatory frequency or luteal progesterone levels. Hence, although dietary fiber may lower circulating estradiol levels and thereby affect systemic estrogen actions – possibly lowering the risk of breast cancer, for example – effects on the reproductive axis itself may involve up-regulation rather than down-regulation.

Environmental pollution

Considerable concern has been raised recently over the potential effects of estrogenic compounds in the environment, particularly synthetic compounds such as polychlorinated biphenyls (PCBs) that can accumulate as residues from pesticide production, on human reproductive biology (Davis and Bradlow 1995). Both increasing breast cancer incidence among women and declining sperm counts among men have been attributed to the actions of these compounds (Carlsen *et al.* 1992; Stoll 1995; Auger *et al.* 1995; Irvine *et al.* 1996). Biologically active forms of PCBs have been isolated from human serum with both estrogenic and antiestrogenic potency (Connor *et al.* 1997; Moore *et al.* 1997; Nasaretnam *et al.* 1997). PCBs have also been found to be capable of stimulating proliferation of breast cancer cells lines (Shekhar *et al.* 1997; Gierthy *et al.* 1997).

Empirical evidence for deleterious effects of environmental estrogens on human fecundity, however, is very weak. Reports of declining sperm counts among men have been plagued by methodological problems which lead many to doubt whether there has in fact been any decline (Bromwich *et al.* 1994; Handelsman 1997). Several recent studies fail to find any

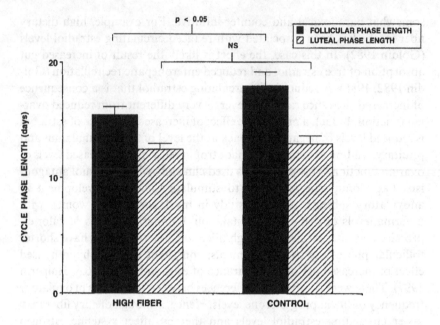

Figure 6.5   Average (±SE) follicular and luteal phase lengths for 11 women with high dietary fiber intake and 15 women with low to normal dietary fiber intake, the diets of the two groups being otherwise comparable in levels of calories, protein, and fat. High fiber consumers have shorter follicular phases than controls (p 0.05). Luteal phase lengths are not significantly different. (Halperin 1997)

evidence of a decline in sperm counts with time (Bujan *et al.* 1996; Fisch and Goluboff 1996; Fisch *et al.* 1996; Rasmussen *et al.* 1997; Handelsman 1997; Lemke *et al.* 1997) or have actually found evidence of an increase (Paulsen *et al.* 1996; Fisch *et al.* 1996).

A different research design to investigate the effects of industrial pollution on urban populations could be based on the comparison of populations which differ in exposures to pollutants while being as homogeneous in other respects as possible. We have identified a possible venue for such research in Poland. One legacy of industrial and environmental policy during the communist regime are levels of industrial pollution in some cities that are among the highest on record, particularly in the coal and steel producing region of Silesia. Yet within the same country cities of comparable size exist with remarkable low levels of industrial pollution (Figure 6.6, Table 6.1). The populations in both settings are genetically and culturally quite homogeneous. It would be a rather straightforward matter to assess ovarian function in age-matched women, or other sensitive biomarkers of reproductive function, from both situations. On the

Figure 6.6   Location of small cities in Poland with markedly different levels of industrial pollution. See Table. 6.1

basis of those results further research could focus on exposures to particular pollutants and begin to describe the nature of the dose response.

Conclusion

As the effects on ovarian function of intermediate variables such as these are elucidated, we are in a better position to articulate useful and testable predictions regarding the effects of urban environments. Most predictions will be conditional in form. For example, we can predict that urban environments, to the extent that they are associated with earlier maturation and larger adult body size, will also be associated with increased ovarian function among adult women and consequently higher risks of breast cancer. We can also predict that the age-pattern of fecundity will be the same in urban as in rural environments. Thus, to the extent that urban environments are associated with increased rates of female participation in the workplace and consequent delays in childbearing, more women may

Table 6.1 *Industrial pollution levels in Polish cities*

City	Population	Particulate Emissions (tons/year)
Konin	82,000	188,000
Tarnowskie Gory	81,500	187,000
Jaworzno	98,500	147,000
Rybnik	143,100	106,000
Tychy	120,000	100,000
Ostroleka	51,000	64,500
Lomza	61,500	4,800
Suwalki	63,900	4,800
Elk	53,800	3,100
Nowy Sacz	80,500	3,400
Krosno	49,500	2,700

Source: Polish Central Office for Statistics (Glowny Urzad Statystyczny, Warsaw), as published in *Ochrona Srodawiska* (1993). See also Figure 6.6.

experience difficulty conceiving in short periods of time leading to an increased demand for infertility services. To the extent that urban environments are associated with different levels of energy intake and expenditure – and here patterns may be very heterogeneous among different socioeconomic categories – associated changes in ovarian function can be predicted. To the extent that energetic stresses are reduced, the contraceptive effect of breastfeeding should be weaker. To the extent that urban environments may be characterised by higher levels of chronic or acute psychosocial stress, or higher levels of industrial pollution, we may be able to predict associated variation in levels of ovarian function. The predictions we make will be highly contingent on the specific features of the urban environment that may be present, and we recognise that heterogeneity within urban environments is likely to be very important. But the predictions themselves will rest on firm foundations born of empirical research.

The preceding examples are meant to illustrate the way in which the methods and models of reproductive ecology can contribute to our understanding of the effect of urban environments on this important aspect of human biology. The examples are certainly not exhaustive, but the same research program could be usefully applied, for example, to male reproductive function, or to fetal loss, or other reproductive endpoints. It may also be that an analogous approach, proceeding backward from dependent variables of interest toward independent variables associated with urban environments, carefully documenting the linkages among intermediate variables in the causal pathway, will prove useful in other areas of human biology as well.

References

Arnot, A. M., Vandekerckhove, P., DeBono, M. A. and Rutherford, A. J. (1995). Follicular volume and number during in-vitro fertilization: association with oocyte developmental capacity and pregnancy rate. *Human Reproduction*, **10**, 256–61.

Auger, J., Kuntsmann, J. M., Czyglik, F. and Jouannet, P. (1995). Decline in semen quality among fertile men in Paris during the past 20 years. *New England Journal of Medicine*, **332**, , 281–5.

Bailey, R. C., Jenike, M. R., Bentley, G. R., Harrigan, A. M. and Ellison, P. T. (1992). The ecology of birth seasonality among agriculturalists in central Africa. *Journal of Biosocial Science*, **24**, 393–412.

Bentley, G. R. (1985). Hunter-gatherer energetics and fertility: a reassessment of the !Kung San. *Human Ecology*, **13**, 79–109.

Bentley, G. R. (1987). Dietary fiber and age at menarche: a reexamination of the evidence. *Annals of Human Biology*, **14**, 69–75.

Bentley, G. R. (1996). How did prehistoric women bear "Man the Hunter"? Reconstructing fertility from the archaeological record. In *Gender and Archaeology*, ed. R. P. Wright, pp. 23–51, Philadelphia: University of Pennsylvania Press..

Bentley, G. R., Goldberg, T. and Jasienska, G. (1993). The fertility of agricultural and non-agricultural traditional societies. *Population Studies*, **47**, 269–81.

Bernstein, L., Henderson, B. E., Hanisch, R., Sullivan-Halley, J. and Ross, R. K. (1994). Physical exercise and reduced risk of breast cancer in young women. *Journal of the National Cancer Institute*, **86**, 1403–8.

Bongaarts, J. and Potter, R. G. (1983). *Fertility, Biology, and Behavior: An Analysis of the Proximate Determinants*. New York: Academic Press.

Bromwich, P., Cohen, J., Stewart, I. and Walker, A. (1994). Decline in sperm counts: an artefact of changed reference range of "normal." *British Medical Journal*, **309**, 19–22.

Buikstra, J. E., Konigsberg, L. W. and Bullington, J. (1986). Fertility and the development of agriculture in the prehistoric midwest. *American Antiquity*, **51**, 528–46.

Bujan, I., Mansat, A., Pontonnier, F. and Mieusset, R. (1996). Time series analysis of sperm concentration in fertile men in Toulouse, France between 1977 and 1992. *British Medical Journal*, **312**, 471–2.

Caldwell, J. C. (1982). *Theory of Fertility Decline*. New York: Academic Press.

Campbell, K. L. (1994). Blood, urine, saliva, and dip-sticks: experiences in Africa, New Guinea, and Boston. *Annals of the New York Academy of Sciences*, **709**, 312–30.

Campbell, K. L. and Wood, J. W. (1988). Fertility in traditional societies. In *Natural Human Fertility and Biological Determinants*, eds. P. Diggory, M. Potts and S. Teper, pp. 39–69. New York: Macmillan.

Carlsen, E., Giwercman, A., Keiding, N. and Skakkebaek, N. E. (1992). Evidence for decreasing quality of semen during past 50 years. *British Medical Journal*, **305**, 609–13.

Cohen, M. N. 1977. *The Food Crisis in Prehistory*. New Haven, CT: Yale University Press.

Cohen, M. N. (1989). *Health and the Rise of Civilization*, New Haven CT: Yale University Press.

Cohen, M. N. and Armelagos, G. J. (1984). *Paleopathology at the Origins of Agriculture*. New York: Academic Press.

Cohen, M. N., Malpass, R. S. and Klein, H. G. (1980). *Biosocial Mechanisms of Population Regulation*. New Haven, CT: Yale University Press.

Connor, K., Ramamoorthy, K., Moore, M., Mustain, M., Chen, I., Safe, S., Zacharewski, T., Gillesby, B., Joyeux, A. and Balaguer, P. (1997). Hydroxylated polychlorinated biphenyls (PCBs) as estrogens and antiestrogens: structure activity relationships. *Toxicology Applied Phamacology*, **145**, 111–23.

Davis, D. L. and Bradlow, H. L. (1995). Can environmental estrogens cause breast cancer? *Scientific American*. Oct. 1995, 167–72.

Davis, K. and Blake, J. (1956). Social structure and fertility: an analytic framework. *Economic Development and Cultural Change*, **4**, 211–35.

Demyttenaere, K., Nijs, P., Koninckx, P. R., Steeno, O. and Evers-Kiebooms, G. (1988). Anxiety and conception rates in donor insemination couples. *Journal of Psychosomatic Obstetrics and Gynecology*, **8**, 175–81.

Dickey, R. P., Olar, T. T., Taylor, S. N., Curole, D. N. and Harrigill, K. (1993). Relationship of biochemical pregnancy to pre-ovulatory endometrial thickness and pattern in patients undergoing ovulation induction. *Human Reproduction*, **8**, 327–30.

Dickson, L. E., MacMahon, B., Cole, P. and Brown, J. B. (1974). Estrogen profiles of oriental and caucasian women in Hawaii. *New England Journal of Medicine*, **291**, 1211–13.

Domar, A. D., Seibel, M. M. and Benson, H. (1990). The mind/body program for infertility: a sequel to aberrant folliculogenesis. *Fertility and Sterility*, **35**, 489–99.

Downey, J. and McKinney, M. (1992). The psychiatric status of women presenting for infertility evaluation. *American Journal of Orthopsychiatry*, **62**, 196–205.

Dubey, A. K., Wang, H. A., Duffy, P. and Penzias, A. S. (1995). The correlation between follicular measurements, oocyte morphology, and fertilization rates in an in vitro fertilization program. *Fertility and Sterility*, **64**, 787–90.

Eaton, S. B., Pike, M. C., Short, R. V., Lee, N. C., Trussell, J., Hatcher, R. A., Wood, J. W., Worthman, C. M., Blurton Jones, N. G., Konner, M. J., Hill, K. R., Bailey, R. and Hurtado, A. M. (1994). Women's reproductive cancers in evolutionary context. *Quarterly Review of Biology*, **69**, 353–67.

Eisner, B. G. (1963). Some psychological differences between fertile and infertile women. *Journal of Clinical Psychiatry*, **19**, 391–5.

Ellison, P. T. (1995). Understanding natural variation in human ovarian function. In Human Reproductive Decisions: Biological and Social Perspectives, ed. R. I. M. Dunbar, pp. 22–51. New York: St. Martin's Press.

Ellison, P. T. (1996). Developmental influences on adult ovarian function. *American Journal of Human Biology*, **8**, 725–34.

Ellison, P. T. (1996). Age and developmental effects on adult ovarian function. In

Variability in Human Fertility: A Biological Anthropological Approach, eds. L. Rosetta and C. N. G. Mascie-Taylor, pp. 69–90. Cambridge: Cambridge University Press.

Ellison, P. T. (1988). Human salivary steroids: methodological considerations and applications in physical anthropology. *Yearbook of Physiological Anthropology*, **31**, 115–32.

Ellison, P. T. and Lager, C. (1986). Moderate recreational running is associated with lowered salivary progesterone profiles in women. *American Journal of Obstetrics and Gynecology*, **154**, 1000–3.

Ellison, P. T., Lipson, S. F., O'Rourke, M. T., Bentley, G. R., Harrigan, A. M., Panter-Brick, C. and Vitzthum, V. J. (1993). Population variation in ovarian function. *Lancet*, **342**, 433–4.

Ellison, P. T., Peacock, N. R. and Lager, C. (1989). Ecology and ovarian function among Lese women of the Ituri Forest, Zaire. *American Journal of Physical Anthropology*, **78**, 519–26.

Ellison, P. T. (1995). Breastfeeding and fertility. In *Breastfeeding: A Biocultural Perspective*, eds., K. A. Dettwyler and P. Stuart-Macadam, Hawthorne NY: Aldine/de Gruyter.

Ellison, P. T. (1988). Human salivary steroids: methodological considerations and applications in physical anthropology. *Yearbook of Physical Anthropology*, **31**, 115–42.

Ellison, P. T. (1990). Human ovarian function and reproductive ecology: new hypotheses. *American Anthropology*, **92**, 933–52.

Ellison, P. T. (1998). Reproductive ecology and reproductive cancers. In *Hormones and Human Health*, eds., C. Panter-Brick and C. Worthman, Cambridge: Cambridge University Press.

Eveleth, P. B. and Tanner, J. M. (1990). *Worldwide variation in human growth*, 2nd edn. Cambridge: Cambridge University Press.

Fisch. H. and Goluboff, E. T. (1996). Geographic variations in sperm counts: a potential cause of bias in studies of semen quality. *Fertility and Sterility*, **65**, 1044–6.

Fisch, H., Goluboff, E. T. and Olson, J. H. (1996). Semen analyses in 1283 men from the United States over a 25-year period: no decline in quality. *Fertility and Sterility*, **65**, 1009–14.

Ford, K., Huffman, S. L., Chowdhury, A. K. M. A., Becker, S., Allen, H. and Menken, J. (1989). Birth-interval dynamics in rural Bangladesh and maternal weight. *Demography*, **26**, 425–37.

Gierthy, J. F., Arcaro, K. F. and Floyd, M. (1997). Assessment of PCB estrogenicity in human breast cancer cell line. *Chemosphere*, **34**, 1495–505.

Goldin, B. R., Adlercreutz, H., Gorbach, S. L., *et al.* (1982). Estrogen excretion patterns and plasma levels in vegetarian and omnivorous women. *New England Journal of Medicine*, **307**, 1542–7.

Goldin, B. R., Aldercreutz, H., Gorbach, S. L., Woods, M. N., Dwyer, J. T., Conlon, T. , Bohn, E. and Gershoff, S. N. (1986). The relationship between estrogen levels and diets of Caucasian American and Oriental immigrant women. *American Journal of Clinical Nutrition*, **44**, 945–53.

Green, B. B., Weiss, N. S. and Daling, J. R. (1988). Risk of ovulatory infertility in relation to body weight. *Fertility and Sterility*, **50**, 721–6.

Halperin, F. (1997). The effect of dietary fiber on ovarian function. A. B. honors

Thesis, Department of Anthropology, Harvard University.

Handelsman, D. J. (1997). Sperm output of healthy men in Australia: magnitude of bias due to self-selected volunteers. *Human Reproduction*, **12**, 2701–5.

Harrison, R. F., O'Moore, A. M., O'Moore, R. R. and McSweeney, J. R. (1981). Stress profiles in normal infertile couples: pharmacological and psychological approaches to therapy. In *Advances in diagnosis and treatment of infertility*, eds. V. Insler and G. Bettendorf, pp. 143–57. New York: Elsevier North Holland.

Henry, L. (1961). Some data on natural fertility. *Eugenics Quarterly*, **8**, 81–91.

Huffman, S. L., Chowdhury, A., Allen, H. and Nahar, L. (1987). Suckling patterns and post-partum amenorrhea in Bangladesh. *Journal of Biosocial Science*, **19**, 171–9.

Irvine, S., Cawood, E. and Richardson, D. (1996). Evidence of deteriorating semen quality in the United Kingdom: birth cohort study in 577 men in Scotland over 11 years. *British Medical Journal*, **312**, 467–71.

Jasienska, G. and Ellison, P. T. (1998). Physical work causes suppression of ovarian function in women. *Proceedings of the Royal Society: Biological Sciences*, **265**, 1847–54.

Lager, C. and Ellison, P. T. (1990). Effect of moderate weight loss on ovarian function assessed by salivary progesterone measurements. *American Journal of Human Biology*, **2**, 303–12.

Lasley, B. L., Mobed, K. and Gold, E. B. (1994). The use of urinary hormonal assessment in human studies. *Annals of the New York Academy of Sciences*, **709**, 299–311.

Lemcke, B., Behre, H. M. and Nieschlag, E. (1997). Frequently subnormal semen profiles of normal volunteers recruited over 17 years. *International Journal of Andrology*, **20**, 144–52.

Lenton, E. A., Gelsthorp, C. H. and Harper, R. (1988). Measurement of progesterone in saliva: assessment of the normal fertile range using spontaneous conception cycles. *Clinical Endocrinolology*, **38**, 637–46.

Lipson, S. F. and Ellison, P. T. (1996). Comparison of salivary steroid profiles in naturally occurring conception and nonconception cycles. *Human Reproduction*, **11**, 2090–7.

Lipson, S. F. and Ellison, P. T. (1992). Normative study of age variation in salivary progesterone profiles. *Journal of Biosocial Science*, **24**, 233–44.

Liu, H. C., Jones, G. S., Jones, H. W. Jr, and Rosenwaks, Z. (1988). Mechanisms and factors of early pregnancy wastage in in vitro fertilization-embryo transfer patients. *Fertility and Sterility*, **50**, 95–101.

Lustig, R. H., Bradlow, H. L. and Fishman, J. (1989). Estrogen metabolism in disorders of nutrition and dietary composition. In *The Menstrual Cycle and its Disorders*. eds. K. M. Pirke, W. Wuutke and U. Schweiger, pp. 119–32. Berlin: Springer-Verlag.

Lustig, R. H., Hershcopf, R. J. and Bradlow, H. L. (1990). The effects of body weight and diet on estrogen metabolism and estrogen-dependent disease. In *Adipose Tissue and Reproduction*. ed. R. E. Frisch, 107–124. Basel: Karger.

MacMahon, B., Cole, P., Brown, J. B., Aoki, K., Lin, T. M., Morgan, R. W. and Woo, N-C. (1971). Oestrogen profiles of Asian and North American women. *Lancet*, **2**, 900–2.

MacMahon, B., Cole, P., Brown, J. B., Aoki, K., Lin, T. M., Morgan, R. W. and Woo, N-C. (1974). Urine estrogen profiles of Asian and North American women. *International Journal of Cancer*, **14**, 161–7.

Mai, F. M., Munday, R. N. and Rump, E. E. (1972). Psychiatric interview comparisons between infertile and fertile couples. *Psychosomatic Medicine*, **34**, 431–40.

Malina, R. M. and Himes, J. H. (1977). Seasonality of births in a rural Zapotec municipio, 1945–1970. *Human Biology*, **49**, 125–37.

Meldrum, D. R. (1993). Female reproductive aging–ovarian and uterine factors. *Fertility and Sterility*, **59**, 1–5.

Moore, M., Mustain, M., Daniel, K., Chen, I., Safe, S., Zacharewski, T., Gillesby, B., Joyeux, A. and Balaguer, P. (1997). Antiestrogenic activity of hydroxylated polychlorinated biphenyl congeners identified in human serum. *Toxicology and Applied Pharmacology*, **142**, 160–8.

Mosher, S. W. (1979). Birth seasonality among peasant cultivators: the interrelationship of workload, diet and fertility. *Human Ecology*, **7**, 151–81.

Nasaretnam, K., Corcoran, D., Dils, R. R. and Darbre, P. (1997). 3,4,3',4'-Tetrachlorobiphenyl acts as an estrogen in vitro and in vivo. *molecular Endocrinology*, **10**, 923–36.

Nijs, P., Koninckx, P. R., Verstraeten, D., Mullens, A. and Nicasy, H. (1984). Psychological factors of female infertility. *European Journal of Obstetrics and Reproductive Biology*, **18**, 375–9.

O'Rourke, M. T. and Ellison, P. T. (1993). Age and prognosis in premenopausal breast cancer. *Lancet*, **342**, 60.

O'Rourke, M. T., Lipson, S. F. and Ellison, P. T. (1996). Ovarian function in the latter half of the reproductive lifespan. *American Journal of Human Biology*, **8**, 751–60.

Omran, A. R. (1971). The epidemiological transition: a theory of the epidemiology of population change. *Millbank Memorial Fund Quarterly*, **49**, 509–38.

Panter-Brick, C. and Ellison, P. T. (1994). Seasonality of workloads and ovarian function in Nepali women. *Annals of the New York Academy of Sciences*, **709**, 234–5.

Panter-Brick, C., Lotstein, D. S. and Ellison, P. T. (1993). Seasonality of reproductive function and weight loss in rural Nepali women. *Human Reproduction*, **8**, 2248–58.

Paulsen, C. A., Berman, N. G. and Wang, C. (1996). Data from men in greater Seattle area reveals no downward trend in semen quality: further evidence that deterioration of semen quality is not geographically uniform. *Fertility and Sterility*, **65**, 887–93.

Peacock, N. R. (1990). Comparative and cross-cultural approaches to the study of human female reproductive failure. In Primate Life History and Evolution, ed. C. J. DeRousseau, pp. 195–220, New York: Wiley-Liss.

Pirke, K. M., Schweiger, U., Laessle, R., Dickhaut, B., Schweiger, M. and Waechtler, M. (1986). Dieting influences the menstrual cycle: vegetarian versus nonvegetarian diet. *Fertility and Sterility*, **46**, 1083–8.

Rasmussen, P. E., Erb, K., Westergaard, L. G. and Laursen, S. B. (1997). No evidence for decreasing semen quality in four birth cohorts of 1,055 Danish men born between 1950 and 1970. *Fertility and Sterility*, 1059–64.

Schenker, J. G., Melrow, D. and Schenker, E. (1992). Stress and human reproduction. *European Journal of Obstetrics, Gynaecology, and Reproductive Biology,* **45**, 1–8.

Schover, L. R., Collins, R. L. and Richards, S. (1992). Psychological aspects of donor insemination: evaluation and follow-up of recipient couples. *Fertility and Sterility*, **57**, 15–28.

Schover, L. R., Greenhalgh, L. F., Richards, S. I. and Collins, R. L. (1994). Psychological screening and the success of donor insemination. *Human Reproduction*, **9**, 176–8.

Schweiger, U., Tuschl, R. J., Platte, P. , Broocks, A., Laessle, R. G. and Pirke, K-M. (1992). Everyday eating behavior and menstrual function in young women. *Fertility and Sterility*, **57**, 771–5.

Shekhar, P. V., Werdell, J. and Basrur, V. S. (1997). Environmental estrogen stimulation of growth and estrogen receptor function in preneoplastic and cancerous human breast cell lines. *Journal of National Cancer Institute*, **89**, 1774–82.

Short, R. (1987). The biological basis for the contraceptive effects of breast feeding. *International Journal of Gynaecology and Obstetrics*, 25 suppl: 207–17.

Stewart, D. R., Overstreet, J. W., Nakajima. S. T. and Lasley, B. L. (1993). Enhanced ovarian steroid secretion before implantation in early human pregnancy. *Journal of Clinical Endocrinology and Metabolism*, **76**, 1470–6.

Stoll, B. A., ed. (1995). *Reducing breast cancer risk in women.* Amsterdam: Kluwer Academic.

Sukalich, S., Lipson, S. F. and Ellison, P. T. (1994). Intra- and interwoman variation in progesterone profiles. *American Journal of Physical Antropology*, **18**, 191.

Tay, C. C. K., Glasier, A. F. and McNeilly, A. S. (1996). Twenty-four hour patterns of prolactin secretion during lactation and the relationship to suckling and the resumption of fertility in breast-feeding women. *Human Reproduction*, **11**, 950–5.

Trichopoulos, D., Yen, S., Brown, J., Cole, P. and MacMahon, B. (1984). Effect of westernization on urine estrogens, frequency of ovulation, and breast cancer risk. *Cancer*, **53**, 187–92.

Wasser, S. K. (1990). Infertility, abortion, and biotechnology: when it's not nice to fool mother nature. *Human Nature*, **1**, 3–25.

Wasser, S. K. (1994). Psychosocial stress and infertility: cause or effect? *Human Nature*, **5**, 293–306.

Wasser, S. K. and Isenberg, D. Y. (1986). Reproductive failure among women: pathology or adaptation. *Journal of Psychosomatic Obstetrics and Gynaecology*, **5**, 153–75.

Wasser, S. K., Sewall, G. and Soules, M. R. (1993). Psychosocial stress as a cause of infertility. *Fertility and Sterility*, 59(3):685–9.

Worthman, C. M., Jenkins, C. L., Stallings, J. F. and Lai, D. (1993). Attenuation of nursing-related ovarian suppression and high fertility in well-nourished, intensively breast-feeding Amele women of lowland Papua New Guinea. *Journal of Biosocial Sciences*, **25**, 425–43.

Worthman, C. M., Stallings, J. F. and Gubernick, D. (1991). Measurement of hormones in blood spots: a non-isotopic assay for prolactin. *American Journal of Physical Anthropology*, Suppl., **12**, 186–7.

Wright, J., Duchesne, C., Sabourin, S., Bissonnette, F., Benot, J. and Girard, Y.

(1991). Psychosocial distress and infertility: men and women respond dif-ferently. *Fertility and Sterility*, **55**, 100–8.

Yoshimura, Y. and Wallach, E. E. (1987). Studies of the mechanism(s) of mam-malian ovulation. *Fertility and Sterility*, **47**, 22–34.

7 Pollution and child health

L. M. SCHELL AND A. D. STARK

Editors' introduction

Contemporary urban environments in industrialised countries that contain features that impact on human biology and health are represented in this essay that focuses on children. Two approaches to human biologic study are demonstrated: (1) starting with a factor of the urban environment and tracing its effects, and (2) starting with a characteristic of urban populations and seeking the causes of it in the urban physicosocial environment. The analysis of the current asthma epidemic is an example of the latter, and it describes the role that other animals, e.g. cockroach and dust mites, may play in triggering asthma sensitisation and attacks. Although pollution is often included among the new environmental features of cities, rarely are animal allergens considered even though they are pollutants in the strict sense of the term. In an example of the first analytic approach, the effects of low level lead pollution on child development are illustrated with new data from a longitudinal study of infant growth. Just as Ellison's approach in the previous chapter illustrates the importance of measuring putative proximate determinants of ovarian function, the illustrations in this chapter emphasise the measurement of environmental features, whether allergens or lead, and move away methodologically from urban–rural comparisons. Even if there are no differences between urban and rural comparison groups in some biologic parameter, it can result from strong but counterbalancing effects. When environmental, social and biological factors are as interwoven as they are in cities, analysis of specific factors is thought to yield more secure and generalisable statements of biocultural relationships.

Pollution is a ubiquitous feature of urban environments in developed countries. Although levels vary within and between populations, few individuals today can be found who do not bear some pollutant burden. Metals are many times more concentrated in human tissues now than 500 years ago. For example, the level of lead may have increased more than 300 fold over the same period (Committee on Measuring Lead in Critical Populations 1993). Polychlorinated biphenyls, pesticides and herbicides, as well as their metabolites, can be detected in samples from humans with no specific occupational exposure (Rogan and Gladen 1985). Urban populations are particularly at risk because many of the largest sources of pollutants, primarily industrial and transportation sources, have been concentrated in cities. Children are at special risk because they are exposed to pollutants at critical times of development, their exposure may be greater because they often have more contact with dirt and dust than adults, and they inhale more air and consume more food and water per unit of body weight than adults (Bearer 1995). Furthermore, effects on children may be lifelong (Landrigan and Carlson 1995) and may affect growth, learning, fertility and morbidity (Rogan 1995; Schell 1991; Waldbott 1978).

Despite the importance of pollution, it is difficult to define precisely. It has no specific referent in nature and its definition is somewhat arbitrary. It usually includes materials and energy such as radiation and sound that are unwanted because of real or supposed detrimental effects on health. Although some materials commonly considered pollutants are produced by natural forces (e.g. particulates from volcanoes, methane from organic waste), the term most often refers to materials or energy that are produced through human activity. Often this definition is broadened to include any environmental threats to the quality of life whether to humans or non-humans. Many of the detrimental effects on humans concern areas of human function that are of traditional interest to human biologists: reproduction, aging and growth as well as exercise, cardiovascular and respiratory physiology. This chapter seeks to show how pollutants are related to certain of these parameters in children, and in so doing, address the methods human biologists employ to study the human biology of urban populations.

Human biologists have approached the study of interactions between the environment and biology from myriad perspectives, but these may be grouped into two categories: (1) studies that begin with a particular outcome, such as measures of respiratory physiology and trace constituent variation to features of the environment; and (2) studies that begin with a feature of the environment, such as a metal, or a dietary or energy expenditure regime, and follow the consequences for areas of human function of interest to human biologists such as reproduction, aging, exercise physiology and growth. Both types of approach have concrete implications for the planning of research and problems in measurement that are likely to occur. To illustrate each approach, this chapter will first examine the impact of a specific pollutant, lead, on child growth, and then examine the causes of a specific disease, asthma, tracing its connection to pollution and other urban factors.

Both approaches require careful definition and measurement of both exposure and outcome. Techniques for exposure measurement can be based on either estimating behavioral regimes that put an individual or group at risk, or directly measuring the body's burden of the pollutant or an immediate effect of it (a biomarker). Pollutants, especially mixtures, and other environmental factors that cannot be measured will be more difficult to relate to specific biological outcomes of interest to human biologists.

Lead in the United States

Among the many pollutants present in urban environments, lead is chosen as an example because it leaves a measurable residue in the body that marks past exposure, and because it affects human function in socially meaningful ways. Concern about lead exposure intensified during the early 1970s in the United States public health community (Lin-Fu 1973). At that time the usual definition of lead poisoning was a blood lead level of 60 μg/dL, with 40 μg/dL being defined as an undue body burden. In 1973 the undue body burden was revised downward by the United States Centers for Disease Control and Prevention (CDC) to 30 μg/dL. A national survey of lead levels conducted through National Health and Nutrition Examination Survey (NHANES) in the late 1970s demonstrated that one in six African-American children,[1] 0.5–5 years of age who lived in the central area of cities with a million or more inhabitants, had a blood lead concentration above the action level of 30 μg/dL (Annest et al. 1982). Both white and non-white American children had consistently higher lead levels in urban places.

Studies showed that the most common proximal sources of lead were dust, air, food and water (Figure 7.1), and that the main route of uptake was through ingestion (Agency for Toxic substances and Disease Registry 1988). In children, acute exposure occurs through ingestion of lead-containing paint chips and chronic exposure from chronic ingestion of minute quantities of lead-containing dust that precipitates from the air onto mundane objects that are put into the mouth. The daily ingested dose is 10–20 times larger than the daily respiratory dose. In the United States, the ultimate sources of lead were traced to emissions from automobiles that used lead-containing gasoline, dilapidated structures that shed lead-based paint dust, and to a lesser extent, food and water. In countries where lead-based paint is uncommon, leaded gasoline is a prominent source of the pollutant. Industrial emissions, from smelters, for example, can contribute substantially to local air pollution regardless of what other sources may be in the environment.

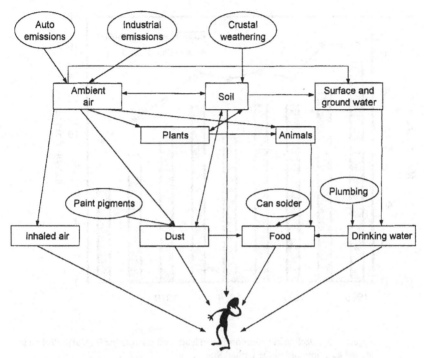

Figure 7.1   Sources and routes of exposure to lead (adapted from Agency for Toxic substances and Disease Registry, 1988)

To protect catalytic converters in automobiles, the United States Clean Air Act mandated a reduction in the use of leaded gasoline. This had the additional effect of greatly reducing lead levels in air and particularly dust. The effect on lead levels in people was considerable (Figure 7.2). Lead levels in the United States declined in tandem with the decline in the use of leaded gasoline (Annest *et al.* 1983). Other government actions such as efforts to remediate lead-contaminated housing, have also contributed. As efforts to reduce lead exposures were enacted through legislation, lead levels in the United States declined (Pirkle *et al.* 1994). Simultaneously, new research suggested that commonly occurring lead levels could be associated with detrimental effects in children (Table 7.1: for review see Agency for Toxic Substances and Disease Registry 1988), and the lead level at which the United States Centers for Disease Control recommended clinical involvement was reduced to 10 μg/dL in 1991.

Data from the NHANES III survey conducted from 1988 to 1991 show a decline in the average blood lead level among children, but the percentage of children classified as lead impacted (above 10 μg/dL) remains substantial (Brody *et al.* 1994). The most recent data from Phase 2 of the

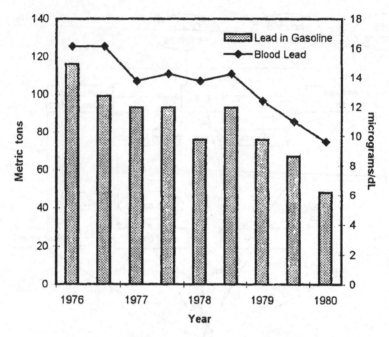

Figure 7.2   Decline in children's blood lead level mirrors the legislated decline in
leaded gasoline sold in the United States.
Source: adapted from Annest *et al.* (1983)

NHANES III survey on lead levels (October 1991 to September 1994),
reveals that differences among social groups defined by ethnicity or race
persist (Morbidity and Mortality Weekly Report 1997). African-American
and Hispanic subpopulations have higher rates of elevated lead levels in
children than whites; over 35% of urban African-American children have
blood lead levels above 10 µg/dL. Within groups, urban–rural gradients
differ: they remain strong for non-whites but are trivial for whites, reveal-
ing the changing nature of the urban environment and the different interactions that socially defined groups have with it. The age of housing is now
an important risk factor, and when combined with the risk associated with
African-American ethnicity, the likelihood of an elevated lead level is
enormous: young African-American children living in housing built before
1946 have approximately ten times the risk of elevated body burdens of
lead compared to whites living in post-1973 housing.

At the same time that levels are at their lowest, new research has pushed
the level at which we might be concerned about lead to a new low level.
Preliminary results from a study showed that maturity at birth, indexed by
the Brazelton Neonatal Assessment Scale, was reduced even for neonates

Table 7.1. *Effects on children of blood lead at different levels*

Lowest effects Pb-B (µg/dL)	Neurological effects	Heme synthesis Effects	Other effects
10–15 (prenatal and postnatal)	Deficits in neurobehavioral development (Bayley and McCarthy Scales) Electrophysiological changes	ALA-D inhibition	Reduced gestational age and weight at birth Reduced size up to age 7–8 years Hearing deficits
15–20		EP elevation	Impaired vitamin D metabolism; Py-5-N inhibition; Increased erythrocyte protoporphyrin production
< 25	Lower IQ, slower reaction time (studied cross-sectionally)		
30	Slowed nerve conduction velocity		
40		Reduced hemoglobin; elevated CP and ALA-U	
70	Peripheral neuropathies	Frank anemia	
80	Encephalopathy		Colic, other GI effects; kidney effects
> 80	Death		

Source: Adapted from Agency for Toxic Substances and Disease Registry (1988).

with lead levels below 10 µg/dL (Emory *et al.* 1997). Questions remain as to whether lead has detrimental effects on child development at the lowest commonly observed levels.

The Albany Pregnancy Infancy Lead Study

To investigate the effects of low levels of lead on child growth and development we initiated the Albany Pregnancy Infancy Lead Study. Growth and development was chosen as a focal health outcome because of its significance to the fields of public health and human biology (Schell 1997). To better understand how lead burdens are acquired, we sought to determine the effect of maternal nutrition during pregnancy on lead transmission to the fetus, and of the infant's diet on the infant's subsequent lead levels. The study has been conducted in two phases: Phase I with recruitment from

1986 to 1992, and Phase II with recruitment from 1992 to 1998. The two phases differ in length of infant follow-up (24 *vs.* 12 months, respectively), the types of developmental outcomes assessed, and the technique used to measure lead in blood. Only Phase I data have been analysed at this time, and the analysis presented here focuses on size at birth.

The study sample was drawn from the population of women attending the Albany County Obstetric Clinic in Albany, New York. Women utilising the clinic are considered to be at high risk for adverse pregnancy outcome, and the clinic population is drawn primarily from the working and lower-class residents of Albany County. Over 95% of all clinic patients are eligible for the Women's and Infant's Supplemental Foods Program, a government program of nutrition supplementation and have incomes below 1.85 times the federally defined poverty limit.

To be included in the sample, a woman had to have been seen at the clinic for prenatal care before the 24th week of her pregnancy. The sample of neonates analysed in this report are the offspring of those women who were recruited between June 16, 1986 and October 30, 1989. Only women of either African-American or white race/ethnicity are included in the analysis owing to the small number of women of other ethnicities in our sample.

Descriptive statistics for the sample of mothers are presented in Table 7.2. On average, mothers are young and multiparous. Most women (73%) are unmarried; 57% of the sample analysed here are African-American, the remainder white, and 72% were unemployed when seen at the clinic for their first prenatal visit. Half of the mothers completed secondary school while 11% have completed one or more years of higher education.

Data were collected from interviews with mothers during pregnancy and from clinic records. Neonatal anthropometry was performed in the hospital nursery by one of the investigators (LMS) and three specially trained pediatric nurses. The examination included length, using a Harpenden neonatometer, head circumference, upper arm circumference, triceps and subscapular skinfolds. Birthweight and chest circumference measurements were taken from hospital records, and to assess maturation at birth, a Dubowitz examination to estimate maturation and gestation length was performed in the nursery.

The lead level of the neonate was determined from an umbilical cord blood sample obtained in the delivery room or in a few cases from a venous sample taken in the nursery. Lead levels in venous and cord bloods did not differ in our sample. Blood lead analyses were performed using the anodic stripping voltametry method which is capable of quantitatively measuring blood lead to a minimum level of 4–5 µg/dL. Exact quantification of lead levels below 5 µg/dL is not possible.

Table 7.2. *Characteristics of mothers and infants, Albany Pregnancy Infancy Lead Study (* n = *130)*

	Mean	SD
Characteristics of the neonate		
Blood lead (μg/dL)	5.7	2.42
Birth weight (gms)	3374	593.7
Dubowitz – gestation age (wks)	39.4	1.60
Subscapular skinfold (mm)	3.9	1.22
Triceps skinfold (mm)	3.5	1.05
Upper arm circumference (cm)	10.1	1.05
Length (cm)	49.6	2.24
Chest circumference (cm)	33.2	2.31
Head circumference (cm)	34.7	1.49
Characteristics of mother		
Age (yrs)	21.7	5.15
Mother's education (yrs)	11.2	1.58
Number of pregnancies	2.59	1.86
Cigarette use (cig/day)[a]	4.8	8.5
Alcohol use (oz/wk)[a]	0.09	0.434
Weight (lbs.)[a]	161.1	36.89
Arm circumference (cm)[a]	28.3	4.86
Triceps skinfold (mm)[a]	20.2	8.41
Weight change 2nd to 3rd trim. (lbs)	13.0	6.87
Height (in)	64.1	2.99

[a] Average of 2nd and 3rd trimester values.

Table 7.3. *Distribution of blood lead levels at birth*

Blood lead[a] μg/dL	N	Trichotomous <5,5-7, >7	Dichotomous <5, ≧5
2	6		
3	19	48	48
4	23	—	—
5	29		
6	10	53	
7	14	—	
8	9		82
9	8		
10	7	29	
11	4		
13	1		

[a] Values below 5 are imputed.

The distribution of neonatal blood lead levels is shown in Table 7.3. Of all neonates, 48 (37%) had blood lead levels below 5 µg/dL. The distribution of blood lead levels above 4 µg/dL follows a log normal distribution. The highest blood lead level was 13 µg/dL.

Because so many cases had non-quantifiable levels of lead in blood, we tested for effects of lead on the neonate's anthropometry in two ways. One approach was to treat lead as a continuous variable with values for cases with non-quantifiable values assigned by an imputation technique. The most appropriate imputation technique is one which would retain the basic shape, mean and variance of the true distribution. Since lead levels usually follow a log-normal distribution, we have used the observable extent of this distribution above 4 µg/dL and estimated the cases to be expected having values of 4, 3, 2 and 1 µg/dL. These values were randomly assigned to the cases with non-quantifiable levels, but in proportion to the number of cases expected from the log-normal distribution. The advantage of this technique is that it uses information about known distributions of lead levels to preserve the mean and variance in lead levels. However, since the assignment of values to particular cases is random, the effect on hypothesis testing is to reduce the strength of whatever true association is present. Several alternative models for imputation have also been tried which involved substituting one value (either 3 or 4) for all non-quantifiable levels. The results from these methods do not differ from the results using the other approaches and will not be described further. After imputation, and prior to all statistical analyses, all lead levels were transformed to their natural logarithm.

The second approach is to treat lead level as a categorical variable dividing the sample into two groups: cases with quantifiable lead level of 5 µg/dL or more, and non-quantifiable cases with lead levels of 4 µg/dL or less. We also have trichotomised the sample by dividing it into a low level group with non-quantifiable lead levels, a moderate group with lead levels of 5, 6 and 7 µg/dL, and a higher lead group with values above 7 µg/dL.

The intent of employing different approaches is to include in our analysis all the cases with non-quantifiable levels since they constituted such a large proportion of the data (37%) and could provide important information about the effect of lead. Consistency of results among analyses using these several techniques is considered strong evidence for a real association of neonatal lead level with the dependent variables under investigation.

Before testing for an effect of lead we adjusted for other influences on neonate anthropometry (Table 7.4). To insure an adequate adjustment model, we included a variable in the model if it is related to the dependent variable as determined by either a significant ($p < 0.10$) bivariate cor-

Table 7.4. *Models of covariates used in multiple regression analysis of infant anthropometry*

Independent variables	Dependent variables[a]				
	BWT & CHSTC	TRI & ARMC	SUBS	LEN	HEADC
Gestation age (wks)	X	X	X	X	X
Race	X	X	X	X	X
Weight gain (2nd to 3rd trim.)	X	X	X	X	X
Weight (3rd trim.)	X	X	X	X	X
Mother's age	X	X	X	X	X
Mother's education[b]	X	X	X	X	X
Number of pregnancies	X	X	X	X	X
Cigarette use[c]	X	X	X	X	X
Alcohol consumption	X	X	X	X	X
Mother's height	X	X	X	X	X
Mother's marital status	X	X	X	X	X
Mother's lean arm circ.[d]		X			X
Mother's triceps skfd. (3rd trim.)		X	X		
Mother's arm circ. (3rd trim.)				X	

[a]BWT: Birthweight and chest circumference
TRI: Triceps skinfold and arm circumference
SUBS: Subscapular skinfold
LEN: Length
ARMC: Arm circumference
HEADC: Head circumference
CHSTC: Chest circumference
[b]Mother's eduction was coded into four categories: one or more years of college, completed high school, begun high school, did not begin high school.
[c]Use in 2nd & 3rd trimesters.
[d]Lean arm circumference = upper arm circumference − (3.14 × triceps skinfold).

relation coefficient, or a significant ($p < 0.20$) role in a multivariate regression analysis. We also added variables that are known from other studies to affect the dependent variables even if they are only moderately related in our study. Thus, the adjustment model contains variables that independently, or in combination with others, are related to the dependent variable. There are seven anthropometric measures of the neonate and so seven models are possible. However, the models are virtually the same. All models include several measures of maternal size as well as several measures of socioeconomic status.

Table 7.5 presents the unstandardised coefficients representing the association of neonatal blood lead level (continuous scaling) with each dependent variable after adjustment for the covariates. All the coefficients relating neonatal lead level to size are negative. Tests of significance

Table 7.5. *Unstandardised coefficients representing the effect of lead on neonatal anthropometry: models using lead as a continuous, a dichotomised and a trichotomised variable*

Dependent variables	Trichotomous categorisation ($<5, 5$-$7, >7$)		Dichotomous categorisation ($<5, >5$)		Continuous	
	b	t	b	t	b	t
Birthweight	−123.4	−2.20*	−159.3	−1.81	−202.6	−2.07*
Subscapular skinfold	−0.476	−3.59***	−0.728	−3.53***	−0.699	−3.03***
Triceps skinfold	−0.345	−2.90**	−0.541	−2.92**	−0.527	−2.55*
Arm circumference	−0.359	−3.40***	−0.474	−2.84**	−0.659	−3.66****
Chest circumference	−0.537	−2.37*	−0.833	−2.38*	−0.899	−2.29**
Length	−0.407	−1.80	−0.437	−1.24	−0.723	−1.87
Head circumference	−0.095	−0.639	−0.0004	−0.002	−0.237	−0.93

*$p < 0.05$, **$p < 0.01$, ***$p < 0.005$, ****$p < 0.0001$

($p < 0.05$) indicate that neonatal blood lead level is significantly associated with reduced birth weight, chest circumference, subscapular and triceps skinfolds, and arm circumference, but not with length or head circumference. For each log unit of neonatal blood lead there is a reduction of 202 grams in birth weight.

A similar pattern of significant differences is observed when lead is scaled as a trichotomous or dichotomous variable (Table 7.5). Differences in regression coefficients reflect differences in the scaling of the lead level (13 levels in the continuous model *vs.* 2 or 3 in the categorical ones). Evidence of an effect of lead is strong given the consistency of results among analyses using different analytic strategies.

The interaction of race and blood lead level was tested by including an interaction term in the multiple regression model after entering neonatal lead level, and it was not statistically significant.

These results are quite comparable to the one other study of a biracial, socioeconomically disadvantaged population in Cincinnati, Ohio (Bornschein *et al.* 1987). An important difference between the studies, however, is that neonatal size is related to maternal blood lead levels in the Cincinnati cohort while the analysis here related neonatal size to neonatal blood lead levels.

In the Albany cohort there is an estimated reduction in birth weight of 202 grams per log unit of neonatal blood lead after controlling for other factors including gestational age. This reduction is well within the range of deficits observed among the Cincinnati longitudinal cohort. The reduction in length at birth in the Albany cohort is 0.7 cm per natural log unit of neonatal blood lead which is similar to the reduction seen in the Cincinnati study (Bornschein *et al.* 1987). In both studies head circumference is not related to the measure of blood lead.

The effects of lead in the Albany cohort primarily concern soft tissue: the skinfolds and arm circumference. The deficit in soft tissue is consistent with the decrease in birth weight. Prenatal growth in soft tissue, particularly adipose tissue, occurs relatively late in prenatal life. Conversely, the rapid growth of head circumference, the dimension least affected in the Albany sample, occurs earlier (Tanner 1989). Rapid growth in length occurs at a time intermediate between the prenatal growth spurts of head circumference and soft tissue deposition, and length is also affected to an intermediate degree. It is possible that lead levels most affect fetal growth late in pregnancy. Although the mechanism of prenatal growth suppression is unknown, endocrinological studies of lead-related postnatal growth failure indicate that inhibited growth hormone secretion may be responsible (Huseman *et al.* 1992).

The results from the study of the Albany cohort are noteworthy because the lead levels themselves are low and such levels are quite common in the general population. Since reduced prenatal growth is associated with poorer postnatal development including physical, behavioral and cognitive development, the growth deficits we have observed suggest that lead exposure, even at low levels, may be associated with poorer child development.

Asthma

Few anthropological studies are oriented towards understanding the causes of specific diseases, focusing instead on the causes of variations within the normal range of function and morphology. However, some diseases represent the extreme of a continuous range of variation rather than a qualitatively different state, and their study can be conducted with the same tools that human biologists usually apply. Hypertension, obesity and growth failure are examples of "diseases" that represent the tail of a distribution of human function that human biologists are fully equipped to study. Also, some diseases are so clearly the result of interaction of social and biological factors that their understanding demands the biosocial approach characteristically applied by human biologists.

Asthma is one of these diseases. Its study combines four features of interest to human biologists. It has a significant environmental component in that its occurrence is linked to certain air pollutants: oxides of sulfur and of nitrogen, ozone, lead, volatile organic compounds, and particulate matter. Also, there is growing evidence for the presence of biological variation in asthma susceptibility, and for subpopulations of susceptibles defined by genetic and familial factors. Furthermore, there appear to be intergenerational effects. Exposure in one generation produces effects that predispose the next generation to additional exposure. This cycle produces intergenerational continuity in the biological characteristics of the affected subpopulation. Finally, sociocultural factors including race or ethnicity influence morbidity, treatment and mortality, leading to suitable applications of biocultural modeling. Because of these characteristics of asthma, a human biology approach may have multiple benefits: its study may result in a greater understanding of the disease itself, and to refinement of theories within human biology that concern the interaction between the physical environment, the social system and genetic variation.

Asthma is now defined as "a chronic lung disease characterised by intermittent inflammation and bronchospasm manifesting in recurrent cough, wheeze, shortness of breath, reduced expiratory flow, exercise intolerance, and respiratory distress" (Kemper 1996:111). Inflammation is the most prominent characteristic, and eosinophils are prominent among the inflammatory cells. They are partly responsible for the hyperreactivity, and ensuing damage to the airway (Koren 1995). Asthma was once commonly viewed as a mild episodic disease, likely of psychosomatic origin, but it is now considered a chronic or persistent abnormality in airway responsiveness that produces intermittent symptoms. It is usually a life-long disease; most asthmatics are diagnosed before 5 years of age. Population rates of asthma are usually based on the rate among persons 5 to 35 years of age because most cases are diagnosed by 5 years of age and there is limited diagnostic accuracy of new cases in individuals greater than 35 years of age.

The prevalence of childhood asthma in the United States and other developed countries has increased markedly in recent years (Woolcock 1991). From 1982 to 1991 the age-adjusted death rate increased by 40% from 13.4 to 18.8 per million in the United States and the self-reported prevalence rate increased by 42% (Morbidity and Mortality Weekly Report 1995). Asthma currently affects 10–12 million persons and costs $6.2 billion annually in the United States, just over half of which is for in-patient, emergency and out-patient care, physician services and medications while the rest is from loss of work and premature death (Boushey and Fahy 1995). In Great Britain asthma now affects one child in seven (Cookson and Moffatt 1997).

Asthma is found in both urban and rural populations, but it is more common in urban ones (Gergen *et al.* 1988), and is often considered an "urban disease" because the risk factors for asthma so often include exposures that are more common or more concentrated in urban environments (Weiss and Sullivan 1994; Weiss *et al.* 1993). A large survey of children in the United States ($n = 15,416$) aged 6–17 years, found these risk factors associated with asthma: African-American ethnicity, male gender, weighing less than 2.5 kg at birth or having spent one or more nights in neonatal intensive care, having a mother who smokes more than 10 cigarettes per day, having a mother who was less than 20 years of age at the child's birth or with less than a secondary school education, not residing with both biological parents, being from a large family and living in poverty (Weitzman *et al.* 1990). These factors are tightly interwoven in urban populations.

Ethnic and socioeconomic differences in prevalence are most striking among children aged 2–5 years when such social factors play large roles in differential diagnosis, treatment and mortality. In the United States, young African-American children have 1.2–2.2 times the prevalence of the disease (depending on the date of comparison and the definition of asthma) compared to white children. Even when an objective indicator of asthma, exercise induced bronchospasm, is used to prevent confounding of socioeconomic status itself with differences in the definition of asthma that are related to socioeconomic status, very well-off schoolchildren have less than half the prevalence compared to very poor children (Ernst *et al.* 1995). In the US asthma mortality is five to seven times greater for African-American children than their white counterparts, and African-American children are twice as likely to be hospitalised for asthma. However, since the difference in prevalence of asthma is far lower, socioeconomic status seems to have a major impact on treatment for asthma that can lead to mortality (Malveaux and Fletcher-Vincent 1995). Thus, asthma is a disease in which differences in treatment that are based on social factors affect mortality.

The biocultural nature of asthma has an additional complexity. Certain urban ambient air pollutants are risk factors for the onset and exacerbation of asthma (Koren 1995), and children of lower socioeconomic status and non-white ethnicity are likely to have greater exposures to these pollutants than other children. In the United States, exposure to a variety of pollutants, including air pollutants, is associated with poverty and non-white ethnic group membership. Ethnicity and poverty often are associated with distinctive patterns of residence, occupation and diet, all of which may increase exposure to asthma-associated factors (Schell and Czerwinski 1998; Czerwinski, this volume).

People in poverty are more likely to live where air quality is poor, and, according to an analysis of recent data by the United States Centers for

Table 7.6. *Sources and effects of pollutants implicated in asthma and other respiratory diseases of children*

Pollutant	Source	Effects
SO$_x$ (Oxides of sulfur including sulfur dioxide	Combustion of fossil fuels containing sulfur	Respiratory irritant, soluable, absorbed in upper airways at rest and lower ones with increasing ventilation
O$_3$ (Ozone)	Produced by sun-driven reactions with nitrogen oxides, and volatile organic compounds that usually are produced by combustion	A potent oxidant
PM (Particulates defined by aerodynamic diameter: PM10 ($\leq$ 10 µm); PM2.5 ($\leq$ 2.5µm)	Fine particles produced by combustion in transportation, industry, energy generation, and from breakdown of materials.	Direct damage to lung determined by location of particle deposition which is related to particle size.
NO^x (Oxides of nitrogen including nitrogen dioxide)	Combustion of fossil fuels, particularly from gasoline and gas stoves	An oxidant which at low levels is weakly associated with lung function in asthmatics

Source: Adapted from Waldbott (1978), Koren (1995).

Disease Control and Prevention, "In 1991, an estimated 6.4 million (63%) of the 10.3 million persons with asthma in the United States resided in areas where at least one National Ambient Air Quality Standard was exceeded." Pollutants may play a preliminary role by sensitising an individual to allergens that are encountered subsequently, or after asthma develops they may trigger asthma attacks. For example, air toxics may increase the likelihood of sensitisation by reducing the integrity of the epithelial barrier and thus easing the penetration of allergens through the pulmonary epithelium (Osebold *et al.* 1988; Yanai *et al.* 1990). Pollutants that can cause bronchospasm in normal non-allergic persons cause bronchospasm at lower levels in asthmatics.

Four urban air pollutants are implicated most often (see Table 7.6 for definitions): oxides of sulfur, oxides of nitrogen, particulates and ozone (Koren 1995). Several studies have reported that children's hospital admissions or emergency room visits were increased in relation to ozone levels (Bates 1995; Thurston *et al.* 1994; Cody *et al.* 1992; White *et al.* 1994). Emergency room visits also have been related to larger particulate pollution (PM10) despite the fact that the levels where the study was conducted never exceeded the current United States standard (Schwartz *et al.* 1993). Fine particulate pollution (PM2.5) has been related to decreased respiratory function of asthmatic children but not normal ones (Koenig *et al.* 1993). Koren (1995) reviewed evidence for effects of sulfur dioxide and found that it can lower respiratory function in asthmatics while having no effect on normal subjects. Among asthmatics it can increase the development of bronchospasm even at very low levels (0.25 ppm), increase airway resistance and decrease FEV1 (forced expiratory volume) values.

Indoor environments may have more air pollution than outdoor ones and contribute significantly to the development of asthma and the severity of symptoms (Cuijpers *et al.* 1995). Environmental tobacco smoke has been strongly associated with the development and exacerbation of asthma symptoms in a number of studies (for review, see Etzel 1995; Leikauf *et al.* 1995). Other common indoor air pollutants include allergens deriving from pets, cockroaches and dust mites. The latter two are strongly implicated as playing a crucial role in stimulating the development of asthma through early sensitisation of the immune system.

The interacting roles of urban factors in the development of asthma and in triggering asthma attacks is based on the underlying immunologic response. Asthma is strongly associated with high production of immunoglobulin E (IgE; Burrows *et al.* 1989). Immunoglobulin E production is influenced by the balance of T helper cells, a class of T lymphocytes required for B lymphocytes to respond to allergens and develop into antibody producing cells. T helper cells produce cytokines, hormone-like peptides, that stimulate the B cells to make antibodies. There are two populations of T helper (Th) cells defined in terms of the cytokines each produces. Many cytokines are produced by both Th1 and Th2 cells, but some cytokines appear to be produced mainly by one or the other Th cell population: Th1 cells producing interleukin 2 (IL-2), tumor necrosis factor (TNF-β) and interferon-γ (IFN-γ), and Th2 cells producing interleukins 4, 5, 6 and 10. Typically there is a balance between the two groups of Th cells: upregulation of one downregulates the other. Immediate allergic reactions are associated with selective activation of Th2 cells and IL-4, and this facilitates the IgE antibody responses and respiratory sensitisation (Koren 1995). This pattern of Th2 cell response may become part of the T cell memory and be reinforced with each reaction.

Much research on asthma is focused on determining how a predominant Th2 pattern of reaction is induced in young children. The familial nature of asthma is well known, but the cause is postulated to be shared environment, genes and their interaction. Family studies show that monozygotic twins have four times the concordance rate of dizygotic twins (19% *vs.* 4.8 %), and the presence of two parents with asthma produces a 49% risk of developing the disease in the offspring (Sarafino and Goldfedder 1995). The latter statistic also is consistent with the theory that there is some contribution from the environment.

A familial pattern also can be produced by congenital or neonatal sensitisation. There is considerable evidence that normal pregnancy involves a shift in maternal immune response toward humoral immunity and away from cell mediated immunity, and that this may include a shift in Th cell balance toward Th2 predominance (Wegmann *et al.* 1993). Wegman and colleagues speculate that unless the infant restores the Th cell balance, the infant may be predisposed to mount a Th2 cell predominant reaction to allergens encountered early in life. There also is some evidence for transplacental leakage of reactivity such that some neonates may be especially predisposed to Th2 predominance. If this is so, early antigen stimulation may begin a pattern of Th2 predominant reactions. Early antigen stimulation is likely if common environmental factors can act as allergens.

Candidates for early stimulation are environmental tobacco smoke (ETS), house dust mites, cockroaches, cat and dog allergens (Etzel, 1995). Chronic exposure to these allergens and irritants may provide early sensitisation or trigger an asthmatic attack. Cockroach allergy in low SES asthmatics is common, ranging from 52% to 78%. Of urban subjects with asthma and positive skin reactivity to cockroach antigen, 91% had immediate positive bronchial reactivity. Recurrent exposure may be responsible for the airway inflammation in asthma. Children allergic to house dust mites when challenged with house dust mite allergen produce large amounts of IL 4, 5, 9 and 13 as well as some IFN-γ, but non-reactors produce no IL 4, 5, 9, 13 but some IFN-γ. Children with heightened sensitivity to dust mite or cockroach allergens due to genetic or familial–environmental predisposition to the Th2 response comprise a subpopulation with greater risk of asthma development and attack. Since children encounter these allergens before the age of 5, the pattern of predominant Th2 reactivity may be set very early (Holt 1994).

The importance of early sensitisation that predisposes towards predominant Th2 reaction is further supported from studies of children who have been stimulated to produce the Th1 cell response. In Japan, where there is compulsory childhood vaccination against tuberculosis, a sample of children with positive delayed hypersensitivity test to tuberculin (due to successful vaccination) had a lower prevalence of asthma, lower IgE levels and less Th2 predominance compared to children without delayed hypersensitivity presumably due to lack of immunologic response to the vaccination (Shirakawa *et al.* 1997). This suggests that early exposure to infection may predispose T cells towards predominantly Th1 profile cytokines that downregulate Th2 cytokine production, reducing the risk of developing elevated levels of IgE and asthma (Cookson and Moffatt 1997). Concordant with this thesis is the observation in Germany that among children in large families, children born later in the birth order have lower rates of asthma than children earlier in birth order (von Mutius *et al.* 1994). Cookson and Moffat (1997) suggest that children born later in the family get more infections at a young age from their older siblings than their older brothers and sisters did at comparable ages since the earlier-born children were infants in a smaller family. They further suggest that the childhood asthma epidemic is partly the result of not having Th1 stimulating infections in early years, rather than a sensitivity to cockroach or dust mite allergen since any allergen will produce a similar result if encountered early in life. Holt (1994) suggests that manipulating the immune system to produce Th1 cytokines could promote a pattern of Th1 cytokine predominance, downregulating Th2 cytokines, thereby decreasing the risk of developing of asthma. One contradiction in this interpretation of the interplay between early infection and asthma onset is the contradictory role played by poverty. On the one hand, data from the United States show that low socioeconomic status is associated with less vaccination, more infections and a higher prevalence of asthma. However, the theory of Holt, Cookson and Moffat suggest that low infection rates and early allergen exposure raise the risk of asthma onset. Clearly the interaction among poverty, early infection, early allergen exposure and asthma requires further investigation.

While association between non-allergen air pollutants and asthma onset and attacks has already been described, at least one airborne pollutant, lead, may also affect cytokine profiles. In rodents, lead enhances B cell differentiation, increases the number of antibody producing cells, reduces Th1 producing cells and enhances Th2 cell numbers (Heo *et al.* 1996). Lead crosses the placenta and children may be born with a low level lead burden. The effect of lead on the development of the immune system is not well known, but existing evidence suggests that early lead exposure may contribute to predominance of a Th2 cytokine profile and to the development of asthma as well.

In summary, asthma is multicausal. Among the prominent risk factors are genetic predisposition, the nature of the stimulating antigen, age, route and duration of exposure, and environmental pollutants. Urban environments combine these factors with low socioeconomic status, and minority group membership to produce elevated rates of asthma compared to rural districts. Attention to the interplay among these biological, environmental and social factors is necessary to sort out the causal pathways and cofactors.

Conclusion

Pollutants affect children through multiple physiological pathways that may influence specific systems as well as general markers of child well-being. The urban environments often contain pollutants at higher concentrations than rural ones, and urban populations include substantial numbers of socioeconomically disadvantaged children. The combination can work synergistically, exposing the most disadvantaged children to the most pollutants.

The examples of lead and asthma described here illustrate two approaches to the study of the human biology of urbanism. They also illustrate the effect of differences in measurement of pollutants on the certainty of the research findings. Past lead exposure produces a long-lasting body burden that can be measured precisely even at low levels. Current studies of the effects of lead on children rely on measuring lead in the individual and, as a result, effects may be observed at low levels. In contrast, past exposure to oxides of nitrogen and sulfur or ozone do not leave measurable body burdens, and current studies of air pollutants and asthma cannot relate variation among children in exposure to asthma onset or symptoms on an individual level. Instead, comparisons are made in which groups with different exposures are compared for asthma prevalence and symptoms, resulting in a less precise statement of the relationship and more uncertainty regarding the true cause of the group differences in asthma rates and symptoms. In general, studies of urban factors and human biology may be more informative if those factors can be measured precisely in individuals rather than in groups.

Note

1. The use of the terms white, non-white, African-American, and Hispanic as ethnic groups is intentionally designed to emphasise the biocultural nature of such designations, and does not mean that any characteristics of such groups are more than transient accompaniments of social position until proven otherwise. Such terms are not explicitly racial but refer to the large body of research documenting variation in physiological parameters and health according to such group identities however they have been defined by government agencies and researchers.

References

Agency for Toxic Substances and Disease Registry (1988). *The Nature and Extent of Lead Poisoning in Children in the United States: A Report to Congress.* Atlanta: US Department of Health and Human Services.
Annest, J. L., Pirkle, J., Makuc, D., Neese, J. W., Bayse, D. D. and Kovar, M. G.

(1983). Chronological trend in blood lead levels between 1976 and 1980. *New England Journal of Medicine*, **308**, 1373–7.

Annest, J. L., Mahaffey, K. R., Cox, D. H. and Roberts, J. (1982). Blood lead levels for persons 6 months–74 years of age: United States, 1976–80. *NCHS Advance Data*, **79**, 1–23.

Bates, D. V. (1995). The effects of air pollution on children. *Environmental Health Perspectives*, **103**, 49–53.

Bearer, C. F. (1995). Environmental health hazards: how children are different from adults. *The Future of Children*, **5**, 11–26.

Bornschein, R. L., Succop, P. A., Dietrich, K. N., Krafft, K., Grote, J., Mitchell, T., Berger, O and Hammond, P. B. (1987). Prenatal Lead Exposure and Pregnancy Outcomes in the Cincinnati Lead Study. In *Heavy Metals in the Environment*, eds. S. E. Lindenburg and T. P. Hutchinson. Edinburgh: CEP Consultants.

Boushey, H. A. and Fahy, J. V. (1995). Basic mechanisms of asthma. *Environmental Health Perspectives*, **103**, 229–33.

Brody, D., Pirkle, J., Kramer, R., Flegal, K., Matte, T. and Gunter, E. (1994). Blood lead levels in the US population. Phase 1 of the Third National Health and Nutrition Examination Survey (NHANES III, 1988–1991). *Journal of the American Medical Association*, **272**, 277–83.

Burrows, B., Martinez, F. D., Halonen, M., Barbee, R. and Cline, M. (1989). Association of asthma with serum IgE levels and skin-test reactivity to allergens. *The New England Journal of Medicine*, **320**, 271–7.

Cody, R. P., Weisel, R. A., Birnbaum, G. and Lioy, P. T. (1992). The effect of ozone associated with summertime photochemical smog and the frequency of asthma visits to the hospital emergency departments. *Environmental Research*, **58**, 184–94.

Committee on Measuring Lead in Critical Populations. (1993). *Measuring Lead Exposure in Infants, Children, and Other Sensitive Populations*. Washington DC. National Academy Press.

Cookson, W. O. C. M. and Moffatt, M. F. (1997). Asthma: an epidemic in the absence of infection? *Science*, **275**, 41–2.

Cuijpers, C. E. J., Swaen, G. M. H., Sturmans, F. and Wouters, E. F. M. (1995). Adverse effects of the indoor environment on respiratory health in primary school children. *Environmental Research*, **68**, 11–23.

Emory, E. K., Pattillo, R. A., Archibold, E., Bayorh, M. A. and Sung, F. (1997). Effects of low level lead exposure and neural behavioral functions in neonates. [Abstract] *Speaker and Poster Abstracts from the 1st Children's Environmental Health: Research, Practice, Prevention and Policy*, 31. Feb. 21–23.

Ernst, P., Demissie, K., Joseph, L., Locher, U. and Becklake, M. R. (1995). Socioeconomic status and indicators of asthma in children. *American Journal of Respiratory and Critical Care Medicine*, **152**, 570–5.

Etzel, R. A. (1995). Indoor air pollution and childhood asthma: effective environmental interventions. *Environmental Health Perspectives*, **103**, 55–8.

Gergen, P. J., Mullally, D. I. and Evans, R. (1988). National survey of prevalence of asthma among children in the United States, 1976 to 1980. *Pediatrics*, **81**, 1–7.

Heo, Y., Parsons, P. and Lawrence, D. (1996). Lead differentially modifies cytokine production in vitro and in vivo. *Toxicology and Applied Pharmacol-*

ogy, **138**, 149–57.

Holt, P. G. (1994). A potential vaccine strategy for asthma and allied atopic diseases during early childhood. *The Lancet*, **344**, 456–8.

Huseman, C., Varma, M. and Angle, C. (1992). Neuroendocrine effects of toxic and low blood lead levels in children. *Pediatrics*, **90**, 186–9.

Kemper, K. J. (1996).Chronic asthma: an update. *Pediatrics in Review*, **17**, 111–18.

Koenig, J. Q., Larson, T. V., Hanley, Q. A., Rebbolledo, V., Dumler, K., Checkoway, H., Wang, S., Lin, D. and Pierson, W. E. (1993). Pulmonary function changes in children associated with particulate matter. *Environmental Research*, **63**, 26–38.

Koren, H. S. (1995). Association between criteria air pollutants and asthma. *Environmental Health Perspectives*, **103**, 235–42.

Landrigan, P. J. and Carlson, J. E. (1995). Environmental policy and children's health. *The Future of Children*, **5**, 34–52.

Leikauf, G. D., Kline, S., Albert, R. E., Baxter, C. S., Bernstein, D. I., Bernstein, J.and Buncher, C. R. (1995). Evaluations of a possible association of urban air toxics and asthma. *Environmental Health Perspectives*, **103**, 253–71.

Lin-Fu, J. S. (1973). Vulnerability of children to lead exposure and toxicity. *New England Journal of Medicine*, **289**, 1229–33.

Malveaux, F. J. and Fletcher-Vincent, S. A. (1995). Environmental risk factors of childhood asthma in urban centers. *Environmental Health Perspectives*, **103**, 59–62.

Morbidity and Mortality Weekly Report (1995). Asthma–United States, 1982–1992. *Morbidity and Mortality Weekly Report*, **43**, 952–5.

Morbidity and Mortality Weekly Report (1997). Update: blood lead levels–United States, 1991–194. *Morbidity and Mortality Weekly Report*, **46**, 141–6.

Osebold, J. W., Zee, Y. C. and Gershwin, L. J. (1988). Enhancement of allergic lung sensitisation in mice by ozone inhalation. *Proceedings of the Society for Experimental Biology and Medicine*, **188**, 259–64.

Pirkle, J., Brody, D., Gunter, E., Kramer, R., Paschal, D., Flegal, K. and Matte, T. (1994). The decline in blood lead levels in the United States. The National Health and Nutrition Examination Surveys (NHANES). *Journal of the American Medical Association*, **272**, 284–91.

Rogan, W. J. (1995). Environmental poisoning of children–lessons from the past. *Environmental Health Perspectives*, **103** (Suppl 6), 19–23.

Rogan, W. J. and Gladen, B. C. (1985). Study of human lactation for effects of environmental contaminants: The North Carolina Breast Milk and Formula Project and some other ideas. *Environmental Health Perspectives*, **60**, 215–21.

Sarafino, E. P. and Goldfedder, J. (1995). Genetic factors in the presence, severity, and triggers of asthma. *Archives of Disease in Childhood*, **73**, 112–16.

Schell, L. M. (1991). Effects of pollutants on human prenatal and postnatal growth: noise, lead, polychlorinated compounds and toxic wastes. *Yearbook of Physical Anthropology*, **34** (Supplement 13 to the American Journal of Physical Anthropology,) 157–88.

Schell, L. M. (1997). Using patterns of child growth and development to assess communitywide effects of low-level exposure to toxic materials. *Toxicology and Industrial Health*, **13**, 373–8.

Schell, L. M. and Czerwinski, S. A. (1998). Environmental health, social inquality and biological differences. In *Human Biology and Social Inequality,* eds. S. S.

Strickland & P. S. Shetty, p. 114–31. Cambridge: Cambridge University Press.
Schwartz, J., Slater, D., Larson, T. V., Pierson, W. E. and Koenig, J. Q. (1993). Particulate air pollution and hospital emergency room visits for asthma in Seattle. *American Review of Respiratory Diseases*, **147**, 826–31.
Shirakawa, T., Enomoto, T., Shimazu, S. and Hopkin, J. (1997). The Inverse Association Between Tuberculin Responses and Atopic Disorder. *Science*, **275**, 77–9.
Tanner, J. M. (1989) *Fetus into Man*. Cambridge, MA: Harvard University Press.
Thurston, G. D., Ito, K., Hayes, C. G., Bates, D. V. and Lippmann, M. (1994). Respiratory hospital admissions and summertime haze air pollution in Toronto, Ontario: consideration of the role of acid aerosols. *Environmental Research*, **65**, 271–90.
von Mutius, E., Martinez, F. D., Fritzsch, C., Nicolai, T., Reitmeir, P. and Thiemann, H. H. (1994). Skin test reactivity and number of siblings. *British Medical Journal*, **308**, 692–5.
Waldbott, G. L. (1978). *Health Effects of Environmental Pollutants*. 2nd edn. St. Louis: The C.V. Mosby Company.
Wegmann, T., Lin, H., Guilbert, L. and Mosmann, T. (1993). Bidirectional cytokine interactions in the maternal–fetal relationship: is successful pregnancy a Th2 phenomenon? *Immunology Today*, **14**, 353–6.
Weiss, K. B., Gergen, P. J. and Wagener, D. K. (1993). Breathing better or wheezing worse? The changing epidemiology of asthma morbidity and mortality. *Annual Review of Public Health*, **14**, 491–513.
Weiss, K. B. and Sullivan, S. D. (1994). Socio-economic burden of asthma, allergy, and other atopic illnesses. *Pediatric Allergy Immonology*, **5**, 7–12.
Weitzman, M., Gortmaker, S. and Sobol, A. (1990). Racial, social, and environmental risks for childhood asthma. *AJDC*, **144**, 1189–93.
White, M. C., Etzel, R. A., Wilcox, W. D. and Lloyd, C. (1994). Exacerbations of childhood asthma and ozone pollution in Atlanta. *Environmental Research*, **65**, 56–68.
Woolcock, A. J. (1991). The problem of asthma worldwide. *European Respiratory Review*, **1**, 246.
Yanai, M., Ohrai, T. and Aikawa, T. (1990). Ozone increases susceptibility to antigen inhalation in allergic dogs. *Journal of Applied Physiology*, **68**, 2267–73.

8 Urbanism and health in industrialised Asia

J. PETERS

Editors' introduction

Southeastern Asia is currently urbanising at a faster rate than any other major geographical region, and four of the ten largest cities in the world are now to be found there. In examining the complexity of factors which have lead to changing patterns of urban health in Asian cities, Peters develops themes that are resonant with those of other authors in this volume, including McMichael (chapter 2), Clark (chapter 3), Di Ferdinando (chapter 5), Schell and Stark (chapter 7), and Parker (chapter 14). Overlapping environments graded by social, economic, cultural, and ethnic factors are implicit in descriptions of unemployment, social instability and insecurity, undernutrition, and infectious disease sitting in close proximity to the emerging middle classes who are now experiencing rising levels of obesity and diseases of affluence. Although this description could equally apply to many cities of the industrialised West, there are some differences between Asia and Europe and the United States. Unlike Johnston and Gordon-Larsen (chapter 10), Peters describes obesity as a problem of the Asian middle classes, and not of the poor. Furthermore, exposure to pollution is perhaps greater than among the North American populations described by Schell and Stark (chapter 7), and Czerwinski (chapter 11). Despite the dismal picture painted of urban life in contemporary Asia, the health benefits must outweigh the drawbacks, if broad health indicators are to be believed; the more urbanised Asian countries have lower infant mortality rates and higher life expectancies, than those with low levels of urbanism. In large part, improved access to health care, education, and empowerment have come with urbanism in Asia, making health improvements possible. However, there are problems in determining the trajectory of urban health and human biology change in Asia, since the pace of urbanisation is dramatically greater than ever it was in the Western countries, while environmental issues have a low priority among Asian nations, and there are still far too few data linking health and urban factors in this region.

Although the absolute level of urbanism in Southeastern Asia, with the exception
of countries such as Japan, South Korea, Singapore and Hong Kong, could be
considered low by world standards, the rate of urbanisation exceeds that of the
world standard.   (Talib and Agus 1992)

Factors promoting urbanisation

Changes in trading patterns

Until the mid-twentieth century, for many countries in Southeastern Asia
trade with the developed countries tended to be limited to extraction of
raw materials, transportation of these to the industrial West for processing, and the selling back of the finished products. Consequently, industrial development in the area was limited and focused in a few major cities
which were also seaports (Chan 1995). The oil crises of the 1970s resulted
in a total re-evaluation of this arrangement as manufacturing costs in the
Western countries escalated. Six million jobs were moved from the advanced economies, representing 20% of all manufacturing jobs in Western
Europe and 8% in America, to the countries which sourced the raw
materials. This created newly industrialised countries, many of which were
in the Southeast Asia Region. As their economies upgraded, transnational
activities and therefore options for work became concentrated in and
around a number of key cities and their supporting hinterlands leading to
the emergence of large urban regions (Chan 1995).

Population movement

With an estimated 124 million unemployed in the rural areas of China and
another 60–100 million surplus rural workers adrift between the villages
and the cities, many of whom are subsisting through low-paid, part-time
jobs (CIA 1997) it is understandable that many villagers are migrating to
the cities to find work and this is a major reason for the increases seen in the
urban population. As an example of the rate of change from rural to urban
and the size of that change: in 1979 China's economy was predominantly
rural with self-supporting local industry. Urban reform was introduced in
1984 because land development was accelerating, and, by 1991, China had
95 cities whose populations were in excess of 1 million and an annual rate
of increase in new urban residents of 12–13 million annually (Chan 1995).
In other countries such as Indonesia, the Philippines, Singapore, and
Malaysia, whilst the percentage urban population in 1983 was low, ranging from 18% to 39%, the average urban growth rate in the previous 10
years had been 4.8%, 4.0%, 3.9%, and 3.5% per year respectively (Badri
1992). In South Korea, in parallel with its rapid economic development,

the population of Seoul increased from 2.5 million in 1960 to 10 million in 1990. Together with Pusan, Seoul now contains half the total population of the country (He *et al.* 1992). To a much lesser extent, given the age structure of many countries in this region, fertility levels have also contributed to net population increases (WHO 1994). Whatever the cause the overall result is that four of the world's most populated cities are now to be found in this region: Tokyo-Yokohama 25 million (largest), Shanghai 13 million (4th), Beijing 12 million (7th), Jakarta 9 million (10th).

Issues associated with the effect of urbanisation on health are less related to the magnitude than to the rate of population movement and the likely ability of urban authorities to cope with the demands of such rapid change. Development of urbanism has been very rapid in some areas although slower in others. But, in most cases, planning has been inadequate, reactive rather than proactive, and in response to the negative consequences of urbanisation. Neither infrastructure, legislation, or local past experience have been available to guide responses to the problems that have arisen through these rural–urban, agricultural–industrial changes and the consequent economic, social, and environmental issues have had an impact upon the health of the urban population.

Urbanism and health

The people who come to the cities looking for work swell the numbers of homeless or exacerbate already overcrowded housing, they contribute to crime rates, and suffer malnutrition, general mental and physical illnesses. However, the health problems that arise in these expanding urban populations cannot necessarily be addressed by health agencies alone as many fall within the remit of other government departments, such as those responsible for sanitation, pollution, work safety, and housing.

Housing and health

Increases in urban population density lead to greater demands on existing land for local housing, employment opportunities, and services. To accommodate the influx of migrant workers farmland around the towns is converted into manufacturing and housing units and as the cost of such land rises, housing developments become taller, ultimately skyscrapers, and more expensive. If suitable housing is unavailable or unaffordable, illegal squatter dwellings develop. A large, and growing proportion of the population, the urban poor, in many Southeast Asian cities live in such slums and squatter settlements, many of which have no basic amenities such as potable water, sanitation, drainage, or waste management facilities. The rubbish dumps will be breeding grounds for pests and the squatter camp inhabitants consequently suffer from excessive exposure to diseases transmitted by insect and rodent vectors. In 1990, there were estimated to be 235,000 squatters around Kuala Lumpur (Talib and Agus 1992). Even if official local housing is available such dwellings may still be overcrowded. In Hong Kong, approximately 50,000 families live in overcrowded accommodation, defined as having a living space of less than 60 sq. ft per person. Living quarters described as "caged homes" are a reality (Ng 1996).

Overcrowding encourages the spread of infectious diseases such as tuberculosis. Indeed, most of the factors which make people vulnerable to such diseases are evident in the newly emerging urban areas in Southeast Asia: poor quality living environments including crowded conditions, poor lighting and ventilation, noise, and lack of facilities for personal hygiene; absence of friends and family, which may lead to lack of social support and inattention to own health and nutrition; long working hours resulting in fatigue and stress and a poor quality working environment, with air pollution and noise. Although the prevalence of tuberculosis had declined in many industrialised countries, levels are increasing once more in some, such as Japan. The Southeastern Region of Asia has a problem. In 1990, there were 8 million new cases of tuberculosis world-wide, with 2.6 million and 2.5 million in the West Pacific and Southeast Asia respectively. Likewise, of the 2.9 million deaths globally, 1.8 million of them were in these two regions (Kochi 1991).

Employment and health

Work is generally available in the Region as unemployment rates are low at around 3.5% or less. There are some notable exceptions, such as China with unemployment rates of 5.2% in urban areas, the Philippines at 9.5% and Vietnam at 25% (CIA 1997). However, not all are able to work, such as those who are physically or mentally ill. Work may also not be available at the right level, with many migrants lacking the necessary skills to enter the higher ranks of the job markets. There is also an increasing level of risk of losing work in this region. For some of those countries of Southeast Asia which have only recently become urbanised the risks of an industry failing can be higher because the companies involved have developed so quickly and may have overextended themselves financially. In South Korea in 1997, wage controls were imposed and people were having to cope with a decline in what was, until recently, a growing economy. Unemployment and short-time work became more prevalent and this resulted in strikes, protests, clashes with the armed forces, and damage to physical and mental health. In the past in China, where industry was basically state run, every one was guaranteed a job for life plus a pension or some form of care at the end of their working life (whatever age that might be), the so-called"Iron rice bowl" phenomenon. But by the end of 1995, 41,000 state run enterprises had failed and suspended production leaving 5.5 million workers and pensioners without salaries or pensions and in the first three months of 1996, more than 25% of China's 370,000 state run companies were loss making (CIA 1997).

A lack of work, the need to move to the urban environment in search of work, and even the loss of work create social instability and insecurity, and can contribute to mental and physical ill-health. Movement into the towns is a move away from access to the extended family and both the physical and mental support it offers. Regrettably, recognition and treatment of mental and psychological health conditions in Southeast Asia is about 20 years behind that in the West. Other issues associated with mental illhealth and its treatment include the cultural unacceptability of using mental health therapist services and the lack of time or money for treatment in those whose total time is devoted to providing sufficient food for self and/or family (personal communication, Dr C. Betson, Hong Kong).

Lifestyle factors and health

For those in work, the higher incomes and consumer orientation of the urban environment plus the development programs underway in urban areas are resulting in rapidly improving socioeconomic conditions. Lifestyles are changing and a new middle class is emerging. In the more urbanised countries of Southeastern Asia, whilst the prevalence of infectious diseases and malnutrition has declined an increase has occurred in the so-called diseases of affluence, such as cardiovascular conditions. In Hong Kong, deaths from malignant neoplasms, heart and cerebrovascular diseases comprise the majority (57%) of total deaths (Department of Health, Hong Kong 1994). Likewise, coronary heart disease is the leading cause of death in urban China, accounting for 51% of deaths (Yao et al. 1993) with age standardised mortality rates for specific circulatory diseases (such as cerebrovascular disease and ischaemic heart disease) higher in urban compared with rural populations (WHO 1994). This pattern of higher cardiovascular mortality and morbidity in urban areas is also seen in other countries in this region, such as Indonesia (Boedhi-Darmojo 1993). A recent study of risk factors for cerebrovascular disease suggests that this is potentially a major health problem for Southeast Asia given the increased prevalence of smokers and those with higher body mass index (BMI) values, and higher cholesterol levels found in some urban populations in the region (INCLEN 1994). Overnutrition and obesity are becoming major problems in many cities (Gross *et al*. 1992). In one Tokyo-Yokohama school which included a dietitian on its staff, 12% of 227 children were considered obese using either a weight/height index (BMI) or upper arm skinfold measurements (Peters *et al*. 1990).

Higher income can result in increased spending power and the opportunity to purchase holidays. Advertisements in a well-known local daily Hong Kong newspaper offered trips to a number of countries in the region: Cambodia, Burma, Thailand, and Vietnam with direct flights, a tourist guide, good food, and a stay in a new hotel. The trips were promoted as a "wonderful holiday tour for men". An out-of-town sex adventure in a region where HIV positivity ranges up to 63% (Waller 1992). It is predicted that by 1999 Asia will have the highest number of new cases of HIV and AIDS world-wide. The infection is also spreading from urban to rural areas and further into Southeast Asia (Mertens and Low 1996) and, given such holidays and the migration of young village girls to the cities, where many survive through prostitution, this is not surprising.

Industrial pollution and health

Other hazards to health associated with urbanisation are those attributable to the man-made urban environments and the man-made pollution of these environments. Rapid industrialisation and subsequent urbanisation without adequate planning controls can result in industrial development in unsuitable places, such as intermixed with housing or by usurping prime farm land. This can impact directly and indirectly on health. For example, the demands of increased urbanisation following industrial development in the China/Hong Kong development corridor has resulted in:

(a) areas of the New Territories, Hong Kong, being converted into open storage spaces and parking lots for which the owner receives a higher income than from farming. The result of this is a loss of farmland, fish ponds, and consequently local sources of food;

(b) changes to land use and damage to natural drainage systems leading to serious flooding and pollution of both land, water, and crops, with, for example, diesel spills (Chan 1995);

(c) the loss of 20% of agricultural land in China since 1957 to economic development and soil erosion. The result has been reduced grain production, insufficient food to feed the country's population, and serious famines in some years with many deaths;

(d) attempts to improve crop production on the remaining land with increased use of both appropriate and inappropriate pesticides, leading to numerous outbreaks of pesticide poisoning among the consumers, as reported in the local daily press;

(e) an inadequate transport infrastructure; roads are congested with motor vehicles, mainly diesel-powered container lorries and delivery trucks, which are a major source of air pollutants: particulates, nitrous oxides, carbon dioxide, carbon monoxide, and volatile hydrocarbons.

In Hong Kong in 1991, vehicles were responsible for 75% of ambient nitrogen dioxide and 51% of the overall emissions of total suspended particulates (with diesel vehicles producing 82%, and 98% respectively) (Environmental Protection Department 1992). The other major source of air pollution is sulphur dioxide produced by burning fossil fuels containing high levels of sulphur. For China, the main source of fuel for cooking and heating is coal briquettes with a high sulphur content. Air pollution, acid rain and the impact of these upon respiratory health and crop production are major problems for many countries in Southeast Asia. Table 8.1 provides a summary of the major air pollutants, mean annual levels recorded in two districts (Kwai Tsing with poor air quality and Southern District with better air quality), in Hong Kong over a three-year period and one-year mean concentrations not to be exceeded. The impact of a government intervention and government legislation in July 1990, limiting the use of fossil fuels containing high levels of sulphur (Hong Kong Government 1990), can be clearly seen. Following the implementation of legislation, sulphur dioxide levels fell by 84% in Kwai Tsing with unsustained falls of 24% and 18% in total and respirable particulates respectively (Environmental Protection Department 1991).

A survey of government out-patient department clinics in Hong Kong in 1990 found that the most common cause of patient visits (66%) was for respiratory complaints (Department of Community Medicine 1991). The respiratory health was studied of approximately 12,500 Hong Kong primary school children aged 8 to 13 years, half of whom lived in Kwai Tsing, a district with high levels of industry and air pollution, and the rest in Southern, a district with negligible industry, cleaner air but similar housing. Using self-completed, internationally standardised questionnaires (MRC 1960, Florey and Leeder 1982), significant differences in prevalence of reported symptoms were found between the two districts. The excess risk to children associated with living in the air polluted district ranged from 11% for phlegm, through 22% for cough or sore throat to 34% for wheeze, after adjustment for socioeconomic factors and exposure to environmental tobacco smoke in the home (Table 8.2) (Peters et al. 1996). For a subgroup of children given a histamine challenge, a higher prevalence of

Table 8.1. *Major air pollutants, mean annual levels in Hong Kong* ($\mu g/m^3$), *1989 to 1991, source of pollutants and one year mean concentration* ($\mu g/m^3$) *not to be exceeded*

District	Southern (less air polluted)					Kwai Tsing (air polluted)				
Pollutant	SO_2	NO_2	TSP	RSP	SO_4in RSP	SO_2	NO_2	TSP	RSP	SO_4in RSP
1989	11*	NA	61*	38*	9.9*	111	36	100	58	12.5
1990	8	21	55	37	5.8	67	39	86	52	8.8
1991	7	25	60	43	6.4	23	44	83	54	7.7

SO_2	Sulphur dioxide	Combustion of high sulphur content fuels (coal, oil) in industrial furnaces and power plants, motor vehicles (mainly diesel)	80
NO_2	Nitrogen dioxide	Combustion of fossil fuels in power plants, gas-fuelled appliances e.g., domestic cookers and heaters, motor vehicles (mainly diesel)	80
TSP	Total suspended particulates	These arise from man-made activities of construction, mining, earth-moving and from incomplete combustion of hydrocarbon fuels such as those used in motor vehicles	80
RSP	Respirable suspended particulates	RSP are defined as suspended particles in air with a nominal aerodynamic diameter of ⩽ 10 micrometers	55
SO_4 in RSP	Sulphated particulates	Some suspended matter consists of sulphuric acid and other sulphates. These result from the combination of SO_2 and cold moist air in the presence of certain metal catalysts	

* Measurements for Southern District were only available from September 1989 onwards.
# levels in April and May 1990 (before implementation of Sulphur Fuel Regulations, July 1990) were 136 $\mu g/m^3$ and 113 $\mu g/m^3$ respectively.

bronchial hyper-reactivity was observed in the children living in the more polluted district (χ^2=7.74, df=3, p=0.05) (Tam *et al.* 1994). Following government legislation limiting the use of fossil fuels containing high levels of sulphur (Hong Kong Government 1990), a follow-up study of children living in the same two districts found no differences in children's respiratory symptoms between districts (Peters *et al.* 1996) and higher declines in bronchial hyper-reactivity in children living in Kwai Tsing (Wong *et al.* 1998). One further point of interest from this study is that there were also excess risks for specific respiratory symptoms in those children exposed to environmental tobacco smoke in the home. Tobacco smoke is a major contaminator of indoor air. These excess risks associated with tobacco smoke exposure were higher than those for the district effect and remained

Table 8.2. *Any grouped respiratory symptoms odds ratios (OR) and 95% confidence intervals (95 CI) adjusted for confounding factors, before and after the air pollution intervention*

Any grouped symptom	Variable		1989 and 1990			1991		
			OR	95% CI	p	OR	95% CI	p
Cough or sore throat	District (KT vs. S)		1.22	1.05–1.42	< 0.05	0.92	0.73–1.15	n.s.
	No. of smoker categories	1	1.20	1.04–1.40	< 0.05	1.23	0.97–1.56	n.s.
		2–4	1.67	1.36–2.06	< 0.001	1.55	1.09–2.21	< 0.05
Phlegm	District (KT vs. s)		1.11	0.96–1.30	n.s.	0.88	0.68–1.13	n.s.
	No. of smoker categories	1	1.33	1.15–1.54	< 0.001	1.26	0.96–1.64	n.s.
		2–4	1.95	1.58–2.40	< 0.001	1.75	1.19–2.56	< 0.01
Wheezing	District (KT vs. S)		1.34	1.09–1.66	< 0.01	1.18	0.89–1.57	n.s.
	No. of smoker categories	1	1.10	0.91–1.33	n.s.	0.94	0.69–1.28	n.s.
		2–4	1.13	0.84–1.51	n.s.	1.70	1.15–2.54	< 0.01

Note: Other variables: gender, age, session of study, housing and father's education were adjusted in the model (results not shown); KT = Kwai Tsing district, S = Southern District, Smoker categories are defined as any of the following who smoke: mother, father, siblings living at home, others living in the family home
Source: Peters *et al*. 1996.

at the same high level, after the improvement seen in air quality (Table 8.2) (Peters *et al.* 1996).

Much of the early industrial development in the Southeastern Asian region was located close to rivers or coastal areas so these became the sites for urban development. But factories may discharge chemicals and other wastes into these rivers or seas either legally or illegally. Minimata Bay, Japan is a classic case of industrial development with limited control, resultant water pollution, and an impact on health. A fertiliser factory discharged methyl mercury into Minimata Bay and between 1953 and 1960, approximately 200 people were diagnosed with neurological disorders. In 1959, the cause of these symptoms was traced back to contaminated fish and shellfish from the bay eaten as part of the local diet. By 1974, 800 cases had occurred, 100 of whom had died, as the factory continued to discharge its waste into the Bay (BMA 1991). Coastal waters are still being contaminated. Those around Indonesia have been found to contain heavy metals: cadmium, mercury, chromium, copper, lead, and zinc (Supriharyono 1992).

Disposal of sewage is also a major problem for urban authorities, especially when there are no formal waste disposal mechanisms in place such as in squatter camps. Less than 10% of sewage in China receives any treatment before discharge onto the land (CIA1997). A study of the sea water around the coasts of Hong Kong, found high levels of anaerobic bacteria from fecal contamination (Boost *et al.* 1992). Similarly, in Hong Kong, swimming at "barely acceptable" beaches incurred higher risks of contracting skin, ear, respiratory, and gastrointestinal illnesses than at "relatively unpolluted" beaches. A quantitative relationship was found between densities of *E coli* in beach water and swimming-associated gastrointestinal symptoms and skin symptoms (Department of Community Medicine 1990).

Health gains

The implications for public health in this region from polluted air, water, and land are broad. They include both direct effects upon health from such as respiratory problems, gastrointestinal infection, sickness, and diarrhoea, and indirect impact through food availability, its quality and safety, job losses through closure of fisheries, shell fisheries and loss of farming land, and recreational loss with closure of beaches. However, if urbanisation is managed proactively and such environmental issues can be controlled, there are also benefits to the public's health from urbanism. Using selected indicators of health and/or socioeconomic development, Table 8.3 illustrates the strong correlation seen between declines in child mortality ($r=-0.658$, $p=0.028$) and increases in life expectancy ($r=0.746$, $p=0.013$) with increasing urbanism. These trends could possibly be explained by the amount spent by individual governments on health care rather than the benefits of urbanism. But assuming that an adjustment can be made for high and low-income countries by using the percentage of government income spent on healthcare in each country, those countries in this region for whom data are available appear to spend a similar proportion of their income, between 0.4% and 1.5%, on health care, irrespective of level of urbanism. There appears to be no relationship between the spending on health and either child mortality ($r=-0.573$, $p=0.179$) or life expectancy ($r=0.608$, $p=0.147$) (Table 8.3). A similar exercise looking at spending on health care adjusted for size of population or population density countrywide also produced no obvious pattern of spending with level of urbanism.

Table 8.3. *Selected indicators for health*

Country	% urban*	Population (m)	Density: Population/sq km country-wide	Child mortality per 1,000 live birth	Life expectancy for those born 1990–95	% Govt. income spent on health 1987
Vietnam	20	64	226	65	63.9	—
Thailand	20	59	116	34	67.1	1.1
China	30	1 304	19	42	72.1	1.4
Indonesia	34	192	106	97	62.7	0.5
Malaysia	53	19	61	29	70.8	1.5
Philippines	54	68	233	69	65.0	0.7
Taiwan	58	21	588	8	—	—
Japan	78	125	333	6	78.8	—
S. Korea	80	44	451	30	70.8	0.4
Hong Kong	95	6	6 080	7	77.6	—
Singapore	100	3	4790	9	74.5	1.3

— not available
Source: Mackay 1993 (modified) * Asiadata 1997.

Of course, the money may be disproportionately spent but overall it would appear that health care provision, as measured by spending on health, or lack of it, does not explain the considerable variation in specific health indices seen between countries in this area of the world. Other associated contributory factors may include those of access to health care, availability and/or eligibility for free health care, affordability of health care, and subsidisation of health care. For subsidies, there are differences across the region, for example, in Hong Kong the government subsidises health care by 97%, in China it is subsidised by 10% and work units pay some of the costs, but migrants in the cities with no jobs are liable for all costs.

However, it has also been said that education, especially that of women, is the key to improvements in health through the uptake of health services such as family planning and immunisation programs. For the countries in this region listed in Table 8.3, there is a significant correlation and positive trend between the proportion of the population who have attained a level of education beyond the primary stage and level of urbanism within the country ($r=0.892, p < 0.001$) (Figure 8.1). One reason may be the potential for more easily accessed education in urban areas. But the full explanation is likely to be much broader and will include such aspects as nutrition, housing, family planning, the availability of work, and the interaction between them, along with access and availability.

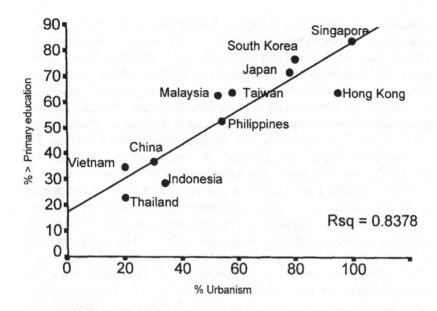

Figure 8.1   The relationship between urbanism and level of education achieved

Conclusions

Almost all countries in Southeast Asia face many dilemmas at present. National interests demand the acceleration of the process of industrialisation but developing industries and the ensuing urbanisation create various health problems. Overall, the health problems in this region as prompted by industrialisation and its resultant urbanisation are not dissimilar to the pattern seen in the West during, and subsequent to, the industrial revolution. But

(a) changes are happening over a much shorter time scale which exacerbates the problem;

(b) changes are happening irrespective of the Western experience;

(c) there is little documented information available on the associated health issues;

(d) the ethos in the general public appears to be one of making money rather than caring, or even being aware, in some instances, of the environment in which the living and working is taking place.

However, there is an awareness and a response by some governments to the problem. For example, in Hong Kong, there is an ongoing extensive housing program started in the 1960s which provides public sector low-cost accommodation for 50% of the population, 46% of the total housing available, and includes a planned infrastructure of supporting clinics, schools, and services (Daryanamu 1996). Legislation is also being enacted, such as that to control ambient sulphur dioxide levels (Hong Kong Government 1990), although enforcement is another issue. What governments need to do is to focus not on the issues of urbanism *per se* but on specific groups who are at risk during the urbanisation process, such as the urban poor, children, and the elderly. To help them in this process, epidemiological studies can provide useful tools for identifying areas of need and for evaluating the effectiveness of initiated preventive measures. It is not easy to establish cause and effect relationships for rural–urban factors and their impact on health. Large-scale well-planned epidemiological population studies are one way in which this problem can be addressed. But this needs to be as part of a multisectoral approach with active participation by regulatory agencies, including those responsible for the environment and for health, both of whom need to continually monitor and collect good quality intelligence, and in partnership with academic research units. The last two steps in the chain are education and legislation where necessary to establish and enforce practice of sound urbanisation policies to benefit the health of the residents.

References

Badri, M. A. (1992). Environmental problems of the tropical city: a brief overview. *Southeast Asian Journal of Tropical Medicine and Public Health*, **23**, Supplement 3, 94–109.

British Medical Association (BMA). (1991). *Hazardous Waste and Human Health*. Oxford: Oxford University Press.

Boedhi-Darmojo, R. (1993). The pattern of cardiovascular disease in Indonesia. *World Health Statistics Quarterly*, **46**, 119–24.

Boost, M. V., Wong, L. P., Wong, P. S. and Ma, K. C. (1992). An investigation into the presence of anaerobic bacteria in coastal waters of Hong Kong. In *The Role of the ASAIHL in Combating Health Hazards of Environmental Pollution*, eds. A. J. Hedley, I. J. Hodgkiss, N. W. M. Ko, T. L. Mottershead, J. Peters, W. W.-S. Yim, pp. 74–81. Hong Kong: University of Hong Kong.

Central Intelligence Agency, Office of the Director (1977). CIA Maps and Publications. *The World Factbook 1995*. 17 March. http://www.odci.gov/cia/publications/95/fact.

Chan, R. C. K. (1995). Urban development strategy in an era of global competition: the case of Hong Kong and South China. In *Globalisation and Regional Development in Southeast Asia and Pacific Rim*, eds. G. Y. Lee and Y. W. Kim, pp. 202–227. Korea: Korea Research Institute for Human Settlements.

Daryanamu, R. (ed.) (1996). *Hong Kong 1995*. Appendix 37:508. Hong Kong: Government Printer.

Department of Community Medicine. Beach Study (1990). *EPD Progress Report* on Phase III. *Epidemiological-Microbiological Studies on Beach Water Pollution*. Hong Kong: Environmental Protection Department.

Department of Community Medicine (1991). *Survey of Government Outpatient Departments. Report to the Department of Health*. Hong Kong: Department of Community Medicine.

Department of Health, Hong Kong (1994). *Annual Report 1993/4*. p. 4. Hong Kong: Government Press.

Environmental Protection Department (1991). *Environment Hong Kong 1991. A Review of 1990*. Hong Kong: Government Printer.

Environmental Protection Department (1992). *Environment Hong Kong 1992. A Review of 1991*. Hong Kong: Government Printer.

Florey, C. du V. and Leeder, S. R. (1982). *Methods for Cohort Studies of Chronic Airflow Limitation*. WHO Regional Publications, European Series No 12. Copenghagen: WHO.

Gross, R., Rasad, A. and Sevenhuysen, G. (1992). Asian pacific workshop on nutrition in the Metropolitan area: a summary. *Southeast Journal of Tropical Medicine and Public Health*, **23**, Supplement 3, 1–2.

He, K. S., Yoon, G. A., Kim, W. K. and Park, O. J. (1992). Urban nutritional problems of Korea. *Southeast Journal of Tropical Medicine and Public Health*. **23**, Supplement 3, 69–76.

Hong Kong Government. (1990). *Air Pollution Control (Fuel Restriction) Regulations 1990*. Air Pollution Control Ordinance (Chapter 311). Gazette 4/1990. Hong Kong: Government Printer.

INCLEN Multicentre Collaborative Group. (1994). Socio-economic status and risk factors for cardiovascular disease: a multicentre collaborative study in the International Clinical Epidemiology Network (INCLEN). *Journal of Clinical Epidemiology*, **47**, 1401–9.

Key Trends and Statistics. (1997). *"Asia Studies Ltd"*. 17 March. ttp://aisadata.pacificnet.com/mkt—data.

Kochi, A. (1991). The global tuberculosis situation and the new control strategy of the World Health Organisation. *Tubercle*, **71**, 1–6.

Mackay, J. (1993). *State of Health Atlas*, pp. 80–7. London: Simon and Schuster.

Mertens, T. E. and Low, B. D. (1996). HIV and AIDS: where is the epidemic going? *Bulletin of the World Health Organisation*, **74**, 121–9.

Medical Research Council. (1960). *Questionnaire on Respiratory Symptoms*. London: MRC.

Ng, K.-C. (1996). Hope for cagemen dashed. *South China Morning Post International Weekly*. 20 January 1996, p. 3.

Peters, J., Tanaka, S., Toyokawa, H., Takayanagi. M., Nishikawa, H. and Kigawa, M. (1990). The choice of measurement in the assessment of obesity. *Japan Journal of Health and Human Ecology*, **56**, 246–7.

Peters, J., Hedley, A. J., Wong, C. M., Lam, T. H., Ong, S. G., Liu, J. and Spiegelhalter, D. J. (1996). Effects of an ambient air pollution intervention and environmental tobacco smoke on children's respiratory health in Hong Kong. *International Journal of Epidemiology*, **25**, 821–8.

Supriharyono. (1992). Heavy metal contamination in coastal waters–a case study along the north coast of central Java, Indonesia. In *The Role of the ASAIHL in Combating Health Hazards of Environmental Pollution*, eds. A. J. Hedley, I. J. Hodgkiss, N.W.M. Ko, T. L. Mottershead, J.Peters and W. W.-S. Yim, pp. 165–169. Hong Kong: University of Hong Kong.

Talib, R. and Agus, M.R. (1992). Social problems and health in urbanisation. *Southeast Asian Journal of Tropical Medicine and Public Health*, **23**, Supplement 3, 84–93.

Tam, A. Y. C., Wong, C. M., Lam, T. H., Ong, S. G., Peters, J. and Hedley, A. J. (1994). Bronchial responsiveness in children exposed to atmospheric pollution in Hong Kong. *Chest*, **106**, 56–60.

Yao, C., Wu, Z. and Wu, Y. (1993). The changing pattern of cardiovascular diseases in China. *World Health Statistics Quarterly*, **46**, 113–18

Waller, A. (1992). A fight on all fronts. *Far Eastern Economic Review*, 13 February, pp. 28–9.

Wong, C. M., Lam, T. H., Peters, J., Hedley, A. J., Ong, S. J., Tam, A. Y. C., Liu, J. and Spiegelhalter, D. J. (1998). Comparison between two districts of the effects of an air pollution intervention on bronchial responsiveness in primary school children in Hong Kong. *Journal of Epidemiology and Community Health*, **52**, 571–8.

WHO. (1994). *World Health Statistics Annual Report*. c4–c7. Geneva: WHO.

Part III
Poverty and health

9 Urban poverty and nutrition in the United Kingdom

E. DOWLER

Editors' introduction

Urbanisation involves clear changes in patterns of subsistence, such that rural subsistence bases are replaced with social and economic structures that involve the commoditisation of work and subsistence. These structures influence the extent of dietary variation and the level of nutritional adequacy experienced by groups and populations in different urban microenvironments. In this chapter, Dowler examines the ways in which socioeconomic factors impact on urban dietary availability in Britain. While the poorest households in Britain spend the highest proportions of their incomes on food, their intakes of most nutrients are lower than among the better-off in society. Furthermore, the greater the level of poverty, the lower is the overall intake of nutrients, including energy and protein. This is in contrast to the United States, where the poor urban community studied by Johnston and Gordon-Larsen (chapter 10) has high intakes of most nutrients, including energy and protein. Despite superficial environmental similarities, poor populations in London and Philadelphia have quite different nutritional experience. The social geography of London and Philadelphia differ in their national and historical bases and political and economic attitudes to food, contributing to these big differences in seemingly similar contexts.

In modern times the human role as hunter, gatherer or grower [of food] has largely been replaced by that of shopper. Fifty years ago most shopping for food was done by walking a short distance to a small local shop which sold national brand goods and local produce; goods were served by the shopkeeper or assistant and would then be delivered by bicycle. Now the retailing industry has moved into an era of mass consumption, the parallel of the mass production that revolutionised industry. Most people buy food from one of a small number of retailing giants that operate nationally and are heavily involved in food production and packaging. The bulk of food purchasing is self-service and the main weekly shop is done by car.   (Piachaud and Webb 1996:1)

175

Introduction

In this chapter I begin to address the complexity of poverty as a feature of contemporary urban life and the consequences for nutrition and health, concentrating on food patterns and indicators of nutrient intake rather than those of physical growth or obesity. I draw on work from a wide range of disciplines, and conclude with a very brief review of attempts in the pubic and private sector to address some of the problems identified.

Others have outlined the wider historical sweep of urban development in the UK; I want to draw attention only to three salient points. The first is that notions of "North-South divide" notwithstanding, differences persist in social and health indicators between the old urban industrial areas, and inner London, and the rest of the UK, particularly southern regions of England (excluding much of inner London) (Dorling and Tomaney 1995). The second is that while urban poverty is characterised by location – the appearance and persistence of "poor places" – its spatial concentration of poverty is none the less relatively low: the rich and poor to some extent live "cheek by jowl." The third point is more pertinent to nutritional consequences of poverty: that food retailing patterns have changed dramatically over quite a short period, with as yet largely underresearched impact on low-income households.

These statements, which might even seem to some extent contradictory, underpin my understanding of urban poverty and its consequences for nutrition and health.

Poverty and deprivation: concepts and measurement

Poverty is notoriously difficult to define and measure consistently in ways that are acceptable to policy-makers and administrators, to members of a given society, or to those who see themselves as constituting "the poor" (the literature is large and summarised elsewhere, e.g. Piachaud 1987; Oppenheim and Harker 1996). Most attempts begin with trying to define a minimal income to enable individuals or households to survive,[1] in a given society. Debate over what constitutes "survival" (e.g. Payne 1992; Smith 1995) and how to cost it, and the need for comparisons over time and place, have been partly responsible for the shift to "relative" poverty definitions: income sufficient for a generally accepted standard of living in a specific society at a specific time. For instance, at the end of 1984 the European Council adopted as a definition of poverty "the Poor shall be taken to mean persons, families and groups of persons whose resources (material, cultural, social) are so limited as to exclude them from the minimum acceptable way of life in the Member State in which they live."

Table 9.1. *Nos. households and individuals receiving Income Support and average amounts (Great Britain, 1995) (in thousands)*

	Households	Individuals	£/week
Total	5,670	9,773	55.6
Aged 60+	1,781	2,112	41.4
Disabled	739	1,306	60.3
Lone parent	1,056	2,983	76.6
Unemployed	1,672	2,710	53.9
Other	422	663	61.8

Source: Department of Social Security (1996).

Such a definition clearly goes beyond meeting basic biological needs and acknowledges human need to engage in customs, practices and expectations of society. This concept of social exclusion,[2] with its implications of disempowerment and low self-esteem, has been important for adminstrators, campaigners and those who experience poverty.

However, despite these perceptions and the recognition of resources other than income, relative poverty is usually operationalised for national and international purposes in terms of income, as those whose income falls below a specified level – usually 50% average disposable per capita income (Oppenheim and Harker 1996). In the UK, the level laid down by law as the minimum amount sufficient to live on is the rate used to calculate entitlement to Income Support, the basic social protection measure. Levels for a family of two adults and two children would in 1997 be about £118 a week – which in fact is roughly the same as 50% of average income (after housing costs). About 14 million people in the UK live in households with incomes below half the national average. Almost 10 million of these people rely on Income Support; the remainder live on low or insecure wages (Department of Social Security 1994: 5, 6).[3] Whether it is appropriate to use numbers claiming Income Support to count the poor is to some extent a moot point (the published statistics in fact refer to them as "Low Income Families"), but since the rates received are intended to meet only basic needs and receipt is means tested, it seems reasonable to use the level as an income cut-off and its receipt as signaling avoidance of poverty (Dowler and Dobson 1997).

As Table 9.1 shows, of those who currently claim Income Support, 22% are aged 60+ yrs, 31% are lone parents and 28% are unemployed (the remainder are disabled or in other categories); the majority of claimants over 60 years or classified as lone parents are women. One in three children lives in a household with incomes below 50% of average income; 58% of lone-parent households and 35% of single pensioner households have incomes below 50% average. Goodman and Webb (1994) showed that unemployment, having children, being a lone parent, and being a pensioner without a private pension, were the main contemporary risk factors for being poor. Similar sociodemographic conditions signal a high risk of poverty throughout Europe (Oppenheim and Harker 1996).

One of the problems with income based measures of poverty is finding appropriate data sets; another is in interpreting what that income will purchase. Although minimal food and shelter needs could theoretically be costed, assumptions are made about physical access and economic and social entitlement. An alternative approach has been to use deprivation measures: to look at the level of key material (or social) conditions experienced. Several different indices have been developed, reflecting the focus and needs of investigators (e.g. Department of the Environment 1983; Jarman 1983; Townsend et al. 1987; Carstairs and Morris 1989). All tend to use census based indicators of household conditions, such as crowding, occupational social class, access to a car, housing tenure and employment status. Sometimes individual demographic characteristics such as proportion of pensioners or under-fives, or ethnicity, are included; sometimes access to health care or basic amenities is also included. Inclusion of different terms, use of different weighting systems or geographical base can lead to different rank orders for geographical areas, and comparisons between indices can be hard to interpret (Goodwin 1995). The consequences for drawing conclusions about urban areas and poverty are discussed below

Sources of data on poverty and nutrition

Work on poverty and deprivation, and their consequences for food and nutrition in low-income countries has used notions of vulnerability (social, economic), entitlement (to food, or to particular social conditions) (e.g. Sen 1981; Drèze and Sen 1989), food security (Maxwell 1996), and latterly, livelihood security (Clay 1997) rather than income based indicators. These approaches have had some impact on work in the UK where food and nutrition is concerned (e.g. Williams and Dowler 1994; Department of Health 1996), and concepts of access and vulnerability have potential for research and intervention. In practice, empirical research in these terms is only just beginning in food and nutrition. Until recently, most national and smaller-scale nutrition surveys used occupationally based social class as socioeconomic indicators; whether people in manual classes constitute the "poor" is a moot point and it is not always clear how those who are unemployed, or economically inactive (many of whom are elderly or lone parents) are classified in such surveys. Recent national surveys have also classified subjects in terms of economic activity or means-tested state benefit receipt (on a "yes/no" basis). However, there is no time element included, nor are these indicators combined with household demographic or locational characteristics: the data sets for commenting specifically on urban poverty and nutrition are limited. Rather, we have to look at who the urban poor are likely to be and infer their nutritional conditions from household type.

The main source of national data on income and food is the National Food Survey (NFS), an annual household food consumption survey of approximately 7,000 households in England, Scotland and Wales. Data on food usage patterns and nutrients/head (collected at household level) are published by broad region, social class, household composition, and for five income categories and two categories of household without earners (>/< £140/week) (Ministry of Agriculture, Fisheries and Food 1996). Some income comparisons can be made, but the level of "low income" here is in fact higher than Income Support rates for most households (a two-adult, two-children family would receive £100–£116 a week; a lone-parent, two-children family £97–£112, depending on the age of the children).

Urban poverty and health

Cities are often seen as synonymous with wealth and poverty, but are characterised by diversity within and between them. The geography of poverty is complex and depends partly on the deprivation indices and the area boundaries. For instance, in an analysis of the 1991 Census, Forrest and Gordon (cited in Goodwin 1995) listed the 20 most and least materially deprived local authority districts. Thirteen of the most materially deprived districts were in inner London, two in outer London, and four were in Liverpool, Knowsley, Birmingham and Kingston-upon-Hull, the latter being the only non-large conurbation. The least deprived areas were characterised by containing small or medium towns. Social deprivation score ranking had some of the same districts, but also periurban estates, high on the list of most deprived. Other research supports this finding (Goodwin 1995).

There has been a concentration of poverty location over time. Green (1994) also used Census data, and compared ranking of "best" and "worst" electoral wards (about 2,000 households) in 1981 and 1991. She showed that poverty was concentrated in inner cities, especially old industrial areas, whereas wealth, though less concentrated, was more likely to be found in rural, suburban or small/medium towns. But she also found that

Table 9.2. Greater London wards: measures of multiple deprivation: highest ranked (five wards) and lowest ranked (five wards)

Ranked ward	Borough	Z-score index	Unemployed (%)	Over-crowded (%)	Not owning home (%)	Not owning car (%)
1. Spitalfields	Tower Hamlets	16.8	32.5	29.8	81.9	73.6
2. Liddle	Southwark	10.9	30.8	12.1	96.4	74.4
3. St Dunstan's	Tower Hamlets	10.8	27.4	16.2	83.8	68.6
4. St Mary's	Tower Hamlets	9.1	22.0	14.7	79.5	68.8
5. Shadwell	Tower hamlets	8.9	25.1	12.7	82.2	66.8
760. Cranham West	Havering	−6.2	5.0	0.4	3.3	13.2
761. Biggin Hill	Bromley	−6.2	4.6	0.9	7.9	7.9
762. Cheam South	Sutton	−6.2	4.1	0.4	7.4	11.0
763. Selsdon	Croydon	−6.3	3.9	0.7	4.6	11.5
764. Woodcote	Sutton	−6.5	5.8	0.3	4.8	5.4

Source: Goodwin 1995: ch. 4.
Data from OPCS (1991), Census of Population (London, HMSO); analysis by Department of Geography, University of Aberystwyth.

although the "poorest" wards tended to be in the "poorest" localities, at least half the "poorest" neighbourhoods lay outside the "poorest" areas (Green 1994). Goodwin (1995) summarises a number of studies which support this finding; he includes outer urban estates of local authority housing in areas of concentrated poverty. These estates were built originally either to service a postwar new industrial development which has now disappeared, or to rehouse families from inner city delapidated ninteenth-century housing. In other words, areas that were intended to counter traditional "urban" evils of unemployment and housing slums in many cases have become the new poverty places.

A characteristic of the last 15 years is the marked polarisation of poverty in the UK (Joseph Rowntree Foundation 1995). Table 9.2 shows evidence from London; data from current research in Aberystwyth ranks London wards using measures of multiple deprivation. The differentials are extreme: up to 74% of households in the poorest wards (in inner London) had no car access, and over 80% rented their housing (home ownership was much promoted by the government during the 1980s as a desirable, reasonable goal for all), whereas fewer than 10% owned neither car nor home in the richest wards. In London as elsewhere, social and economic changes have led to the growth in "no-earner" households and neighborhoods, and corresponding growth in "dual-earner" households and neighborhoods (Green 1994). Polarisation of housing and service provision has tended to match employment trends, so that local authority housing is increasingly tenanted by the single elderly, the sick or disabled, the unemployed and lone-parent households, with a higher proportion of minority ethnic groups among each (Noble and Smith 1994; Joseph Rowntree Foundation 1995; Goodwin 1995). Of course, there is nothing intrinsic about being unemployed, a lone parent, a pensioner or disabled that inevitably leads to poverty; it depends on what alternative income sources are available through social protection, or rates of re-employment. What is clear is that neither has worked to prevent people in these circumstances from falling into poverty in the UK over the last two decades, whichever definitions or measurements are used.

Turning to measures of health, there is a large and growing literature on the relationship between poverty or deprivation and differential morbidity and premature mortality in the UK, as in other industrialised societies (e.g. among many, Crombie *et al.* 1989; Eames *et al.* 1993; Davey Smith *et al.* 1994), and most of these health effects have been seen in the urban areas and periurban estates (Drever and Whitehead 1995; Charlton 1996). These differentials have widened as income inequality and deprivation increased (e.g. McCarron *et al.* 1994; Philimore *et al.* 1994): that is, as individuals have become poorer, and as societies have become more unequal, the risk of illness and early death for those who are poor has increased relative to those who are not poor (Wilkinson 1992).

Much of this work has used electoral ward as the geographic base, and has led to area-based targeting for intervention by the health and other sectors. Sloggett and Joshi (1994) recently examined the relationship between premature mortality and deprivation measured at the individual level, and found a similar relationship to other studies: the most materially deprived, whether by employment status, tenure, car access or occupational social class, died youngest. In fact, Sloggett and Joshi's work showed that car access and housing tenure, the income proxy indicators used above, were powerful predictors of mortality, with a reduction in life expectancy from age 25 of almost 5 years for those in the worst circumstances. But they also argued that the poorest areas had the worst health indices because there was a concentration of the poorest people – that deprivation attaches to the household rather than to the locality (i.e. to people rather than geography), despite the characteristics of urban deprivation which contribute to poor health – housing quality, service access, transport, etc.

Consequences of poverty for food and nutrition

Nutrition is clearly an input to individual health, as well as an outcome measure in terms of diet quality, growth or body size. Despite the role of garden plots, allotments or other means for people to grow food in cities, people need money for food and somewhere to buy it. Nutritional measurements are outcome indicators of a household's access to food, determined in part by the money a household or individual allocates to food expenditure, which itself depends on how much money they have and what priority they can place on food as opposed to other demands of the domestic economy. Access also depends on the shops people can reach, the price of foods, and the range of food commodities available. Government information and commercial advertising and promotion can influence choice of commodity and food usage patterns, which are also affected by people's work (paid or voluntary) and leisure activities.

There are a number of ways to investigate the impact of poverty on food and nutrition. We could look at the percentage of income spent on food, and ask whether the amounts spent would be sufficient to purchase diets adequate for health. In the UK, households in the lowest income decile spend the highest proportion of income on food (26% vs. 15%), although they spend less than richer households in absolute terms (in 1991, £21 a week rather than £73; households with weekly incomes below £80 spent about £1 a week on fruit; households with weekly incomes above £550 spent £3.60 or more on fruit a week (Central Statistical Office 1992). Work by the Budget Standards Unit suggests that a "modest-but-adequate" diet at 1992 prices would have cost £19.48 a week for a single pensioner, and £39.41 for a lone-parent household (Nelson *et al.* 1993). (Adjusting the food prices for a "low-cost-but-adequate" budget reduces the weekly amount to £18.65 and £38.40 respectively.) Clearly, people living on Income Support could not afford to spend these theoretical amounts to obtain a low-cost but healthy diet.

An alternative method would be to look at what people living on low incomes, or classified as "poor," actually consume in terms of foods or nutrients. The NFS data on low-income households and those without earners show that nutrient intakes are more likely to be below the reference levels[4] in the lowest income or non-earner households, than in richer households. The NFS also shows that vitamin C, folate, iron, zinc and magnesium are much more likely to be below reference levels in households with more than three children, or headed by a lone parent (Ministry of Agriculture, Fisheries and Food 1996). National data of nutrient intakes measured on individuals (by weighed dietary intakes or food frequency questionnaires) are available from a number of surveys (e.g. Department of Health 1989; Braddon *et al.* 1988; Gregory *et al.* 1990; Bolton-Smith *et al.* 1991; Gregory *et al.* 1995). Table 9.3 shows data on women's nutrient intakes from the Nutritional Survey of British Adults: women who were unemployed, or in households claiming benefit, or in lower/manual social classes, had significantly lower intakes of many vitamins (especially vitamins A, C and E) and minerals (especially iron) than women not in these categories. The findings were similar for men. The Pre-School Child Survey showed that children from manual social classes, or households claiming benefits, or from lone-parent families, had lower intakes and blood levels of most vitamins and minerals than those not in these circumstances, even when intakes were adjusted for differences in energy intake – the children were not just eating less food. In the survey on schoolchildren, children who received free school meals (and were therefore from households claiming benefits) had lower vitamin and mineral intakes than those not from benefit households.

Findings such as these from large surveys are comparable to those from smaller surveys looking at nutrient intakes in different age groups, and in households of different socioeconomic conditions, summarised in Craig and Dowler (1997). We recently surveyed nutrient intakes in lone-parent households in London and, in common with others, found those who lived on income support had lower nutrient intakes than those who did not. We

Table 9.3. *British women: mean daily intakes of energy and nutrients*

Nutrient	All women (n = 1110)	Unemployed (n = 57)	Benefit recipients (n = 153)	Soc class IV & V (n = 222)	Lone mothers (n = 76)
Energy (kcals)	1,680	1,640	1,560	1,580	1,580
Protein (g)	62.0	56.5	55.6	57.3	56.3
Fat (g)	73.5	71.9	67.7	69.6	69.9
Fibre (g)	18.6	17.7	16.8	16.9	16.1
Iron (mg)	12.3	10.4	11.8	11.6	11.5
Calcium (mg)	730	642	636	660	605
Zinc (mg)	8.4	7.5	7.4	7.7	7.5
Folate (µg)	219	190	192	196	199
Vitamin A (ret equiv µg)	1,488	1,003	1,328	1,217	1,206
Vitamin C (mg)	73.1	72.1	55.4	55.8	55.8
Vitamin E (mg)	8.6	7.1	7.5	8.2	7.3

Source: Gregory et al. 1990.

also tried to look at the cumulative effects of living on a low income and constructed a deprivation index reasonably similar to those constructed from census data (few lone parents in the study had access to a car). Table 9.4 shows that nutrient intakes were very much less likely to be adequate (in terms of percentage reference intakes) in the long-term unemployed, who lived in local authority housing on means-tested benefits, particularly where automatic deductions were made from those benefits for rent or fuel debt recovery (Dowler and Calvert 1995). Those living in the worst deprivation had about half the nutrient intakes of parents not in such circumstances. This finding was largely independent of smoking, and was also not seen to such an extent in the children's diets: parents seemed to protect their children from the worst consequences of poverty in relation to food.

These surveys show that members of the poorest households, who were likely to be unemployed, lone parents or pensioners, consistently have worse diets in terms of nutrients consonant with good health, than those from better-off households. The surveys mentioned also include data on dietary patterns, and the story is similar: poorer households consume fewer foods recommended for health and have a much less diverse food base. They eat monotonous diets with little variation. There is no evidence that poorer people do not know what constitutes an appropriate diet for health; qualitative surveys have continually shown they do not have enough money to purchase it or cannot get to decent shops (Cole-Hamilton and Lang 1986; Health Education Authority 1989; Dobson et al. 1994).

Our survey findings also contributed to work on budget management on

Table 9.4. *UK lone parents: adequacy of nutrient, fat and NSP intakes, by the material poverty index**

Nutrient (se)	Poverty index = 0 $n = 59$	Poverty index = 1 $n = 35$	Poverty index = 2 $n = 24$	p Value ANOVA
Protein % RNI	158 (6.2)	143 (6.5)	127 (7.6)	0.0121
Total fat g	83 (3.5)	77 (4.2)	70 (5.1)	n.s.
Iron % RNI	90 (4.5)	66 (4.3)	56 (4.6)	< 0.0001
Calcium % RNI	111 (5.6)	93 (5.7)	83 (7.6)	0.008
NSP % EAR	66 (4.6)	54 (6.9)	44 (8.1)	0.0004
Zinc % RNI	122 (5.9)	106 (6.0)	96 (7.7)	0.0197
Folate % RNI	114 (6.4)	83 (4.3)	76 (6.5)	< 0.0001
Vitamin C % RNI	149	101	74	0.0022
Vitamin A (ret.eq.) % RNI	129	99	79	0.0275
Vitamin E % safe intake	188	168	148	n.s.

* *Poverty Index*: those in categories A or B, or in both.
Category A = those in local authority or private rented housing, with no job, no holiday, for more than 1 year.
Category B = those with rent/fuel automatically deducted or on a key meter.
Poverty index:
0 = not in either category
1 = in one category (A or B) only
2 = in both categories (A and B)
NSP = non-starch polysaccharide; ret. equiv. = retinol equivalent; n.s. = not significant.
Figures given are arithmetic mean of % RNI (reference nutrient intake), % EAR (estimated average requirement) or % safe level; except fat, for which g/day are given. Standard errors are given in brackets, taken from ONEWAY or ANOVA. For vitamins A, C and E, the geometric mean is shown and no standard error can be presented.

low incomes (e.g. Kempson 1996): people develop careful strategies to manage tight budgets, and are often very skilled at controlling expenditure and shopping around. Food expenditure is what many reduce to avoid or reduce indebtedness or meet essential demands (Dobson *et al.* 1994). However, about one in five claimants has money taken off their benefit to repay rent or fuel arrears; many others use card-key meters, through which they also pay arrears. These findings of severe nutrient reduction contribute to challenging the adequacy of social provision for long-term living (Kempson 1996) and highlight the health consequences to individuals and society of the failure to provide sufficient resources.

There is another, fundamental aspect to the circumstances of urban living for the poor, and that is physical access to food: shops, transport and personal mobility. As the discussion above has made clear, poor urban people tend to live in inner cities and/or large local authority estates; in recent years food shops have struggled to survive in such places, partly because people there spend less and partly because of the massive changes in food retailing. Street markets and small, specialised high street food shops are disappearing, although discount operators have to some extent filled this retail gap, offering low prices on a limited product range. There has been a fourfold increase in superstores (> 25,000 sq.ft), mostly located outside town centres and designed primarily for car access (Department of Health 1996). By 1994–5, large supermarkets had captured about 70% of average total food expenditure, from about 50% in 1991 (Piachaud and Webb 1996). The poorest in cities and towns do not have cars (Oppenheim and Harker 1996) and public transport to better shopping centres is often inadequate. Over 4 million adults in Britain have some degree of motor disability, of whom 3 million are aged over 60 years, which means at minimum they cannot walk more than 400 yards without severe discomfort (Piachaud and Webb 1996). Poorer people probably spend more time on basic food shopping than others, because they spend more time making cost comparisons (Dobson et al. 1994). Yet those who can get to, choose to, or afford to shop in large supermarkets generally benefit from lower food prices and more choice, particularly if they can purchase bulk sizes (Leather 1996). Piachaud and Webb (1996) found that food in small shops (corner shops, convenience stores, independent small supermarkets) cost on average 24% more than the same items in large supermarkets or discounters – about 10% of average low income. Using data for the cheapest versions of foods, food in small shops cost 60% more than large stores – about 25% of average low income. Other researchers have shown that foods currently recommended for a healthy diet (such as wholemeal bread, leaner meat, fresh fruit and vegetables) not only cost more than cheap filling foods (which are not always "healthy") but they also cost more in the shops where poorer people live than where richer people shop; Morton 1988; Mooney 1990; Dowler and Rushton 1994).

These facts seem to contradict the findings by health statisticians and others that poor health outcomes or higher mortality relates to people rather than places. The answer clearly lies between the two: households need sufficient income to pay for rent, local taxes and utilities (heat, light, water) as well as basic domestic needs. When they don't have enough money, occasionally or regularly, food is where most people cut back: they eat different food (no fruit, fewer vegetables), less food (crackers or jam sandwiches), or they go without. These strategies have health consequences. But in addition, poor people have to pay more than richer people for basic food, and usually have considerably less choice in commodity and quality. Their housing may cost more to heat; their water may be more expensive; the schools and health facilities may be less adequately funded and staffed than those of richer people. Their control over their physical and economic circumstances is likely to be considerably reduced; social exclusion contributes to ill-health and higher mortality. All of these conditions seem to be exacerbated in inner cities and run-down industrial areas; none is inevitable.

A recent project team to the Nutrition Task Force in the British Department of Health addressed the problems of food and low income, outlining potential responses needed from the public and private sectors. Their report (Department of Health 1996) called for a national network of local projects and initiatives on food and low income, and for the creation of local public/private sector food partnerships, especially in areas of multiple disadvantage, to regenerate local food economies. There are some promising beginnings, but it is hard to feel confident that the food, or health, sectors alone can make a serious contribution to reducing differentials in life expectancy of five years (Sloggett and Joshi 1994), or mortality differentials of 100% (Charlton 1996).

I began with a quotation about food shopping in the 1990s; I finish with a different one, about the discipline of poverty:

> It is in this context [of wide range of choice and information about new items] that we refer to the discipline of poverty... when food shopping can no longer be experienced as a relatively relaxed activity it is no longer possible to respond to the routine messages on television and in supermarkets. Instead, shopping is severely constrained as the tight budget limits choice of shopping outlets and food items. The pressure is to cut back... (Dobson *et al.* 1994:35)

Notes

1. Although we should note that one of the earliest to define and measure poverty in modern times, Seebohm Rowntree, in fact identified the poor visually and on a relative poverty basis (comparing the living conditions of working-class people in York with living conditions conventionally recognised and approved). The poverty line with which his name is associated, which was constructed by costing a minimal survival diet and estimating additional living needs (rent, clothes, etc.), was in fact used to separate people already identified as poor into those whose income was insufficient to purchase basic survival necessities, and those whose income was sufficient but who were unable so to do for other reasons (not necessarily inefficiency) (Veit-Wilson 1986).
2. "Exclusion processes are dynamic and multidimensional in nature. They are linked not only to unemployment and/or to low income, but also to housing conditions, levels of education and opportunities, health, discrimination, citizenship and integration into the local community" (European Commission 1994).
3. Among those living on low wages in the UK are many who currently have no entitlement to national insurance or occupational pensions; as a consequence they receive no maternity or sick pay, or state old-age pension: they are likely to

be reliant on Income Support when they can no longer earn a wage. The value of benefit levels in relation to real costs of living becomes critical when people live on them for long periods.

4. For groups of people, the lower their mean nutrient intake is in relation to the reference intake, the less likely all members of the group are to be eating enough of that nutrient to avoid ill-health. The probability of deficiency increases as the percentage of reference value achieved decreases (Department of Health 1991).

References

Bolton-Smith, C., Smith, W. C. S., Woodward, M. and Tunstall-Pedoe, H. (1991). Nutrient intakes in different social class groups: results from the Scottish Heart Health Study. *British Journal of Nutrition*, **65**, 321–5.

Braddon, F. E. M., Wadsworth, M. E. J., Davies, J. M. C. and Cripps, H. A. (1988). Social and regional differences in food and alcohol consumption and their measurement in a national birth cohort. *Journal of Epidemiology and Community Health*, **42**, 341–9.

Carstairs, V. and Morris, R. (1989). Deprivation and mortality: an alternative to social class? *Community Medicine*, **11**, 210–19.

Central Statistical Office (1992). *Family Spending: A report on the 1991 Family Expenditure Survey*. London: HM Stationery Office.

Charlton, J. (1996). Which areas are healthiest? *Population Trends*, **83**, 17–24.

Clay, E. J. (1997). Food security: a status review of the literature. Research Report to the UK Overseas Development Administration (ESCOR no R5911). London: Overseas Development Institute.

Cole Hamilton, I. and Lang, T. (1986). *Tightening Belts: A Report on the Impact of Poverty on Food*. London: London Food Commission.

Craig, G. and Dowler, E. (1997). Let them eat cake! Poverty, hunger and the UK state. In *First World Hunger: Food Security and Welfare Politics*, ed. G. Riches, pp. 108–33. London: Macmillan.

Crombie, I. K., Kencicer, M. B., Smith, W. C. S. and Tunstall-Pedoe, H. D. (1989). Unemployment, socioenvironmental factors, and coronary heart disease in Scotland. *British Heart Journal*, **61**, 172–7.

Davey Smith, G., Blane, D. and Bartley, M. (1994). Explanations for socioeconomic differentials in mortality: evidence from Britain and elsewhere. *European Journal of Public Health*, **4**, 131–44.

Department of the Environment (1983). Urban Deprivation, Information Note No. 2, Inner Cities Directorate. London: Department of the Environment.

Department of Health (1989). *The Diets of British Schoolchildren*. London: HM Stationery Office.

Department of Health (1991). *Dietary Reference Values for Food Energy and Nutrients for the United Kingdom*. Report of the Panel on Dietary Reference Values of the Committee on Medical Aspects of Food Policy, Department of Health. Report on Health & Social Subjects No. 41. London: HM Stationery Office.

Department of Health (1996). *Low Income, Food, Nutrition and Health: Strategies for Improvement*. A Report from the Low Income Project Team to the Nutrition Task Force. London: Department of Health. Obtainable from PO Box 410, Wetherby, LS23 7LN, UK; fax ++44 1937 845 381.

Department of Social Security (1994). *Households Below Average Income: A Statistical Analysis, 1979–1991/2 and 1979–1992/3.* (Revised edn 1995.) Government Statistical Service, London: HM Stationery Office.

Department of Social Security (1996). *Social Security Statistics 1996.* London: HM Stationery Office.

Dobson, B., Beardsworth, A., Keil, T. and Walker, R. (1994). *Diet, Choice and Poverty: social, cultural and nutritional aspects of food consumption among low income families.* London: Family Policy Studies Centre.

Dorling, D. and Tomaney, J. (1995). Poverty in the old industrial regions: a comparative view. In *Off the Map: The Social Geography of Povety in the UK,* ed. C. Philo, pp. 103–22. London: Child Poverty Action Group.

Dowler, E. and Calvert, C. (1995). *Nutrition and Diet in Lone-parent Families in London.* London: Family Policy Studies Centre.

Dowler, E. and Dobson, B. (1997). Nutrition and poverty in Europe: an overview. *Proceedings of the Nutrition Society,* **56**, 51–62.

Dowler, E. and Rushton, C. (1994). *Diet and Poverty in the UK: Contemporary Research Methods and Current Experience: a Review.* Working Paper for Committee on Medical Aspects of Food Policy and the Nutrition Task Force, Department of Health. Also published (with annotated bibliography) as *Department of Public Health and Policy, Publication no. 11.* London: London School of Hygiene and Tropical Medicine.

Drever, J. and Whitehead, M. (1995). Mortality in regions and Local Authority Districts in the 1990s: explaining the relationship with deprivation. *Population Trends,* **82**, 19–26.

Drèze, J. and Sen, A. (1989). *Hunger and Public Action.* Oxford: Clarendon Press.

Eames, M., Ben-Shlomo, Y. and Marmot, M. G. (1993). Social deprivation and premature mortality: regional comparison across England. *British Medical Journal,* **307**, 1097–102.

European Commission (1994). *European Social Policy: A Way Forward for the Union. A White Paper.* Luxemburg: Directorate-General for Employment, Industrial Relations and Social Affairs.

Goodman, A. and Webb, S. (1994). *For Richer, For Poorer: The Changing Distribution of Income in the UK, 1961–91.* London: Institute for Fiscal Studies.

Goodwin, M. (1995). Poverty in the city: "you can raise your voice but who is listening?" In *Off the Map: the social geography of povety in the UK,* ed. C. Philo, pp. 65–82. London: Child Poverty Action Group.

Green, A. (1994). *The Geography of Poverty and Wealth, 1981–1991.* Findings: Social Policy Research 55. York: Joseph Rowntree Foundation.

Gregory, J., Foster, K., Tyler, H. and Wiseman, M. (1990). *The Dietary and Nutritional Survey of British Adults.* London: HM Stationery Office.

Gregory, J. R., Collins, D. L., Davies, P. S. W., Hughes, J. M. and Clarke, P. C. (1995). *National Diet and Nutrition Survey: children aged 1.5 to 4.5 years.* London: HM Stationery Office.

Health Education Authority (1989). *Diet, Nutrition and "Healthy Eating" in Low Income Groups.* London: Health Education Authority.

Jarman, B. (1983). Identification of underpriviledged areas. *British Medical Journal,* **286**, 1705–9.

Joseph Rowntree Foundation (1995). *Inquiry into Income and Wealth,* vol. 1 (chaired by Sir Peter Barclay); vol. 2: *A Summary of the Evidence* (by John

190   E. Dowler

Hills). York: Joseph Rowntree Foundation.

Kempson, E. (1996). *Life on a Low Income*. York: Joseph Rowntree Foundation.

Leather, S. (1996.) *The Making of Modern Malnutrition. An Overview of Food Poverty in the UK*. The Caroline Walker Lecture 1996. London: The Caroline Walker Trust (6 Aldridge Road Villas, London W11 1BP).

McCarron, P. G., Davey Smith, G., and Wormersley, J. J. (1994). Deprivation and mortality in Glasgow: changes from 1980–1992. *British Medical Journal*, **309**, 1481–2.

Maxwell, S. (1996). Food security: a post-modernist perspective. *Food Policy*, **21** (2), 155–70.

Ministry of Agriculture, Fisheries and Food (1996). *National Food Survey, 1995. Annual Report on Food Expenditure, Consumption and Nutrient Intakes*. London: HM Stationery Office.

Mooney, C. (1990). Cost and availability of healthy food choices in a London health district. *Journal of Human Nutrition and Dietetics*, **3**, 111–20.

Morton, S. (1988). *Local Shops Food Survey*. Manchester: Manchester City Council.

Noble, M and Smith, G. (1994). *Increasing Polarization Between Better-off and Poorer Neighbourhoods in Oldham and in Oxford*. Findings: Social Policy Research, 56. York: Joseph Rowntree Foundation.

Nelson, M., Mayer, A. M. and Manley, P. (1993). The food budget. In *Budget Standards for the United Kingdom*, ed. J. Bradshaw, pp. 35–63. Aldershot: Avebury.

Oppenheim, C. and Harker, L. (1996). *Poverty: The Facts*. (Revised 3rd edn.) London: Child Poverty Action Group.

Payne, P. R. (1992). Assessing undernutrition: the need for a reconceptualization. In *Nutrition and Poverty*, ed. S. R. Osmani, pp. 49–96. Oxford: Clarendon Press.

Phillimore, P., Beattie, A. and Townsend, P. (1994). Widening inequalities of households in northern England 1981–91. *British Medical Journal*, **308**, 1125–8.

Piachaud, D. (1987). Problems in the definition and measurement of poverty. *Journal of Social Policy*, **16**(2), 147–64.

Piachaud, D. and Webb, J. (1996). *The Price of Food: Missing Out On Mass Consumption*. Suntory and Toyota International Centre for Economics and Related Disciplines, London: London School of Economics.

Sen, A. S. (1981). *Poverty and Famines*. Oxford: Oxford University Press.

Sloggett, A. and Joshi, H. (1994). Higher mortality in deprived areas: community or personal disadvantage? *British Medical Journal*, **309**, 1470–4.

Smith, D. (1995). The Social Construction of Dietary Standards: The British Medical Association – Ministry of Health Advisory Committee on Nutrition Report of 1934. In *Eating Agendas: Food and Nutrition as Social Problems*, eds. D. Maurer and J. Sobal, pp. 279–303. New York: Aldine de Gruyter.

Townsend, P. with Corrigan, P. and Kowarzik, U. (1987). *Poverty and Labour in London*: Interim Report of a Centenary Study. London: Low Pay Unit.

Veit-Wilson, J. H. (1986). Paradigms of poverty: a rehabilitation of B.S. Rowntree. *Journal of Social Policy*, **15**(1), 69–99.

Wilkinson, R. G. (1992). Income distribution and life expectancy. *British Medical Journal*, **304**, 165–8.

Williams, C. and Dowler, E. (1994). *Identifying Successful Local Projects and Initiatives in Diet and Low Income: A Review of the Issues.* Working Paper No. 1 for the Nutrition Task Force Low Income Project Team (secretariat, Nutrition Division, Department of Health).

10 Poverty, nutrition and obesity in the USA

F. E. JOHNSTON AND P. GORDON-LARSEN

Editors' introduction

This chapter starts with the observation that obesity is highly prevalent among the urban poor in the United States (US), a situation that is quite opposite from that in developing countries where urban poverty is associated with undernutrition. Urban poverty in the US is associated with a cluster of nutrition-related health consequences and biological effects of traditional interest to human biologists. Using a microlevel, biocultural approach similar to that employed by Huss-Ashmore and Behrman in chapter 4, Johnston and Gordon-Larsen add a new dimension termed "communal participatory action research," a special form of intervention. It combines traditional university missions: research, teaching and service, which in this instance is to the neighboring community. This approach removes the study community from the role of specimens or subjects, and through research collaboration returns education and knowledge. The community involved in the microlevel analysis and intervention are the teenage pupils of the Turner School which serves a poverty-stricken neighborhood in a Philadelphia ghetto. The proximity of an elite university and an impoverished ghetto highlights the extremely patchy nature of urban social geography and its biological consequences.

In both developed and lesser-developed countries, poverty is universally recognised as a major risk factor for a broad spectrum of diseases and other conditions which adversely affect health. Some of these correlates are generalisable across regions, social class and ethnicity, while others are more specific to local conditions. But regardless of the scope of any one disease or condition, the poor are characterised by higher levels of morbidity and mortality than are found among their more affluent counterparts (Kumanyika 1993; Olvera-Ezzell et al. 1994).

Across the urban sectors of the USA, the health risks associated with poverty are especially striking. Among Chicago neighborhoods, male life expectancy at birth ranged from 54 to 77 years in 1988–93, with economic

192

inequality a significant predictor (Wilson and Daly 1997). In a study of asthma mortality in Philadelphia the death rate increased between 1978 and 1991, despite this being a period during which the concentrations of major air pollutants declined (Lang and Polansky 1994). The rates were highest in the census tracts with the highest percentages of poor people and minority residents. And researchers have demonstrated numerous other adverse health outcomes among the urban poor, e.g. ranging from exposure to lead (Rothenberg *et al.* 1996; Czerwinski, this volume) to general health behavior (Byrd *et al.* 1996; Carlisle *et al.* 1996; Sikkema *et al.* 1996).

The nutriture of the urban poor is a matter of increasing concern both to biomedical researchers and health workers as the extent of nutritional problems becomes more apparent. In a study of over 500 non-institutionalised African-American, elderly St. Louis residents, Miller *et al.* (1996) recorded high levels of nutritional risk associated with dietary intake and practices, income and dental problems.

Of the nutrition-related conditions which are associated with urban poverty, the increasing prevalence of obesity, especially among minority groups and particularly those of African-American and Hispanic ancestry (Kumanyika 1993; Must *et al.* 1994; Kimm *et al.* 1996) has become a major public health concern. The prevalence of overweight and obesity is relatively high among all age and sex groups, but the frequency of obesity among the young is of particular importance. Some of the highest levels reported exist in the cities of the USA, with values of 20–30% now seen routinely among teenagers and younger adults.

The deleterious effects of obesity are not as firmly established among the young as they are among adults. At the very least, while the risk of short-term mortality among obese adolescents is quite low, teenage obesity is itself a clinically significant risk factor for adult obesity (Must 1996).

However, more direct correlates of obesity among teenagers are suggested by the literature. In a recent study of inner city children and youth, Douglas *et al.* (1996) have reported obesity to be an independent risk factor for hypercholesterolemia, such that the sensitivity of obesity was better as a screening tool than was a positive family history.

Tershakovec *et al.* (1994) studied behavioral correlates of obesity among inner city African-American elementary school children using a battery of psychological measures. Obese girls had a significantly higher "sex problems" score. In addition, the proportion of obese children placed in special education or remedial class settings was twice that for non-obese children. The authors interpret their results as "suggesting subtle behavior differences in obese children."

National data on the nutritional status of low-income American groups are consistent with the research cited above. The most recent report on nutritional monitoring in the USA (FASEB 1995) concludes that "the risk of nutrition-related health problems is high among people with low incomes." The following problems are mentioned specifically:

- anemia
- low birth weight
- lower likelihood of breastfeeding
- high serum total cholesterol
- increased prevalence of hypertension
- lower intakes of calcium, iron and folate
- greater likelihood of experiencing food insufficiency
- lower awareness of diet–health relationships
- increased weight-for-height

Thus urban residence, poverty and obesity constitute a complex of attributes which is associated with a higher probability of adverse health outcomes and which, while previously considered relevant mainly among older adults, now is known to reach deeply into increasingly younger age groups. As a result it becomes even more important to analyse and hopefully to disentangle the network of factors which, through a complex set of interactions, are associated with obesity.

Overweight and obesity among West Philadelphia teenagers

Excessive weight-for-height is a serious problem among young Philadelphia teenagers from economically disadvantaged families. Furthermore, the prevalence of obesity is on the rise. Figure 10.1 presents the mean body mass index (BMI) of 548 11–14 year old African-American youth measured in the 1990s. For comparison we have included the means of a national sample of 838 youth from NHANES III (FASEB 1995). Consistent with the observations of a higher prevalence of overweight among the urban poor, greater values are exhibited by males and females in the urban sample at all ages and in both males and females. Furthermore the differences tend to increase across the age range.

There has been a significant increase in the prevalence of obesity among the African-American population in (at least) recent decades. The increase has been described by Gordon-Larsen (1997) and Gordon-Larsen *et al.* (1997), who studied three samples ranging in age from 11 to 15 years that had been measured in the 1960s, 1970s, and 1990s. A threefold increase in the prevalence of obesity in males and a fourfold increase in females was reported in this 30-year period. (Obesity was defined by BMI and triceps skinfolds above the NHANES-I 95th percentile.) Using only the BMI as

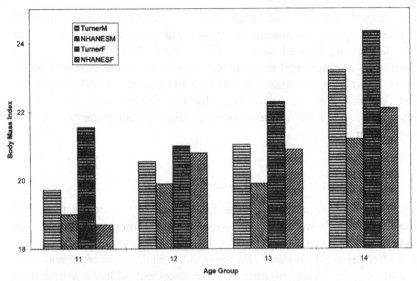

Figure 10.1   Mean body mass index of economically disadvantaged
Philadelphia school students compared to US national data (NHANES)

the criterion, the prevalence of overweight (1990s) was 17.6% based on the
95th percentile and 34.9% based on the 85th percentile of NHANES-I.
These findings exceed the prevalence estimates derived from the first phase
of NHANES-III, 10.9% based on the 95th percentile and 22% based on the
85th percentile (NHES-II and III) (Troiano 1995).

It is particularly troublesome that, despite a heightened recognition of
the public health importance of obesity, a significant increase over the
years in research devoted to the topic, and the application of numerous
interventions designed to reduce it, the problem of overweight and obesity
continues. It is abundantly clear that new approaches must be developed
and new programs utilised. Those groups most at risk must become active
participants in a joint process of health-seeking behavior that is based on
sound research and utilises new and creative techniques that promote
engagement with, and ownership of, programs that are implemented
(Glantz *et al.* 1990). This is especially true with those who are
socioeconomically disadvantaged and who live at the margins of their
societies. Among such groups obesity may not be the problem but rather a
symptom of a broader malaise (Michelson *et al.* 1979). In such circumstances, overweight and obesity cannot be dealt with in isolation, but rather as
part of efforts to help people gain control of their own lives, with nutrition
and health serving as one focus.

The remainder of this chapter focuses on the Turner Nutritional Awareness Project (TNAP), a comprehensive program of service and research carried out primarily among 11–15 year olds from the John B. Turner Middle School, located in West Philadelphia. The TNAP utilises the techniques of communal participatory action research as part of a broad and comprehensive partnership between the University of Pennsylvania and the community within which it is located. The results of this work are compared to US national data as well as particular components of the US population.

The Turner Nutritional Awareness Project

The Turner Nutritional Awareness Project (TNAP) is one component of the partnership between the University of Pennsylvania and the surrounding West Philadelphia community. Its aim is to combine the traditional but often separate missions of a university – teaching, research and community service – through constructive engagement with the neighboring community (Harkavy *et al.* 1996). More specifically, the purposes of the TNAP are the following.

- to describe and analyse the nutritional ecosystem of the Turner community
- to enhance the ability of Turner students and their families to make informed decision regarding their food habits
- to work to increase the quality of the diet and the nutritional status of the Turner community
- to establish community directed, sustainable structures to maintain an awareness of the relationships between diet and health
- to monitor and evaluate dietary quality and nutritional status of the community across time
- to disseminate the results to the public policy and scholarly communities
- to enhance the educational experiences for University of Pennsylvania students through the development of a curriculum which emphasises problem-solving and academically based community service.

The TNAP has been in operation at the Turner School since 1992. Turner is a middle school of the School District of Philadelphia located in West Philadelphia, in the urban periphery that surrounds the central, largely commercial, core of the city. "Row housing" is the typical residential type, built between 1890 and 1920, with individual units family owned or rented. The neighborhood is in an area of high crime and violence and displays significant environmental deterioration (Schwartz *et al.* 1994; Center For Community Partnerships 1995). In characterising changes over the past several decades, the Philadelphia Planning Commission has listed several persistent trends, including:

- loss of middle class
- property deterioration and abandonment
- deteriorating infrastructure
- declining commercial strips

The Turner School serves an economically disadvantaged population. Incomes are low; using federal cut-off values, 84% of the families are categorised as being in poverty. As such, all students are eligible for free breakfast and lunch under federal funding though the majority do not take advantage of the opportunity. School enrollment is 99% African-American.

As part of their anthropology/human biology curriculum, University of Pennsylvania students (Penn) may opt to carry out a set of nutritionally related activities at the Turner School. These activities include teaching nutrition and working with small groups of Turner students on projects aimed at increasing knowledge and enhancing attitudes about the relationships between nutrition and health. Penn students also collect anthropometric data and 24-hour dietary recalls, using standard techniques, for all procedures. Training methods and reliability data have been previously published (Johnston and Hallock 1994). Nutrient values have been obtained from dietary records using Nutritionist III and IV software (N-squared Computing, Salem, Oregon).

Table 10.1 presents the sample size for the analysis reported here. Anthropometric data were collected on 548 individuals between 11 and 14 years with dietary recalls obtained from 418 over the same period. Given the inability of a single 24-hour recall to reflect adequately the habitual diet of a person, the diets and their nutrient values are not analysed at the individual level. However, taken collectively, they characterise the intakes of the sample as a whole. For the same reason, no attempt has been made to link anthropometry to nutrient intake.

Nutritional status of TNAP participants

The heights and weights obtained since 1992 were converted to NCHS z-scores and percentiles using the CDC EpiInfo program. Figure 10.2 gives the distributions of the percentile values of height and weight for the sample. Since there was no sex difference in height or weight z-scores,

Table 10.1. *Sample size by age group and sex, Turner Nutritional Awareness Project*

Age Group	Males	Females	Total
11	61	84	145
12	94	70	164
13	72	87	159
14	50	30	80
Total	277	271	548

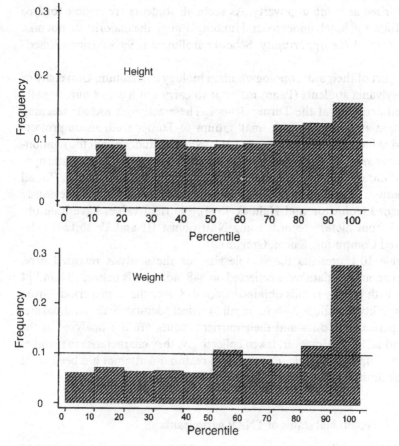

Figure 10.2   Distribution of heights and weights of Turner school students by NCHS centile group

males and females are combined. The horizontal line in the figure indicates the 10% that would be expected in each decile if the distribution Turner heights and weights was the same as that of the US population.

The distribution of both height and weight indicates a greater-than-expected number in the higher centile groups. Both distributions are significantly different from the expected at probabilities < 0.0001. However the excess among higher centile groups is greater for weight than for height. The mean height z-score is 0.23 while the mean weight z-score, at 0.68, is three times as great. Furthermore, the weights of 159, or 27% of the sample, are above the 90th centile. These observations are in accordance with the data presented above showing a significant degree of overweight relative to the USA reference data.

The obesity status of TNAP participants was assessed using, as the criterion, the 90th percentile value for African-Americans seen in NHANES (Frisancho 1990). Figure 10.3 presents the percentage obesity by sex and age group. Separate distributions are plotted for males and females, however the differences are not significant. Overall, of the 556 individuals measured, 27% of the males and 24% of the females were assessed as obese relative to their African-American peers nationally.

Figure 10.3 also shows that there is a clear increase in the percentage obesity with advancing age. Among students over 13 years of age, 22% are obese while among 14-year-olds, the frequency is 39%. Given the age range covered by these data, and the age at menarche – 12.0 years – (Gordon-Larsen 1997) the increase with age in the frequency of obesity cannot be interpreted as a transient circum-pubertal phenomenon. Rather it suggests that the high frequency is likely to be maintained into the later stages of adolescence as well as into the adult years.

Table 10.2 presents the mean BMI by sex and age for the TNAP sample, the US population as a whole, and US Native Americans, the latter a group characterised by high levels of obesity. The TNAP and Native American samples have consistently higher means than the US population throughout the age range and in both males and females. The means of the TNAP and Native American youth do not differ markedly. Native American BMI's are higher at 11, 12 and 13 years of age, but lower at 14.

Dietary intake

The diets of the TNAP sample were assessed from 24-hour recalls collected by University of Pennsylvania undergraduate students. Recalls were scheduled so as to provide information on consumption during the week as well as on weekends. Given the significant skewness of the nutrient data, the medians were generally used as measures of central tendency. Descrip-

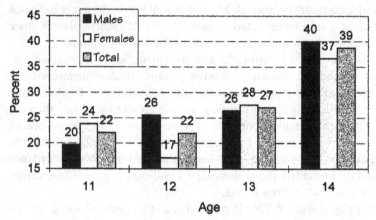

Figure 10.3   Percentage obesity in 275 male and 271 female TNAP participants by sex and age

tion and analysis intakes have been expressed in absolute units as well as the percentage of the recommended dietary allowance (RDA). Since for most nutrients the RDA's are allowances, and not recommended intakes, they overestimate the physiological requirements of most nutrients. We have followed the convention of using 68% of the RDA as the criterion for adequacy (Gibson 1990) for most nutrients. However, this is not appropriate in all cases, in particular dietary energy intake and fat consumption. The intakes are not to be interpreted as necessarily representing the habitual intakes of individual children. Rather they provide a picture of the aggregate daily intakes. In the absence of systematic error, the values should be reliable collective estimates of the intakes of the sample.

Table 10.3 presents the median intakes of selected nutrients as calculated from 24-hour recalls for the 227 males and 231 females. The table gives the nutrient intakes for Turner students, for the US national non-Hispanic black population, 12–15 years (NHB), and for 12–15 year olds of the US population as a whole. Figure 10.4 presents nutrient intakes of the TNAP sample, expressed relative to US national data (FASEB 1995). Turner School females are consuming considerably more dietary energy and nutrients than the national population with some values exceeding 150% of the national data (vitamin C and sodium). The median caloric intake is also quite high: 139%. In fact all nutrients are higher than the US data except for the intake of vitamin E, which is only 46%. On the other hand, Turner males have intakes which cluster around 100%, though there is considerable variation. The vitamin C intake is 127% and that of energy, 108%. Vitamin E is also low in males, at 41% of the national median.

Even though there is a sex difference in the consumption of nutrients

Table 10.2. *Mean body mass index of TNAP, Native American, and US youth by sex and age*

Age	TNAP		US population		Native Americans	
	M	F	M	F	M	F
11	19.7	21.6	19.0	18.7	21.7	21.3
12	20.6	21.0	19.9	20.8	22.1	21.4
13	21.0	22.3	19.9	20.9	22.9	23.6
14	23.2	24.3	21.2	22.1	22.7	23.4

Source: US data from Federated Societies of Experimental Biology (1995).

Table 10.3. *Median daily intake of selected nutrients of Turner School students, US non-Hispanic blacks (NHB), and the US population,[a] 12–15 years*

Nutrient	Males			Females		
	Turner	NHB	USA[a]	Turner	NHB	USA
Energy (kcal)	2,685	2,307	2,486	2,493	1,927	1,799
% energy from fat	34.9	35.2	32.7	37.9	37.7	33.8
% energy from saturated fat	11.5	12.4	12.4	12.1	12.1	12.8
Protein (g)	93.4	72	82	82.2	63	58
Folate (mcg)	201.5	199	299	197	176	181
Calcium (mg)	804	717	1,053	754	613	685
Iron (mg)	15.6	13.23	14.93	13.8	9.76	10.17
Zinc (mg)	10.6	8.91	11.42	9.5	8.53	8.16
Sodium (mg)	4,025	3,287	3,654	4,026	2,872	2,608

[a]US data from Federation of American Societies of Experimental Biology (1995).

Table 10.4. *Proportion of 24-hour recalls of teenagers with percentage of calories from fat less and greater than 30%*

% kcal from fat	Turner School		US Food Stamp Program	
	Males	Females	Males	Females
< 30	25.1%	15.4%	14.0%	17.9%
> 30	74.9	84.6	86.0	82.1

Source: US data from Federation of American Societies of Experimental Biology, 1995.

relative to national data, the patterns of the two sexes are quite similar. In fact the correlation between the male and female percentages, across the nutrients, is 0.96. In other words, the dietary patterns show a quantitative difference between the sexes, with females consuming more than males

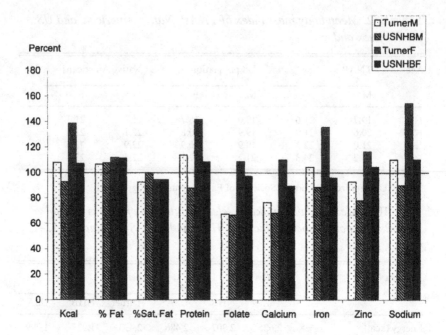

Figure 10.4   Median one-day intakes of selected nutrients among Turner School students (11–15 years of age) and US non-Hispanic black youth (12–15 years of age) as percentage of NHANES population values for 12–15 year olds (FASEB 1995); M = males, F = females

relative to national data. However, again relative to national data, there is no qualitative difference between the sexes.

Table 10.3 presents the median intakes of selected nutrients as calculated from 24-hour recalls of 227 males and 231 females. The table gives the nutrient intakes for Turner students, for the US national non-Hispanic black population, 12–15 years (NHB), and for 12–15 year olds of the US population as a whole. Figure 10.4 presents nutrient intakes of the TNAP sample, expressed relative to US national data (FASEB 1995). Turner School females are consuming considerably more dietary energy and nutrients than the national population. The median caloric intake is 139% of the national data. In fact intakes of all nutrients are higher than the US data and are either higher than. or similar to, the national sample of their non-Hispanic peers. On the other hand, Turner males have intakes that cluster around 100%, though with considerable variation.

The fat intakes of the two African-American groups are given in Table 10.4. While the intakes of fat among the Turner youth are higher than the recommended 30% of total energy, they are not higher than those reported for samples of similar socioeconomic status. As shown in the table, among participants in the US national Food Stamp Program, regardless of ethnicity, 85% of males and 82% of females, 12–19 years of age, had fat intakes that were greater than 30% of their calories. This is a lower proportion than was found among Turner students (68%). These data suggest that the intakes of fats of Turner Middle School students, while higher than recommended values, are not notably greater than reported national data, for youth of comparable socioeconomic status.

Integration of dietary and nutritional status data

The picture presented to date by the Turner Nutritional Awareness Project is one of a community whose teenagers are characterised by a high prevalence of overweight and obesity. By 14 years of age, 40% of the individuals have BMI's which place them into the obese category. It seems clear that the rates of obesity which characterise urban inner-city communities of low SES have their genesis in the years of early adolescence. It is especially significant that the prevalence rises steadily from 11 to 14 years, almost doubling in percentage across this relatively short age range.

The aggregate dietary intake reveals consumption of nutrients which deviate significantly from that recommended for a healthy lifestyle. Caloric intakes are higher in both sexes, markedly so in females. Likewise the intakes of sodium and fat are higher, even when corrected for energy consumption. And finally, minerals, folate and vitamin E are lower than national data after correcting for total energy. Vitamin E intake is especially deficient in both sexes, at only 40% of national data.

It must be noted that in Turner School females, relative to national data, intakes of calcium, iron, zinc, and folate – all of particular concern among low-income groups – are higher than the national data. However, this is achieved not by a balanced diet, but by a high consumption of calories (40% greater).

Strategies for change

The poverty and social marginalisation that characterises many urban neighborhoods imposes a heavy burden upon their residents. This burden has been characterised by Bloom (1964) as constituting a "powerful environment," one that exerts its influence on all members of the community. And while nutritional problems in the USA today are not so obvious as in the lesser developed world, the cycle of urban poverty and its detrimental effects on health have been well documented (e.g. Karp 1993). The Turner Nutritional Awareness Project is part of the University of Pennsylvania's

Center for Community Partnerships. The Center is a structural entity which comprises a comprehensive set of programs and projects that seek to actualise the mission of large research university: to advance and transmit knowledge in order to improve human welfare (Harkavy *et al.* 1996). Thus the TNAP is a broad-based intervention carried out within an academic and research setting and utilising the theory and methods of communal action research (Gustavsen 1996; Toulmin 1996). The nutritional problem is not just one of food, but, as is almost always the case, rooted in the social and economic conditions which confront the community. No single strategy is ever likely to be wholly effective in such a setting (Wilson and Ramphele 1989). Rather, a multiple strategy effort is required as most likely engage all groups that are involved regardless of role, to insure a continuing self-sustaining involvement, to develop a sense of ownership, and ultimately to produce lasting effects.

The conceptual basis for the TNAP encompasses these ideas, such that it is a dynamic, evolving program, with new ideas introduced on a regular basis and driven by the dedication and creativity of all who are involved. Its components include – but are not limited to – the following intiatives:

- the development and implementation of a curriculum in nutrition and health taught by university students and Turner School teachers
- a flexible nutrition textbook written largely by university students with input from Turner School students
- a range of nutrition-centered extra curricular activities carried out on Saturdays and during the summer months
- an increased emphasis on the role of physical activity in health through a set of activities
- a fruit and vegetable stand, managed and directed jointly by university and Turner School students, that sells to students, staff, and members of the community
- the development of a nutrition-centered focus throughout the school

Fruits 'R Us (and Vegetables Too!) – an example of a school-based, child-centered program

One example of a component of the TNAP designed to enhance the nutritional status of the community served by the Turner Middle School is Fruits 'R Us (and Vegetables Too!). This intervention was conceived and implemented by two University of Pennsylvania undergraduates in 1995 (Dubowitz 1996) and planned in collaboration with a small group of university students, along with 60 Turner School students and their teachers. The school students (grade 6, 11–12 years of age) determined collectively the most practical and healthy foods to sell as well as preparing the fruits and other foods, staffing the store, managing the finances and records, and advertising its presence, all under the guidance of university students. Penn students also provided instruction in sanitation and hygiene, in store management and in nutrition and health. The store is open at specified times after school and on the weekends, with the cost of procuring the various items coming from money taken in sales, subsidised when needed. As university and Turner students have moved on, their places have been taken by others so that the school store has been in operation for some two and a half years.

Evaluation of the impact of Fruits 'R Us has been preliminary and based on the first year of operation. Dubowitz (1996) carried out a study of student and staff visits to, and purchases from, the store. Two groups were selected for analysis and comparison. The first group consisted of the 60 students who had participated in the planning of the store (involved) and the second of 132 students from grades 6-8 (11–14 years of age) who had not participated (non-involved). The data collected involved responses to questionnaires about the school store. Dubowitz' evaluation identified three positive sets of outcomes:

- knowledge of the store; among the non-involved students, 86% (114 of 132) knew of the store's existence
- visits to the store; as would be expected, 87% of involved children (53 of 61) had visited the store; of those not involved in the planning and implementation, only 42% (54) had visited; however, among non-involved children who had visited the store, 28% had visited more than once, and more than 11% had visited more than six times
- variety of fruits and vegetables consumed; popular items – apples, oranges – were as expected purchased frequently; many less familiar items, such as kiwi fruit, strawberries, and kumquats, were purchased as frequently as familiar ones; thus Fruits 'R Us exposed students to a wider variety of foods than they otherwise might have known.

Figure 10.5 presents the number of visits to Fruits 'R Us by students who were surveyed, according to their degree of involvement. The importance of involvement as a factor in visits to the store can be seen in the distributions. Also evident is the role of involvement in the frequency of visits. One might expect that children involved in the formulation of the project would visit once, out of curiosity. However, over 75% of involved

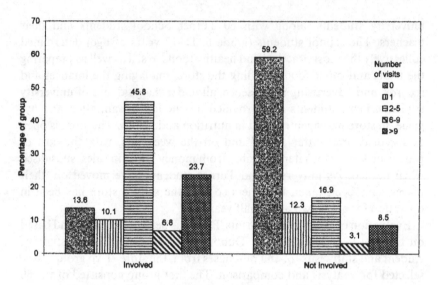

Figure 10.5   Visits of Turner students to Fruits-R Us school store by
involvement in planning, year 1

students had visited more than once, and almost 25% had visited ten times
or more.

The evaluation of the school store at the Turner Middle School suggests
that the consumption of fruits and vegetables can be increased in teenagers
by the development of programs that emphasise significant involvement
from the outset. Based on these data, the involvement is most effective
when the children themselves are empowered, when they participate in the
decisions that are made, when their roles in the project are more than
cursory, when they are not simply the passive recipients of a message but
instead are actively engaged in structuring and delivering the intended
message.

Discussion

Urban poverty is associated with a broad spectrum of social, political and
health-related problems. The stresses generated by these problems are seen
in the lives of those who confront them on a daily basis and are recorded as
a range of undesirable outcomes, among which are related to diet (nutrient
excesses and inadequacies) and to adiposity (overweight and obesity).

Strategies designed to ameliorate these conditions are broadly conceived
and form the basis for the social programs that are developed and applied
by public and private sectors. Partly because of the intractability of the
urban ecosystem and partly because of the nature of the programs themsel-

ves, society's ability to solve the problems of poverty, social stratification and marginalisation have been far less successful than was desired.

One factor in the degree of success of programmatic interventions is the extent to which the recipients are involved in the process as participants, and not just as passive recipients or as subjects of research. This chapter has outlined the nutritional problems associated with urban poverty among young teenagers living in the inner city. It has described a comprehensive project, the Turner Nutritional Awareness Project, which aims to help those teenagers help themselves by empowering them to take more active control of their nutrition.

Acknowledgments

This work was supported in part by the Ford Foundation, the W. K. Kellogg Foundation, the Nassau Fund, and the University of Pennsylvania Center for Community Partnerships. We acknowledge the contributions of Ira Harkavy, Marie Bogle, Cory Bowman, Tamara Dubowitz, Dan Gerber, Jason Hallock, and the students of the University of Pennsylvania and the John B. Turner Middle School.

References

Bloom, B. S. (1964). *Stability and Change in Human Characteristics.* New York: John Wiley.

Byrd, T. L., Mullen, P. D., Selwyn, B. J. and Lorimor, R. (1996). Initiation of prenatal care by low-income Hispanic women in Houston. *Public Health Reports,* **111**, 536–40.

Carlisle, D. M., Leake, B. D., Brook, R. H. and Shapiro, M. F. (1996). The effect of race and ethnicity on the use of selected health care procedures: a comparison of south central Los Angeles and the remainder of Los Angeles county. *Journal of Health Care for the Poor & Underserved,* **7**, 308–22.

Center For Community Partnerships (1995). *Spruce Hill Community Renewal Plan.* Philadelphia: University of Pennsylvania.

Douglas, M. B., Birrer, R. B., Medidi, S. and Schlussel, Y. R. (1996). Obese children should be screened for hypercholesterolemia. *Journal of Health Care for the Poor & Underserved,* **7**, 24–35.

Dubowitz, T. (1996). *Fruits 'R Us (and Vegetables Too!) – A School-Based, Child-Centered Community Health Project: The Strategic Implementation and Quantitative Analysis of Its First Year in Action.* Philadelphia: University of Pa Honors Thesis in Anthropology.

Federation of American Societies of Experimental Biology (1995). *Third Report on Nutrition Monitoring and Related Research,* vols. 1 & 2. Washington DC: US Government Printing Office.

Frisancho, A. R. (1990). *Anthropometric Standards for the Assessment of Growth and Nutritional Status.* Ann Arbor, MI: University of Michigan Press.

Gibson, R. S. (1990). *Principles of Nutritional Assessment*. Oxford: Oxford University Press.

Glanz, K., Lewis, F. M. and Rimer, B. K. (1990). *Health Behavior and Health Education: Theory Research, and Practice*. San Francisco: Jossey-Bass.

Gordon-Larsen, P. (1997). *Ecology of Obesity in West Philadelphia Adolescents*. Ann Arbor, MI: UMI Dissertation Services.

Gordon-Larsen, P., Zemel, B. S. and Johnston, F. E. (1997). Secular changes in stature, weight, fatness, overweight, and obesity in urban African-American adolescents from the mid-1950s to the mid-1990s. *American Journal of Human Biology*, 9, 675–88.

Gustavsen, B. (1996). Is theory useful? *Concepts and Transformations*, 1, 63–78.

Harkavy, I., Johnston, F. E. and Puckett, J. L. (1996). The University of Pennsylvania's Center For Community Partnerships as an organizational innovation for advancing action research. *Concepts and Transformation*, 1, 15–29.

Johnston, F. E. and Hallock, R. J. (1994). Physical growth, nutritional status, and dietary intake of African-American middle school students from Philadelphia. *American Journal of Human Biology*, 6, 741–7.

Karp, R. J. (ed.) (1993). *Malnourished Children in the United States. Caught in the Web of Poverty*. New York: Springer.

Kimm, S., Obarzanek, E., Barton, B. A., Aston, C. E., Similo, S. L., Morrison, J. A., Sabry, Z. I., Schreiber, G. B. and McMahon, R. P. (1996). Race, socioeconomic-status, and obesity in 9-year-old to 10-year-old girls–the NHLBI Growth and Health Study. *Annals of Epidemiology*, 6, 266–75.

Kumanyika, S. K. (1993). Special issues regarding obesity in minority populations. *Annals of Internal Medicine*, 119, 650–4.

Lang, D. M. and Polansky, M. (1994). Patterns of asthma mortality in Philadelphia from 1969 to 1991. *New England Journal of Medicine*, 331, 1542–6.

Michelson, W., Levine, S. V. and Spina, A.-R. (eds.) (1979). *The Child in the City: Changes and Challenges*. Toronto: University of Toronto Press.

Miller, D. K., Carter, M. E., Sigmund, R. H., Smith, J. Q., Miller, J. P., Bentley, J. A., McDonald, K., Coe, R. M. and Morley, J. E. (1996). Nutritional risk in inner-city-dwelling older black Americans. *Journal of the American Geriatric Society*, 44, 959–62.

Must, A., Gortmaker, S. L. and Dietz, W. H. (1994). Risk factors for obesity in young adults: Hispanics, African Americans and Whites in the transition years, age 16–28 years. *Biomedicine & Pharmacotherapy*, 48, 143–56.

Must, A. (1996). Morbidity and mortality associated with elevated body weight in children and adolescents. *American Journal of Clinical Nutrition*, 63(3 Supplement), 445S–447S.

Olvera-Ezzell, N., Power, T. G., Cousins, J. H., Guerra, A. M. and Trujillo, M. (1994). The development of health knowledge in low-income Mexican–American children. *Child Development*, 65 (2 Spec. No.), 416–27.

Rothenberg, S. J., Williams, F. A. Jr., Delrahim, S., Khan, F., Kraft, M., Lu, M., Manalo, M., Sanchez, M. and Wooten, D. J. (1996). Blood lead levels in children in south central Los Angeles. *Archives of Environmental Health*, 51, 383–8.

Schwartz, D., Grisso, J., Miles, J., Wrshner, A. and Sutton, R. (1994). A longitudinal study of injury morbidity in an African-American population. *Journal of the Amererican Medical Association*, 271, 755–60.

Sikkema, K. J., Heckman, T. G., Kelly, J. A., Anderson, E. S., Winett, R.A., Solomon, L. J., Wagstaff, D. A., Roffman, R.A., Perry, M. J., Cargill, V., Crumble, D. A., Fuqua, R. W., Norman, A. D. and Mercer, M. B. (1996). HIV risk behaviors among women living in low-income, inner-city housing developments. *American Journal of Public Health,* **86,** 1123–8.

Tershakovec, A. M., Weller, S. C. and Gallagher, P. R. (1994). Obesity, school performance and behaviour of black, urban elementary school children. *International Journal of Obesity & Related Metabolic Disorders,* **18,** 323–7.

Toulmin, S. (1996). Is action research really 'research'? *Concepts and Transformations,* **1,** 51–62.

Troiano, R. P., Flegal, K. M., Kuczmarski, R. J., Campbell, S. M. and Johnson, C. L. (1995). Overweight prevalence and trends for children and adolescents. The National Health and Nutrition Examination Surveys, 1963 to 1991. *Archives of Pediatric Adolescent Medicine,* **149,** 1085–91.

Wilson, F. and Ramphele, M. (1989). *Uprooting Poverty: The South African Challenge.* Cape Town: David Philip.

Wilson, M. and Daly, M. (1997). Life expectancy, economic inequality, homicide, and reproductive timing in Chicago neighbourhoods. *British Medical Journal,* **314,** 1271–4.

11 Nutritional status and its health consequences among low-income urban pregnant women: diet and environmental toxicants

S. CZERWINSKI

Editors' introduction

This chapter develops the theme that cities are characterised by multiple overlapping microenvironments. Johnston and Gordon-Larsen (chapter 10) showed that the nutritional status of poor urban adolescents differs from the nutritional status of US adolescents nationally. Czerwinski shows that the diets of poor urban gravidae also differ from national reference values. Whereas many investigators relate diet directly to health outcomes such as growth, obesity and cardiovascular function, Czerwinski's analysis reveals that the interactions between diet, nutrition and pollutants are complex. Many people are exposed to pollutants through the foods they eat, such that being well nourished can also mean having a higher pollutant burden. Using new data from the Albany Pregnancy–Infancy Lead Study, he shows that the urban poor experience greater exposure to toxicants such as lead, and their dietary patterns may expose them to more food-borne toxicants. Furthermore, dietary items that might moderate toxicant absorption may be less commonly consumed by the urban poor. Pollutants, as relatively new factors in urban environments, interact with one of the oldest, social stratification, to produce a greater health risk for the urban poor.

Introduction

This chapter will examine the relationship between diet and exposure to environmental toxicants among poor pregnant women living in urban environments. Poor urban populations are often exposed to greater amounts of environmental toxicants, and are consequently at increased risk of certain types of infirmity (Schell 1997). Examining how features of the environment structure exposure to health risks can aid in understanding some of the influences on human biological variation. For example,

210

toxicological insults to the fetus during prenatal development may have long-lasting health consequences (Scott 1979; Morgane *et al.* 1992). The malformation causing teratogenic effects of thalidomide when adminis- tered during certain critical periods of pregnancy are now well known (Lenz 1968). The same is true for alcohol consumption during pregnancy and the suite of behavioral and developmental effects that are known as fetal alcohol syndrome (FAS) (Clarren *et al.* 1978). This chapter will examine exposure to environmental toxicants from diet and will also look at nutritional factors that modify toxicant metabolism.

There are three foci to the present chapter. The first is that diet, in addition to being a source of energy and sustenance, is a significant source of exposure to environmental toxicants. Second, diet can be seen as a modifier of susceptibility to environmental toxicants by affecting toxicant metabolism. And third, there are certain characteristics of the urban diet that make urban-dwellers predisposed or more susceptible to the toxic effects of metals. A case study in which these interactions operate with respect to environmental lead pollution is presented using data collected from the first phase of the Albany Pregnancy–Infancy Lead Study.

Xenobiotics in the human diet

Human beings consume a variety of different substances in their normal diet. Although most of these substances are necessary for proper function, many can have adverse effects on growth, development and cognitive function (Riley and Vorhees 1986). In order to gain an understanding of the kinds of substances the average American diet contains the United States Food and Drug Administration (USFDA) periodically reviews the content of typical diets in the United States (Pennington and Gunderson 1987). Since 1961, the FDA Total Diet Study has periodically tested for various pesticide residues, contaminants and nutrient elements in foods, investigating temporal trends and changes among specific age and sex groups. These studies have been useful in identifying toxicants and pos- sible threats to public health found in the human diet.

In the US, exposure to environmental lead from diet differs by age group (Gunderson 1988). Data from the US FDA Total Diet Survey from the years 1982–84 show the average intake of lead in the human diet for various age groups (Figure 11.1). Lead intake is highest among males in the 14–16 and 25–30 year age groups. Although these levels do not exceed the FAO/WHO tolerable limit of 429 µg/day for adults, it is interesting to note that the tolerable levels set for infants and children are much lower than for adults at 32 and 46 µg/day respectively. More recently, there is substantial evidence that suggests that tolerance levels for adults may not

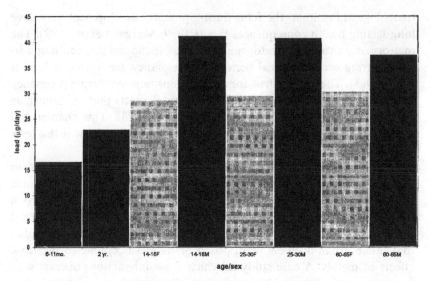

Figure 11.1   Daily intake of lead in the US FDA total Diet Study, 1982–1984.
Source: Gunderson (1988)

be applicable to pregnant women, given the increased vulnerability of the fetus during organogenesis. Tolerable intake levels for women, therefore, should include levels for pregnant and lactating women which are lower and more in line with the tolerance levels of children, because of the additional health risks (Reigart 1995).

Lead levels in food in the US have been decreasing over the past ten years largely due to removal of lead from petroleum and from more stringent processing of food (Gunderson 1995). In other areas of the world, lead intake remains considerably higher than in the US. Figure 11.2 shows intake of lead in several major cities of the world from two major sources, diet and air (Krol *et al.* 1996). Diets in these countries have considerably higher levels of lead when compared with diets in the US. This figure also shows the significantly greater intake of lead from diet compared with respiratory intake in all of the areas surveyed. It is important to note that these differences vary substantially depending upon the level of pollution in the environment and various social and economic factors that influence the sources of lead in the population. Lead exposure also comes from other sources as well, such as through ingestion of leaded paint or dust (Charney *et al.* 1983). These sources are not accounted for in total diet studies.

In addition to heavy metals, there are many other potential contaminants in the human diet. Table 11.1 displays the presence of pesticide residues in various fruits and vegetables sampled by the US Environmental

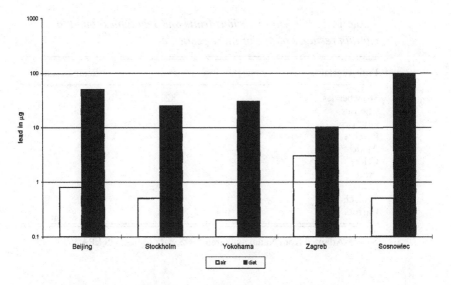

Figure 11.2   Exposure to lead following inhalation or dietary intake.
Source: Krol *et al.* (1996)

Protection Agency (Nadakavukaren 1995). The majority of all fruits sampled contained at least one pesticide residue and in many cases more than one. The fruits that had residues most commonly include strawberries, cherries and oranges. The most prevalent pesticide residues were dichlorodiphenyl dichloroethylene (DDE), malathion, diedrin and pentachlorophenol (Gunderson 1988). Prompted by public awareness, the US EPA passed the Food Quality Protection Act in 1996 to set new standards of tolerable pesticide levels in food. The legislation requires explicit determination of tolerances which are safe for children, which include an additional safety factor (of up to ten-fold) to account for individual variability in susceptibility (Goldman 1995; Fenner-Crisp 1995).

In many areas of the world, food storage containers and food preparation are a large source of exposure to lead contaminants. In Mexico, reliance on lead glazed ceramicware has become a significant health threat (Wallace *et al.* 1985). Figure 11.3 demonstrates the relationship between ceramic use and lead levels in a sample of 98 postpartum mothers in Mexico City. Blood lead levels increase in a dose-response relationship with increased use of lead glazed ceramics. There are many factors including the temperature, acidity of the food and the duration of storage which influence the leaching of lead from ceramic vessels, but in populations that rely on ceramic vessels risk of lead exposure is directly related to frequency of ceramic use.

Certain categories of food can be identified as high in contaminants. For

Table 11.1. *Percent of various fruits and vegetables found to contain residues of one or more pesticide*

Fruit/vegetable	(%)	Samples tested
Strawberries	82	168
Cherries	80	90
Oranges	80	237
Peaches	79	246
Apples	78	542
Celery	75	114
Pears	73	328
Lettuce	68	201
Spinach	54	163
Carrots	50	252

Source: Adapted from Nadakavukaren, 1995; data source: US EPA 1990–1992.

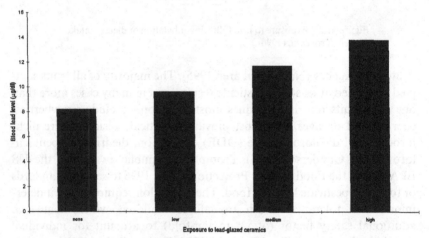

Figure 11.3   Blood lead level by exposure to lead-glazed ceramics among postpartum women residing in Mexico City.
Source: Hernandez-Avila *et al.* (1996)

example, fish have been shown to contain a wide variety of toxicants, including hexachlorobenzene, polychlorinated dibenzo-*p*-dioxins, dieldrin, lindane, chlordane, cadmium, mirex, mercury and polychlorinated biphenyls (PCBs) (Lonky *et al.* 1996). The consumption of contaminated fish has been linked to a number of adverse health effects in pregnant women (Rogan *et al.* 1986; Yamaguchi *et al.* 1971). These effects include higher incidence of low birthweight and cognitive deficits in children. Of these contaminants, mercury and PCBs have been most widely studied.

The most severe occurrence of mercury poisoning happened in Japan between the years of 1953 and 1961 in the coastal fishing town of Min-amata. The townspeople who relied on local subsistence fishing were exposed to very high levels of methyl mercury from eating contaminated fish from the local bay. Doses were so high as to cause severe brain damage and developmental deficits in children exposed *in utero*. These effects have persisted into adulthood (Harada 1995). Although this was an isolated incident, exposure to contaminated fish continues in other settings such as among families that rely on subsistence fishing. Poor populations and ethnic minorities are most likely to eat fish caught in contaminated waters (Reigart 1995). In general, exposure to methyl mercury is likely to occur at much lower levels, but chronically. *In utero* exposure to lower levels of mercury than seen in Minamata has been linked to many more subtle cognitive and behavioral delays (Clarkson 1992).

Similarly, the greatest source of PCB exposure also comes from contaminated fish consumption. Studies investigating high levels of in utero PCB exposure show a higher incidence of low birthweight, reduced gestational age and cognitive impairment during infancy (Jacobson *et al.* 1984, 1996). PCBs are stored in fat and bioaccumulate at higher trophic levels of the food chain, so that large predatory fish can contain unusually high levels of the toxicant (Reigart 1995). A study by Schwartz *et al.* (1983) found maternal serum PCB levels to increase as a function of fish consumption in a sample of 242 pregnant women residing in Western Michigan (Figure 11.4). Other studies have shown similar relationships with PCB levels and fish intake (Fensterheim 1993).

It is evident from the preceding description that diet can be a significant source of toxicant exposure in human populations. The potential teratogenic effects of toxicant exposure are especially critical and damaging during prenatal development (Lenz 1968), when rapid physical and neurological development increases sensitivity to insult. The risk of developmental consequences diminishes with age for most environmental toxicants.

Although most of the studies mentioned have focused on a single toxicant, it is evident that populations are unknowingly exposed to a multitude of toxicants through their diet and from food preparation. The possibility of many toxicants working synergistically to adversely affect human health and reproductive outcomes cannot be overlooked, and data on multiple toxicant exposures is becoming more available. Studies in Baltimore have shown that poor inner-city children are simultaneously exposed to several toxic chemicals. Children with elevated blood lead levels are also more likely to be exposed to higher levels of benzene and environmental tobacco smoke (Weaver *et al.* 1996a,b).

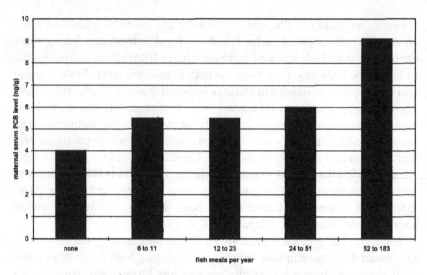

Figure 11.4   Relationship of fish meals to maternal serum PCB level.
Source: Adapted from Schwartz *et al.* (1983)

Diet as a mediator for toxicant metabolism

Diet is also important in mediating toxicant metabolism (Mahaffey 1974). Lippmann (1992) has demonstrated a general path model of toxicant exposure and how adverse effects are manifest (Figure 11.5). Toxicants are initially produced and dispersed into the environment. They are then distributed throughout the microenvironment and come into contact with humans through food, air, water or physical contact. There are a number of endogenous factors that can influence the metabolism of a toxicant, including their *rate of uptake, bioavailability* and *rates of elimination, accumulation and transformation*. Factors that can alter toxicant uptake and elimination include age, sex, genetic variation, states of physical stress (such as pregnancy and disease) and diet (Mahaffey 1985). These factors modulate the biological dose within the organism. Finally, chronic exposure to toxicants leads to functional effects that ultimately result in adverse health consequences such as growth deficits, neurological impairment or cancer.

The interaction of diet and environmental lead

The following discussion will focus on lead since it is probably the best understood of all environmental toxicants. Table 11.2 displays many of the nutrients that can influence the metabolism of lead. Most of what is known

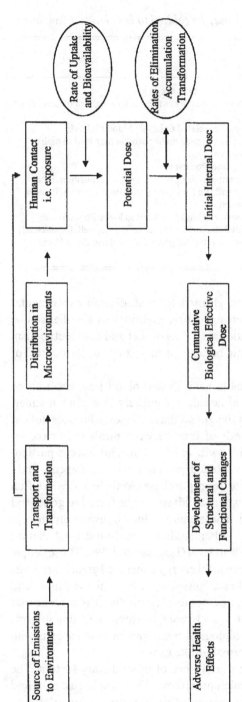

Figure 11.5

Table 11.2. *Dietary factors which may be related to lead levels in humans*

	Direction of effect	Explanation
Total energy	(+/−)	Fasting increases lead uptake in gut
		Higher calorie intake may correspond to higher Pb intake from food
Fat	(+)	Higher fat intakes related to higher blood Pb levels
Protein	(+)	Aging literature suggests higher protein may lead to higher rates of bone loss
Vitamin D	(−)	Influences Ca and Pb absorption
Calcium	(−)	Pb and Ca compete for uptake in the gut, deposition in bone
Iron	(−)	Pb competes with Fe for binding sites on rbc, also influences heme formation
Phosphorus	(−)	Higher phosphorus intakes related to lower Pb levels
Copper	(+/−)	Closely related to Fe metabolism (effect not well documented)
Zinc	(−)	Pb competes for GI absorption, also replaces Zn in Heme formation

about lead metabolism has been gleaned from studies on experimental animals (Miller *et al.* 1989), where exposure, environment and diet can be controlled. In practice, observational studies of diet and lead metabolism in humans have been problematic because of limitations in the control of these experimental variables.

Experimental studies in humans have shown blood lead levels to be related to total energy intake, although not entirely in a clear manner. Fasting increases lead uptake in the gut so that a person who does not eat regular meals or has long periods of time between meals may have increased absorption of lead (Rabinowitz *et al.* 1980). This issue is particularly important in some disadvantaged populations who depend upon relief cheques or food stamps for food purchase (Mahaffey 1990). Often there are gaps between cheques where children may be forced to go several days without a substantial meal. These children may be especially susceptible to increased lead uptake. Other studies have found total dietary intake to be positively related to lead level (Lucas *et al.* 1996). Those people who consumed more food or more total energy consumed greater amounts of lead in their normal diet and as a consequence have higher blood lead levels. Although these appear to be conflicting results, it is evident that fasting is not a normal state, and, in such cases, research on fasting subjects may not demonstrate the same biological relationships as in subjects who are ingesting diets within the normal intake range.

Other studies have noted the significance of other dietary factors. The most important nutrients are calcium, vitamin D, phosphorus, iron and zinc (Miller *et al.* 1989). Higher intakes of these nutrients are related to lower blood lead levels and decreased lead absorption in laboratory animals. The influence of copper, however, remains unclear due to conflicting results (Cerklewski and Forbes 1977; Lauder and Petering 1975). Still, many of these studies are also confounded by the metabolic interaction of many of these nutrients. For example, vitamin D status is closely related to calcium absorption as well as lead absorption (Smith *et al.* 1978). This makes interpreting results in human populations especially difficult.

Some recent human health studies have demonstrated the benefits of certain dietary factors. Hernandez-Avila and colleagues (1996) found a protective effect of milk on blood lead levels in their study of postpartum women. Women with weekly milk consumption have significantly lower blood lead levels compared to women with only monthly consumption. This study however did not control for many of the other dietary factors that may influence lead levels.

West *et al.* (1994) examining 349 pregnant African-American women in Washington DC found the use of vitamin or mineral supplements to be related to lead levels. Women who used vitamin or mineral supplements during all three trimesters of pregnancy had significantly lower mean blood lead levels over the course of pregnancy than those who did not. They also noted significant positive correlations between serum calcium and phosphorus and lead, as well as significant negative correlations with serum vitamin E, vitamin C and lead.

Diet among the urban poor: increased risk for lead toxicity

Most studies of dietary intake among poor urban populations have yielded similar results. Urban populations in the US have high intakes of fat, saturated fat and cholesterol; while having low mineral intakes (Johnston and Hallock 1994). These characteristics may make urban-dwellers at risk for increased lead uptake and absorption, as well as increased retention (Mahaffey 1990). At present, there is little information on the diets of pregnant women from poor urban populations in the US. Of the few studies that have attempted to characterise these diets all have shown similar results. Table 11.3 displays mean nutrient levels for several selected nutrients from three population studies. The data from Johnson *et al.* (1994) are a sample of 322 poor African-American women living in inner city Washington DC attending the Howard University Medical Clinic. The study by Borrud *et al.* (1993) is a sample of 114 poor pregnant women of varied ethnicities from the 1985 and 1986 USDA food intake survey. And finally, Brennan *et al.* (1983) studied 22 pregnant African-American women of lower socioeconomic status. All three studies found similarly low intakes of iron, calcium and zinc. Protein intake in all three studies was

Table 11.3. *Comparison of dietary intakes among similar populations (expressed as percentage of US RDA)*

	US RDA 1989	Johnson *et al.* 1994	Borrud *et al.* 1993	Brennan *et al.* 1983
Energy	2500 kcal	94.3	77.9	74.9
Protein	60 g	162.0	127.1	123.7
Vitamin C	70 mg	223.5	145.0	124.3
Folate	400 µg	58.5	65.7	43.8
Calcium	1200 mg	77.8	80.3	65.8
Phosphorus	1200 mg	113.2	110.8	95.6
Iron	30 mg	48.9	43.0	36.7
Zinc	15 mg	68.9	73.0	58.0

considerably greater than 100% of the US RDA, suggesting a higher risk of lead exposure.

Block and Abrams (1993) comparing women of reproductive age of different SES groups found dietary differences between poor and better-off women of reproductive age, using data from the 1986 Continuing Survey of Food Intakes of Individuals (Figure 11.6). Poor women compared to their more affluent peers were more likely to have lower intakes of all essential minerals including calcium, iron, zinc and folate also suggesting greater risk of lead exposure.

The Albany Pregnancy Infancy Lead Study

Data from the first phase of the Albany Pregnancy Infancy Lead Study (APILS) (1986–92) also corroborates the preceding observations. The APILS study population (n=316) includes women of lower socioeconomic status living in the city of Albany, NY, USA. The sample is 42% white, 53% African-American and 5% Hispanic. The average age of mothers is 21.8 years; 77% of the mothers are unmarried and 48% have not finished secondary school. Mothers were enrolled before 24 weeks of pregnancy and followed until term; 24-hour dietary recalls were collected during each trimester of pregnancy. Of the 316 pregnant women enrolled in the first phase of data collection, 246 have complete dietary data for at least two trimesters of pregnancy. Dietary intakes did not differ significantly by trimester so data over the course of pregnancy were averaged; 35% of the 246 participants had three dietary recalls over pregnancy, the remaining 65% had only two. Nutrient values were converted to US Recommended Dietary Allowances for pregnant women according to National Research Council (1989) recommendations. The distribution of many nutrients is highly skewed so mean and median intakes are presented as measures of

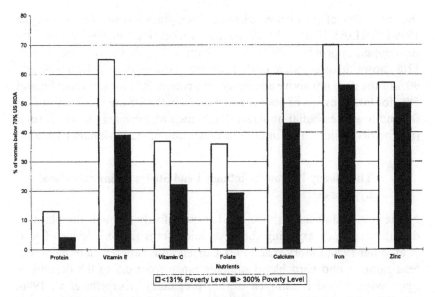

Figure 11.6   Proportion of women aged 19–50 years with intakes below 70% of the 1989 US RDA in four consecutive days.
Source: Block and Abrams (1993)

Table 11.4. *Dietary component as a percentage of US Recommended Dietary Allowance APILS Phase 1*

Dietary component	Total sample			African-Americans		Whites	
	Median	Mean	SD	Mean	SD	Mean	SD
Fat	102	114	57	118	50	112	40
Saturated fat	117	130	65	132	59	130	46
Protein	149	157	71	160	60	156	48
Iron	41	47	29	46	22	49	29
Copper	47	52	25	52	26	54	23
Phosphorus	121	128	48	128	53	132	42
Vitamin D	56	63	36	58	35	70	39
Calcium	82	90	42	84	42	100	42
Zinc	69	72	28	73	30	73	26
Energy	86	89	29	91	32	88	25
$n =$	246			129		105	

central tendency (Table 11.4). Women in our sample have relatively high saturated fat and protein intakes and low intakes of iron, copper, vitamin D and zinc. There are significant differences between African-Americans and whites in calcium and vitamin D intake with African-Americans having considerably lower intakes compared to whites. Figure 11.7 shows the prevalence of diets below adequacy (adequacy is defined as 67% of the 1989 US RDA). Nearly 90% of the diets are below adequacy for iron, 82% for copper, approximately 60% for vitamin D and 50% for zinc. Figure 11.8 shows the prevalence of diets above 100% of the US RDA. Nearly 90% of the diets are above adequacy for protein, 70% for saturated fat and 58% for fat. The diets of poor urban women in the US are relatively high in fat and low in essential minerals. Diets such as these may be considered risk factors for increased uptake and retention of lead (Mahaffey 1990).

The Albany Pregnancy Infancy Lead Study: testing nutritional hypothesis

Numerous studies have shown lead to pass freely through the placenta during pregnancy exposing the developing fetus to toxic levels of lead (Carpenter 1974; Emory et al. 1997). Correlations between maternal serum lead samples and cord blood samples range from 0.5 to 0.8 depending upon when blood is sampled during pregnancy (Korpella et al. 1986; Lauwerys et al. 1978). Many researchers have speculated that lead levels during pregnancy may be partly influenced by the release, during the latter stages of gestation, of lead stored in the maternal skeleton (Silbergeld 1991; Thompson et al. 1985). Dietary factors may influence both this release and the subsequent transport of lead to the fetus.

With data from the first phase of APILS (1986–92) the relationship between maternal diet during pregnancy and blood lead levels in the cord at birth was examined. Table 11.5 displays the results of bivariate logistic regressions predicting the risk of elevated blood lead level (defined as levels $\geqslant 10\ \mu g/dl$) at birth. The US Centers for Disease Control and Prevention have established a $10\ \mu g/dl$ concentration in whole blood as the guideline for lead toxicity in children (Centers for Disease Control 1991). Nutrient intakes have been transformed by natural logarithm to reduce skewness and to minimise the effects of extreme outliers. Of these nutrients, fat, saturated fat, protein and total energy are positively related to elevated blood lead levels ($p < 0.05$). Zinc intake was only moderately related ($p=0.09$). Odds ratios appear elevated because the independent variable is on a log scale so that every increase of one log unit yields a many-fold increase in risk. For example, for every log unit increase in fat intake one can expect a 4.93 increase in risk of having an elevated blood lead level. If log units are converted back into RDAs one can see that someone consuming a diet that is 100% of the US RDA for fat is 4.93 times more likely to deliver a child with an elevated blood lead level than someone consuming a diet that is 37% of the US RDA.

Acknowledging that quantity of consumption may be a confounding

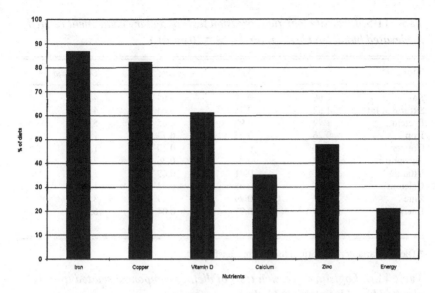

Figure 11.7   Prevalence of diets below 67% of US RDA APILS Phase 1

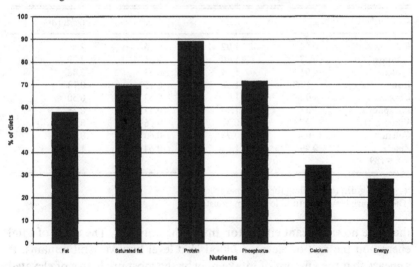

Figure 11.8   Prevalence of diets above 100% US RDA APILS Phase 1

variable in this analysis, a secondary analysis was performed. When total energy intake was controlled for in the logistic model, the effects seen in the bivariate analysis are reduced (Table 11.6). None of the variables are significantly related to elevated blood lead level. By controlling for total energy intake, differences in total food consumption are controlled individually. Thus, in our analysis, if total food consumption is held equal,

Table 11.5. *Bivariate logistic regression tests of dietary component related to elevated blood lead level at birth (> = 10 μg/dL)*

Nutrient*	B	SE	Sig.	Odds ratio
Fat	1.60	0.79	0.04	4.93
Saturated fat	1.52	0.75	0.04	4.57
Protein	1.79	0.92	0.05	5.99
Iron	0.26	0.73	0.73	1.30
Copper	0.62	0.74	0.40	1.86
Phosphorus	0.93	0.82	0.26	2.53
Vitamin D	0.20	0.41	0.62	1.21
Calcium	0.55	0.63	0.38	1.73
Zinc	1.34	0.79	0.09	3.82
Energy	1.90	0.93	0.04	6.68
$n = 189$				

* Dietary variables are natural logarithm transformed.

Table 11.6. *Logistic regression tests of dietary component related to elevated blood lead level at birth[a]*

Nutrient[b]	B	SE	Sig.	Odds ratio
Fat	0.84	1.92	0.66	2.31
Saturated fat	0.87	1.30	0.51	2.39
Protein	0.91	1.44	0.53	2.48
Iron	−1.69	1.18	0.15	0.18
Copper	−0.69	1.07	0.52	0.50
Phosphorus	−0.97	1.33	0.47	0.38
Vitamin D	−0.11	0.37	0.76	0.90
Calcium	−0.41	0.78	0.60	0.66
Zinc	0.49	1.04	0.64	1.63
$n = 189$				

[a](> = 10 μg/dL) (controlling for total energy intake)
[b]Dietary variables are natural logarithm transformed

there are no significant effects for any of the nutrients. The effect of total energy on the risk of elevated blood lead level however still remains. It appears that the amount of intake may be the most predictive of elevated blood lead level. Others have noted this relationship in diets among preschool children (Lucas *et al.* 1996). One possible explanation is that higher intakes of food lead to higher possible intakes of contaminants in the diet. Thus those who eat greater quantities of food may also be eating more lead. Further analyses of these data are planned using structural equations modeling to examine nutrient interactions, while partialling out effects of individual nutrients.

There are several limitations in our study that preclude making more broad generalisations. The 24-hour dietary recall instrument used in our analysis is most reliable when there are at least three observations. In our sample, nearly 65% of the sample had dietary information for only two trimesters of pregnancy. Furthermore, lead levels in our sample are generally low, with little variability, making it difficult to distinguish significant associations. In addition, dietary factors may be more influential on blood lead levels during certain periods. Measuring diet over the entire course of pregnancy may not detect changes in diet during shorter time intervals. These changes may be more influential on blood lead levels. For example, release of lead from the skeleton may increase in the absence of adequate dietary calcium during the later stages of pregnancy when mineralisation of the bones occurs in the fetus. Thus calcium intake during the third trimester of gestation may be a better predictor of blood lead levels at birth than diet over the entire pregnancy. The use of 24-hour dietary recalls spaced at broad intervals precluded a more precise point estimate of dietary intake. The results of this analysis are nevertheless suggestive of certain dietary influences on lead levels during pregnancy.

Summary

Diet can be a source of exposure to contaminants especially among poor urban populations. Numerous contaminants may be found in common diets, including lead, PCBs, mercury and pesticides. The possibility of interaction among toxicants is also acknowledged. Furthermore, diet can act as a mediator of the pharmacokinetics of toxicants once ingested. Diets low in particular nutrients can increase a person's toxicant burden. Urban populations may be at increased risk due to higher exposure to toxicants in the environment in association with dietary inadequacies which make them particularly vulnerable to the toxic effects of pollutants.

References

Barltrop, D. (1969). Transfer of lead to the human foetus. In *Mineral Metabolism in Pediatrics*, D. Barltrop and W. L. Burland, Philadelphia, PA: FA Davis Company.

Block, G. and Abrams, B. (1993). Vitamin and mineral status of women of childbearing potential. In *Maternal Nutrition and Pregnancy Outcome*, eds. C. L. Keen, A. Bendich, and C. C. Willhite. *Annals of the New York Academy of Science*, **678**, 244–54.

Brennan, R. E., Kohrs, M. B., Nordstrom, J. W., Sauvage, J. P. and Shank, R. E. (1983). Nutrient intake of low-income pregnant women: laboratory analysis of foods consumed. *Journal of the American Dietetic Association*, **83**(5), 546–50.

Borrud, L., Krebs-Smith, S. M., Friedman, L. and Guenther, P. M. (1993). Food and nutrient intakes of pregnant and lactating women in the United States. *Journal of Nutrition Education*, **25**(4), 176–85.

Carpenter, S. J. (1974). Placental permeability of lead. *Environmental Health Perspectives*, **82**, 129–31.

Centers for Disease Control (CDC) (1991). *Preventing lead poisoning in young children.* Atlanta: US Department of Health and Human Services.

Cerklewski, F. L. and Forbes, R. M. (1977). Influence of dietary copper on lead toxicity in the young rat. *Journal of Nutrition*, **107**, 143–6.

Charney, E., Kessler, B., Farfel, M. and Jackson, D. (1983). Childhood lead poisoning: a controlled trial of the effect of dust-control measures on blood lead levels. *New England Journal of Medicine*, **309**, 1089–93.

Clarkson, T. W. (1992). Mercury: major issues in environmental health. *Environmental Health Perspectives*, **100**, pp.31–8.

Clarren, S. K. and Smith, D. W. (1978). The fetal alcohol syndrome. *New England Journal of Medicine*, **298**(19).

Emory, E. K., Pattillo, R. A., Bayorh, M. A. and Sung, F. (1997). Effects of low level lead exposure and neural behavioral functions in neonates. Presented at *The First National Research Conference on Children's Environmental Health: Research, Practice, Prevention, Policy*. Abstract. Washington, DC. February 21–23.

Fenner-Crisp, P. A. (1995). Pesticides – The NAS report: how can recommendations be implemented? *Environmental Health Perspectives*, **103** (Supplement 6), 159–62.

Fensterheim, R. J. (1993). Documenting temporal trends of polychlorinated biphenyls in the environment. *Regulatory Toxicology and Pharmocology*, **18**, 181–201.

Goldman, L. R. (1995). Children – unique and vulnerable. Environmental risks facing children and recommendations for response. *Environmental Health Perspectives*, **103** (Supplement 6), 13–18.

Gunderson, E. L. (1988). FDA Total Diet Study, April 1982–April 1984, dietary intakes of pesticides, selected elements, and other chemicals. *Journal of the Association of Official Analytical Chemists*, **71**(6), 1200–9.

Gunderson, E. L. (1995). FDA Total Diet Study: July 1986–April 1991, dietary intakes of pesticides, selected elements, and other chemicals. *Journal of AOAC International*, **78**(6), 1353–63.

Harada, M. (1995). Minamata disease: methylmercury poisoning in Japan caused by environmental pollution. *Critical Reviews in Toxicology*, **25**(1), 1–24.

Hernandez-Avila, M. H., Gonzalez-Cossio, T., Palazuelos, E., Romieu, I., Aro, A., Fishbein, E., Peterson, K. E. and Hu, H. (1996). Dietary and environmental determinants of blood and bone lead levels in lactating postpartum women living in Mexico City. *Environmental Health Perspectives*, **104**, 1076–82.

Jacobson, J. L., Fein, G., Schwartz, P. M. and Dowler, J. K. (1984). Prenatal exposure to an environmental toxin: a test of the multiple effects model. *Developmental Psychology*, **20**, 523–32.

Jacobson, J. L. and Jacobson, S. W. (1996). Intellectual impairment in children exposed to polychlorinated biphenyls in utero. *New England Journal of Medicine*, **335**, 783–9.

Johnston, F. E. and Hallock, R. J. (1994). Physical growth, nutritional status, and dietary intake of African-American middle school students from Philadelphia. *American Journal of Human Biology*, **6**(6), 741–8.

Johnson, A. A., Knight, E. M., Edwards, C. H., Oyemade, U. J., Cole, O. J., Westney, O. E., Westney, L. S., Laryea, H. and Jones, S. (1994). Dietary intakes, anthropometric measurements, and pregnancy outcomes. *Journal of Nutrition*, **124**, 936S–942S.

Korpella, H., Loueniva, R., Yrjanheikki, E. and Kauppila, A. (1986). Lead and cadmium concentrations in maternal and umbilical cord blood, amniotic fluid, placenta, and amniotic membranes. *American Journal of Obstetrics and Gynecology*, **155**, 1086–9.

Krol, B., Knapek, R. and Jedrzejczak, A. (1996). Total exposure to lead and cadmium. Presented at the *International Conference of Environmental Pollution and Child Health: Critical Needs and Issues for Central and Eastern Europe*, Sosnowiec, Poland, May 8–10.

Lauder, D. S. and *Petering, H. G. (1975). Protective value of dietary copper and iron against some toxic effects of lead in rats. Environmental Health Perspectives*, **12**, 77–80.

Lauwerys, R. *et al.* (1978). Placental transfer of lead, mercury, cadmium, and carbon monoxide. *Environmental Research*, **15**, 278–89.

Lenz, W. (1968). Timetable of human organogenesis as a tool in detecting teratogenic effects. In *Memoirs of the Twelfth International Congress of Pediatrics, Mexico City*, vol. 1, pp. 294–295. Mexico DF: Impresiones Modernas, S.A.

Lippman, M. (1992). *Environmental Toxicants: Human Exposures and Their Health Effects*. New York: Van Norstrand Reinhold.

Lonky, E., Reihman, J., Darvill, T., Mather, J. and Daley, H. (1996). Neonatal behavioral assessment scale performance in humans influenced by maternal consumption of environmentally contaminated Lake Ontario fish. *Journal of Great Lakes Research*, **22**(2), 198–212.

Lucas, S. R., Sexton, M. and Langenberg, P. (1996). Relationship between blood lead and nutritional factors in preschool children: a cross-sectional study. *Pediatrics*, **97**, 74–8.

Mahaffey, K. R. (1974). Nutritional factors and susceptibility to lead toxicity. *Environmental Health Perspectives*, May 107–12.

Mahaffey, K. R. (1990). Environmental lead toxicity: nutrition as a component of intervention. *Environmental Health Perspectives*, **89**, 75–8.

Miller, G. D., Massaro, T. F. and Massaro, E. J. (1989). Interactions between lead and essential elements: a review. *Neurotoxicology*, **11**, 99–120.

Morgane, P. J., Austin-LaFrance, R. J., Bronzino, J. D., Tonkiss, J. and Galler, J. R. (1992). Malnutrition and the developing central nervous system. In *The Vulnerable Brain and Environmental Risks*, eds. R. L. Isaacson, and K. F. Jensen. New York: Plenum Press.

Nadakavukaren, A. (1995). *Our Global Environment: A Health Perspective*. 4th edn. Prospect Heights: Waveland Press.

National Research Council (1989). *Recommended Dietary Allowances*. 10th edn. Washington, DC: National Academy Press.

Pennington, J. A. T. and Gunderson, E. L. (1987). History of the Food and Drug Administration's Total Diet Study: 1961–1987. *Journal of the Association of Official Analytical Chemists*, **70**(5), 772–82.

228   S. Czerwinski

Rabinowitz, M. B., Kopple, J. D. and Wetherill, G. W. (1980). Effects of food intake and fasting on gastrointestinal lead absorption in humans. *American Journal of Clinical Nutrition*, 33, 1784–8.

Reigart, J. R. (1995). How should federal policy reflect recent research in the area of intrauterine exposure to environmental hazards. *Environmental Health Perspectives*, 103(Supplement 6), 143–5.

Riley, E. P. and Vorhees, C. V. (1986). *Handbook of Behavioral Teratology*. New York: Plenum Publishing Corp.

Rogan, W. J., Gladen, B. C., McKinney, J., Carreras, N., Hardy, P., Thullen, J., Tingelstad, J. and Tully, M. (1986). Neonatal effects of transplacental exposure to PCBs and DDE. *Journal of Pediatrics*, 109, 335–41.

Schell, L. M. (1997). Culture as a stressor: a revised model of biocultural interaction. *American Journal of Physical Anthropology*, 102, 67–77.

Schwartz, P. M., Jacobson, S. W., Fein, G., Jacobson, J. L. and Price, H. A. (1983). Lake Michigan fish consumption as a source of polychlorinated biphenyls in human cord serum, maternal serum, and milk. *American Journal of Public Health*, 73, 293–6.

Scott, J. P. (1979). Critical periods in organizational processes. In *Human Growth: Neurobiology and Nutrition*, vol. 3, eds. F. Falkner and J. Tanner. pp. 223–41. New York: Plenum Press.

Silbergeld, E. K.(1991). Lead in bone: implications for toxicology during pregnancy and lactation. *Environmental Health Perspectives*, 91, 63–70.

Smith, C. M., DeLuca, H. F., Tanaka, Y. and Mahaffey, K. R. (1978). Stimulation of lead absorption by vitamin D administration. *Journal of Nutrition*, 111, 1321–9.

Thompson, G. N., Robertson, E. F. and Fitzgerald, S. (1985). Lead mobilization during pregnancy. *The Medical Journal of Australia*, 143, 131.

Wallace, D. M., Kalman, D. A. and Bird, T. D. (1985). Hazardous lead release from glazed dinnerware: a cautionary note. *Science Total Environment*, 44, 289–92.

Weaver, V. M., Davoli, C. T., Murphy, S. E., Sunyer, J., Heller, P. J., Colosimo, S. G. and Groopman, J. D. (1996a). Environmental tobacco smoke exposure in inner-city children. *Cancer Epidemiology: Biomarkers and Prevalence*, 5(2), 135–7.

Weaver, V. M., Davoli, C. T., Heller, P. J., Fitzwilliam, A., Peters, H. L., Sunyer, J., Murphy, S. E., Goldstein, G. W. and Groopman, J. D. (1996b). Benzene exposure, assessed by urinary trans, trans-muconic acid, in urban children with elevated blood lead levels. *Environmental Health Perspectives*, 104(3), 318–23.

West, W. J., Knight, E. M., Edwards, C. H., Manning, M. and Spurlock, B. (1994). Maternal low level lead and pregnancy outcomes. *Journal of Nutrition*, 124, 981S–986S.

Yamaguchi, A., Yoshimura, T. and Kuratsune, M. (1971). A survey of pregnant women having consumed rice oil contaminated with chlorobiphenyls and their babies. *Fukuoka Acta Medica*, 62, 117–22.

Part IV
Behavior and stress

12 *Urbanism and psychosocial stress*

T. M. POLLARD

Editors' introduction

For centuries crowding and noise have been considered unpleasant stresses of urban life. Despite the long recognised connection between stress and urbanism, and the more recently revealed connections between stress and health, the role of urban stress in human biology and health is only now becoming clear. Pollard depicts the main methodological difficulties that have hindered the scientific study of stress and human biology: the great variety and number of potential stresses that could be investigated, and the difficulty of measuring human biological reactions without stimulating additional stress. Pollard shows that the former problem can be addressed with a biocultural approach that integrates "emic" cultural constructs of stress with "etic" biopsychological ones. The areas of greatest concern and investigation are work, home, neighborhood and travel in the city, with important cross-cutting themes of perceived control, and role balancing (i.e. domestic and occupational role balancing as performed by women). The development of new techniques to non-invasively measure stress-relevant hormones and neurotransmitters has greatly facilitated observation of physiological responses to stressors in these domains. Now the prospects are favorable for understanding the impact of daily hassles in urban life on physiological systems of interest to human biologists, and comparing measured relationships in different urban centres.

Introduction

Academics writing about cities have often described life in urban environments as stressful, applying the term in particular to the daily hassles of crowding, noise and commuting (e.g. Krupat 1985; Walmsley 1988). In their book *Urban Stress*, Glass and Singer (1972) wrote "Life in the city is an endless round of obstacles, conflict, inconveniences and bureaucratic routine. The urban dweller is confronted daily with noise, litter, air pollution and overcrowding." Others have focused on the social structure of urban life, which is characterised by occupational specialisation, with large

231

numbers of people engaged in non-manual work outside the primary production sector, a pattern which typically brings a marked division between the microenvironments in which people work, live and conduct their leisure activities (Harrison 1973). As James (1991) notes, these microenvironments provide stimuli novel to our species in evolutionary terms and it is therefore not unexpected that we can appear to have difficulty adapting to them.

There is, then, a general consensus in the literature that those who live or work in cities experience more stress than others and that these experiences increase the burden of mental and physical sickness in urbanites. Such a proposition is difficult, if not impossible, to test, largely because of the problem posed by trying to measure stress. Stress is a word gaining global acceptance but it is very hard to define. In the academic literature it has variously been used to describe stimuli which might affect an individual (now usually known as stressors), the behavioral or physiological response to such stimuli, or the perception of an imbalance between external demands and the ability to cope with them. Sometimes only aversive stimuli, responses or perceptions are considered relevant, but others prefer to consider the term stress to include pleasant experiences which bring about a set of physiological responses commonly associated with stress (e.g. Selye 1956); the potential for circularity is obvious. Here I will use stress to describe only aversive experiences, without trying to define it more precisely.

This wooliness in the use of term makes finding a measure of stress very difficult. In many ways the most useful and only definitively defendable position when working with humans may be to accept that stress is a culturally constructed and necessarily subjective phenomenon, and to measure it by asking people how stressed they feel. If we accept this proposition it is not appropriate to seek an objective measure of stress. In any case, while there are physiological responses commonly associated with stress, not only can these same responses be seen in situations which most would not regard as stressful, but they do not seem to be simply related to experiences of stressors or perceptions of stress in people (Pollard 1995; Toates 1995).

Rather than attempt to measure stress, it is surely in the best interests of human biologists, who have a legitimate interest in understanding biological variation in physiological systems generally considered to be affected by stress, to seek measurable correlates of such variation. These may be objective features of the urban physical or social environment, such as mental workload, or subjective, but measurable, perceptions by people of that environment, such as perceived workload. We can be guided in our choice of independent variables by consideration of features of the environment which are commonly considered to be stressors. For example, feelings of lack of control have been highlighted as an important factor in generating stress and some human biologists have tested for the existence of links between perceptions of lack of control and hormone levels. Such an approach also has advantages in allowing a more sophisticated approach to environmental variation than a simple consideration of the qualitative and troublesome dichotomy implied by attempts to make comparisons between rural and urban areas.

The aims of this chapter are therefore to examine more closely some specific features of modern urban life which are purported to be stressful, and to show how far we have succeeded in delineating their effects on human biology. James has undertaken pioneering work with respect to catecholamine and blood pressure variation and the results of his work provide a useful framework within which to discuss the effects of potential urban stressors on these outcomes. James *et al.* (1985) found that Western Samoan men living in the capital had higher urinary adrenaline excretion rates than villagers. They suggest the special demands faced by those undergoing lifestyle change associated with urbanisation and modernisation are responsible. Subsequently, James (1991) has presented evidence that both adrenaline secretion rates and blood pressure vary as New Yorkers move through their daily microenvironments. Adrenaline and blood pressure levels are typically highest when people are at work and lower at home (James 1991; Pickering *et al.* 1991). Using these results as a starting point I will consider biological variation which we can loosely label as linked to stress, first in people living in urbanising contexts and then more generally in those living in urban areas. For the latter I will look in more detail at research conducted in the two principal urban microenvironments (work and home), and at the daily transition between the two. I will begin, however, with a brief summary of the physiological changes linked to stress and outline how they may act to cause disease.

The physiology of stress

Laboratory studies in humans and other species have identified two main axes which respond when individuals are placed in situations considered stressful (Figure 12.1). The most commonly measured components of these responses are the catecholamines (principally adrenaline, but also noradrenaline) and the corticosteroid, cortisol. The sympathetic-adrenal-medullary axis can respond within a few seconds, whereas the hormonally mediated pituitary-adrenal-cortical axis is slower to react. Kirschbaum and Hellhammer (1989) state that levels of cortisol in saliva appear to peak 20–30 minutes after the onset of laboratory stress conditions. Blood pres-

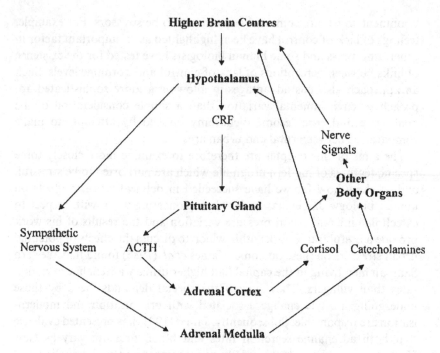

Figure 12.1   Diagram showing pathways from the brain to the adrenal systems involved in stress, showing the sympathetic-adrenal-medullary and pituitary-adrenal-cortical axes. ACTH = adrenocorticotropic hormone, CRF = corticotropin releasing factor.
Source: Redrawn from Frankenhaeuser (1989)

sure is another commonly assessed response.

For those interested in understanding how the so-called stress hormones respond to the real-life demands of urban environments the methodological challenges involved in assessing these responses are rather different to those posed by laboratory studies. The greatest obstacle is the need to measure variation in people going about their normal lives without affecting their day-to-day routine, and, in particular, without inducing feelings of stress. In laboratories, serum hormone levels are usually assessed in blood taken through an indwelling catheter inserted by venipuncture, but this method cannot be used in everyday life. Repeated venipuncture is also unsuitable since it induces feelings of anxiety in most people. Human biologists have made use of less invasive techniques, assaying hormones in urine and saliva (see Baum and Grunberg (1995) for more details). There are a number of difficulties to be tackled in trying to relate variation in urinary and salivary hormones to environmental demands. These are caused mainly by circadian rhythms in hormone levels, difficulties with timing collections for urine in particular (excretion rates over a number of hours are pooled as urine accumulates in the bladder and may be related to the rate of excretion of urine), and the need for correction for other influences on hormone levels, such as physical activity, caffeine and alcohol consumption and smoking. Discussions are provided in Ellison (1988), Kirschbaum and Hellhammer (1989) and Pollard (1995, 1997). Ambulatory blood pressure monitors have proved useful, although self-measurement with automated meters at particular times during the day can be less troublesome to the subject. James (1991) discusses measurement of blood pressure in urban environments and Krantz and Falconer (1995) review methods for the assessment of a number of cardiovascular responses, including blood pressure and heart rate. This type of data is normally used to investigate changes over time within individuals. Resting blood pressure or sometimes overnight or 24-hour urinary hormone levels can also be measured if the researcher is more interested in differences between individuals than in daily fluctuations.

The main reason for studying variation in catecholamines, cortisol and blood pressure is that such variation is likely to have important consequences for health. The best understood mechanisms are outlined in Figure 12.2. The clearest links are between raised catecholamine and blood pressure levels and cardiovascular disease, as outlined below, but many other diseases have been linked to increases in stress hormone levels. For example, the stimulation of blood glucose levels and temporary insulin resistance brought about by increases in the levels of the catecholamines and cortisol may have a role in the onset of diabetes mellitus (Nonogaki and Iguchi 1997). As indicated in Figure 12.2, other proposed mechanisms include the effects of elevated levels of cortisol on cardiovascular disease, cancers and infectious disease, via its suppression of the immune system.

There are a number of mechanisms by which chronically raised catecholamine and blood pressure levels are likely to increase the risk of cardiovascular disease. In the important volume *Man in urban environments* (Harrison and Gibson 1976), Carruthers outlined the effects of elevated catecholamine secretion, suggesting that adrenaline and noradrenaline cause free fatty acids to be released and that the subsequent rise in cholesterol levels would enhance the rate of atherosclerosis and therefore development of cardiovascular disease. Subsequent laboratory work has supported the existence of such a mechanism (e.g. McCann *et al.* 1995). It is also thought that chronic repetition of the adrenaline response may lead to enhanced sympathetic tone, which is implicated in cardiovascular disease, particularly hypertension (Julius 1996).

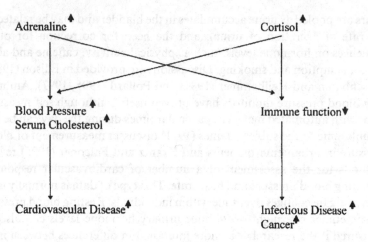

Figure 12.2   Major hypothesised mechanisms linking physiological responses to stress with disease.
Source: Redrawn from Pollard (1997)

Urbanisation and lifestyle change

Limited cross-sectional data have indicated that people living in more urbanised environments typically excrete higher levels of urinary adrenaline than those engaged in traditional subsistence farming work, despite the lower levels of physical activity characteristic of urban jobs. James *et al.* (1985) compared urinary adrenaline excretion rates in young Western Samoan men living in villages or the capital, Apia. Men living in the city, whether non-manual workers, students or laborers, excreted higher levels of urinary adrenaline than villagers on average (Figure 12.3), although the laborers and villagers had similar overnight excretion rates. Similarly Jenner *et al.* (1987) found that Tokelauans living on Fakaofo had higher levels of urinary adrenaline than residents of Nukunonu, a nearby island less influenced by the wage economy. Still higher levels were excreted by Tokelauan migrants to New Zealand living in the urban center of Wellington (this difference was not significant when the results were corrected for excretion of creatinine, but it was not clear whether such a correction, sometimes made to allow for differences in kidney function, was appropriate in this case). In both these studies the results were explained largely with respect to working behavior, with the suggestion that engagement in paid work, as opposed to traditional subsistence work, was more stressful. However, more detailed examination of the particular determinants of variation in adrenaline levels is needed before such a conclusion can be firmly drawn.

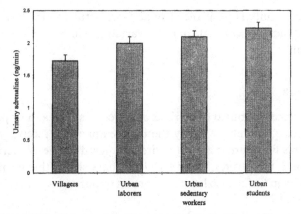

Figure 12.3　Urinary adrenaline excretion rates in Western Samoan men,
adjusted for urine volume, age and adiposity.
Source: Data from James *et al.* (1985)

Dressler (1995) has noted the importance of improving our understanding of the apparent link between lifestyle change associated with urbanisation and increasing risk of cardiovascular disease. He makes the valuable recommendation that we use ethnographic information to allow identification of specific, measurable features of the social environments of people undergoing such change. Using this method he has himself identified a mismatch between occupational or educational status, and lifestyle, as a source of stress amongst those living in urbanising contexts in poorer countries. He suggests that as consumer goods become symbols of status in developing societies, people seek to acquire them and to maintain a status which has not been "truly achieved" and that this creates uncertainty and vigilance in social interaction, and stress. In a number of studies he has related lifestyle incongruity to measures of risk for cardiovascular diseases. For example, in urban Brazil individual measures of lifestyle incongruity were related to high-density lipoprotein levels, such that those with higher lifestyle incongruity had lower levels of this beneficial fraction of cholesterol (Dressler *et al.* 1993). In American Samoa, lifestyle incongruity was associated with blood pressure, but to different degrees in older and younger household heads and in men and women (Dressler and Bindon 1997).

Dressler's model has gained great popularity as a psychosocial explanation for part of the increased risk of cardiovascular disease associated with urbanisation (clearly changes in diet and physical activity levels are also important). Further work is needed to test whether the proximate physiological changes we might expect to account for this increased risk, especially raised adrenaline levels, can be identified, and also to test whether lifestyle incongruity is itself the strongest predictor of increased risk, or whether it is merely a correlate of other variables which are more powerful determinants.

Work

The importance attributed to work as a source of stress, and its perceived economic costs, is highlighted by the huge emphasis it receives in stress research. This literature is almost entirely centered on the Western industrialised world. However, Lin and Lai (1995) have shown that people in the city of Tianjin in China also locate their most important sources of stress in their work.

As noted above, the general perception that stress is greater at work than at home appears to be reflected in higher adrenaline and blood pressure levels in the workplace in city dwellers in New York (James 1991). Others have found similar results. For example, Mulders *et al.* (1982) showed that urinary adrenaline levels were markedly higher in bus drivers (of unspecified sex) on work days than they were on work-free days. Pollard *et al.* (1996) sampled people from the general population of Oxford and found that urinary adrenaline levels were significantly higher on work-day afternoons than on a Sunday afternoon for men. The same was not true for women, despite the fact that men and women reported very similar workloads and changes in mood. There is, thus, some suggestion of a sex difference in physiological responsivity to stress (Matthews 1989). There were no differences related to working and work-free days in urinary cortisol levels in men or women, either in this study (Pollard *et al.* 1996) or an earlier one (Pollard *et al.* 1992). Earlier and larger studies conducted in Oxfordshire focused on the rural area of Otmoor. However, differences in lifestyle between city and village residents are relatively small in the UK. Thus it is not surprising that similar results were seen, with urinary adrenaline levels higher in non-manual workers on work days than on weekend days off work (Jenner *et al.* 1980). In this case the same was true for women (Harrison *et al.* 1981).

Researchers have attempted to identify sources of stress at work. The most widely cited model focuses on the interaction between demands at work and the worker's ability to exert control over his or her work activities and schedule. Karasek (1979) hypothesised that the jobs likely to cause most stress, which he termed high strain jobs, would be ones in which a heavy workload was undertaken in an environment offering little opportunity for control, such as assembly line or some secretarial work. Some, although not all, subsequent studies have shown that, at least in industrialised contexts, people who have high strain jobs do indeed tend to report more mental distress (negative feelings often considered equivalent to feelings of stress) (Karasek 1979; Xie 1996). Some studies have also shown that they have higher blood pressure (Theorell *et al.* 1988; Schnall *et al.* 1990) and more cardiovascular disease (Karasek *et al.* 1981, 1988), even after controlling for likely confounding factors associated with a low socioeconomic status, such as higher body mass index and smoking. With respect to adrenaline secretion, some evidence has been reported that a high workload in the absence of feelings of control leads to elevated levels, but demand, as measured by actual or perceived workload and time pressure, alone appears to predict adrenaline levels more successfully, particularly in men (Frankenhaeuser and Gardell 1976; Lundberg *et al.* 1989; Pollard *et al.* 1996). In one of the few relevant studies, no evidence was found that cortisol levels increased with job strain, or with increased demand or reduced control considered separately (Pollard *et al.* 1996).

An increasingly common cause of worry associated with work is job insecurity, which is a particular threat to well-being at times of high unemployment and when, as has been happening over the past decade, many organisations are 'downsizing', restructuring or merging (Ashford *et al.* 1989). However, there have been few demonstrations of consequences for human biology, partly because of the difficulties of anticipating job insecurity in a study design. Schnall *et al.* (1992) measured blood pressure in New Yorkers and their sample happened to include a group working for a firm which laid people off during the study. They found little effect on blood pressure, with the exception of elevated diastolic blood pressure in people who were measured on the day that their department was sold to another employer. More clear-cut effects were seen in male Swedish shipyard workers in the city of Malmö. These workers were assessed twice, the second time during the closure of the shipyard, and showed a greater increase in serum cholesterol levels between these times than did a control group (Mattiasson *et al.* 1990). There is some evidence (reviewed in Panter-Brick and Pollard 1999) that the acute experience of anticipation of job loss is associated with an elevation in cortisol levels.

Unemployment itself can be considered to be a stressor characteristic of urban society. Unemployment is generally an unwelcome experience for emotional as well as financial reasons. Jahoda (1942) pointed to the loss of benefits of employment, such as having a time structure to the day, maintenance of self-esteem and social support and respect from others, as some of the major costs of unemployment. Unemployment is clearly associated with ill-health (Cameron 1995), but the degree to which stress is responsible for this effect is unclear. A number of studies have examined endocrine variation associated with unemployment in an attempt to identify the effects of stress, but there is little evidence that levels are unusual compared with those of people in work (Panter-Brick and Pollard 1999).

Clearly, work plays an important role in the lives of nearly all people living in industrialised societies, whether they live in cities or not. Many will live in villages but work in cities or towns, but even the few who live and work in villages are tied into the urban economy. On the other hand, in poorer countries it is much more likely that paid employment will be a characteristic of urban locations and less prevalent outside them. Studies have shown that people have higher adrenaline and blood pressure levels at work than they do at home. The key features of work with respect to stress appear to be work demand and control, which show expected relationships with resting blood pressure, but not so clearly with immediate responses within individuals.

Home

Stress associated with living conditions has received less attention, especially with respect to biological outcomes. Here I discuss aspects of homes and homelessness which might cause stress and, where it is available, draw on evidence with respect to the effects of these stressors on the body.

Quality of housing varies a great deal across cities, but many people live at high densities and without basic amenities or in poorly maintained structures and in noisy locations. In the 1930s, Spring Rice (1939) asked working-class women in Britain about the aspects of their housing which they considered affected their health. The women complained most about overcrowding, poor structural conditions in their homes (e.g. damp) and neighborhood noise. These same themes recur in the subsequent literature.

Louis Wirth (1938) identified high population density living as a special challenge of city life, bringing individuals into a large number of social encounters with strangers every day. Crowding within the home may also be important. Gabe and Williams (1993) have reported a strong J-shaped relationship between crowding in the home and psychological distress among women in London, with much the highest level of distress reported by those in the most crowded homes (1.5 or more people per room), although high levels were also reported by those living in the least crowded conditions (less than 0.5 people per room), even after correcting for measures of socioeconomic status. Other research has suggested that city location may be more important than internal density, with people living in inner-city estates likely to show higher levels of distress than those living in the suburbs (McCarthy et al. 1985). Krupat (1985) notes that high

density living "creates problems of co-ordination among people, can reduce people's ability to control their environment, and requires active
coping," but also emphasises that people are capable of adapting to such
challenges, both at the individual and societal levels. He points out that in
Japan cultural norms and internal design allow people to live at higher
densities than most Westerners would find comfortable.

In cities, high-density housing often takes the form of flats in high-rise
buildings and several studies have suggested that life in such circumstances
can be stressful. For example, Fanning (1967) found that families living in
flats in Germany reported more morbidity than families in houses, and
speculated that stress may have contributed to this difference. Gillis (1977)
studied families in a Canadian city and found that women, although not
men, showed more stress the higher up they lived. Freeman (1993) describes the interaction between poverty and high-rise living: "those who
were already disadvantaged may find themselves trapped in dirty, dangerous, noisy and incomprehensible living conditions, which can certainly
be described as 'stressful'." The effects of stress generated by such difficult
living conditions on human biology do not appear to have been investigated.

Urban homes are more likely to be in noisy locations than rural homes.
Many studies have shown that tolerance of noise depends heavily on the
connotation attached to it. For example, one early study manipulated
residents' perceptions of airport noise (Jonsson and Sorensen 1973 cited in
Krupat 1985). The results showed that residents who were told that local
officials were concerned about airport noise were much less likely to be
disturbed by it than those told that officials did not think the noise was a
problem, again suggesting that perceptions of control are very important.
In other studies airport noise has been shown to affect human biology and
health. Schell (1991, 1996) has reviewed studies indicating that children
living near airports are likely to be born early and at low weights and to
continue to show growth retardation. He has suggested that the mechanism for these effects may be via the autonomic nervous system and endocrine systems. Pine *et al.* (1996) back the suggestion that emotional problems in childhood may affect growth, perhaps via abnormalities in growth
hormone secretion. The commonest cause of complaints about noise in
England was neighbors (Department of the Environment 1990), and road
traffic noise is also a major concern. However, there are few data available
to address the effects of these sources for human biology.

The very poorest city dwellers may not have any kind of home of their
own. Clearly this will have profound consequences for health in a number
of ways. Evidence that the sympathetic-adrenal-medullary axis may be
affected is provided by Dimsdale *et al.* (1994), who showed that beta-

adrenergic receptor density was reduced in homeless men who reported high levels of life stress. In richer countries, otherwise homeless people are generally found places to live under social welfare schemes. Conway (1993) reports high levels of perceived stress amongst women living in temporary bed-and-breakfast accommodation in which they have been placed by local authorities. Boredom, isolation and loneliness were considered important causes of this stress.

Again, therefore, perceptions of lack of control appear crucial, whether they are caused by crowding or unpreventable noise. There are some indications that such noise affects growth, but otherwise the home has been neglected as an arena for research on stress-related physiological outcomes.

Reconciling home and work

We have seen that the workplace and the home bring different sets of stressors. Many people, particularly women, also experience problems in combining their roles in these two microenvironments. It is still generally true, across most cultures, that women undertake the bulk of household work and childcare (Doyal 1995) and some studies have suggested that fulfilling a mother's role is for many women the primary source of stress (Barnett and Baruch 1987). Thus, for many women work in the home may be particularly stressful and this is likely to be especially true for those who combine paid employment with responsibility for household duties. Traditional patterns whereby women are expected to undertake maintenance of the home and family have been slow to change. Aryee (1993) notes the large burden of work for married women in urban Singapore, where women are also usually responsible for helping children to cope in a highly competitive educational system. Among dual-earner couples he found greater levels of burnout in women than in their husbands. In Sweden the same pattern was identified by Lundberg et al. (1994); women in full-time work reported higher levels of work overload (combining workplace and household responsibilities) and of stress than men. Walters (1993) interviewed women in the Canadian city of Hamilton about the causes of stress in their lives and found that women spoke most about the difficulties of combining the many demands they faced, usually involving family responsibilities. Women living in Tunis reported similar problems (Hays and Zouari 1995). Women are also increasingly likely to undertake their paid work within the home, bringing another potential stressor into the household. There is a general trend towards homeworking in industrial societies, but it is even more common amongst urban women, in particular, in poorer countries (Tinker 1997).

While James's work suggests that the home may generally be a place of refuge after a hard day's work, not all those he studied showed low blood pressure at home. Some working women gave higher ratings of stress for their time at home than at work and these women showed less difference between their work and home blood pressures than did women more stressed by work. Married women with children showed higher blood pressure at home than single women who reported their greatest stress came from work (James *et al.* 1996). Frankenhaeuser *et al.* (1989) measured blood pressure in men and women working at the same company in Sweden. They found that while male managers' blood pressure dropped after work, that of female managers remained on the same level as during the day and their noradrenaline levels actually increased after work, in contrast to the men.

It appears, therefore, that the combination of demands from work and the home are reflected in the maintenance of high catecholamine and blood pressure levels after work in some women. It will be interesting to see whether such patterns change over the coming years.

Travel in the city

Many city dwellers do not live close to their workplace and have to commute each day; such travel is commonly cited as a source of stress by those writing about the city. Krupat (1985) summarises work conducted with commuters in Irvine, California, which showed that both distance traveled and time taken predicted the degree of inconvenience reported as well as feelings of tension on arrival at work. Recent reports of "road rage" in cities such as Los Angeles are also usually linked to problems of traffic congestion and such events bring their own threat to human health. In studies of daily variation in blood pressure, levels during "transportation" are typically raised nearly as much as they are at work (Pickering 1988).

Studies on professional drivers also provide evidence of stress-related physiological responses. For instance, Evans and Carrère (1991) found that self-reported control in male bus drivers in Los Angeles had an inverse relationship with urinary adrenaline excretion rate change from a morning baseline level. There was also a significant positive relationship between exposure to traffic congestion and adrenaline level, and the authors suggested that traffic congestion decreased perceived control and therefore led to hormone elevation. Similarly, Gardell (1987) attributed increased urinary adrenaline levels in Swedish bus drivers to greater perceived time pressure created by the demands of remaining on schedule and dealing with traffic congestion.

Travelling on public transport as a passenger is also commonly identified as a stressful experience. Swedish work has shown that urinary adrenaline levels were higher after than before a train journey to work and, furthermore, that people who joined the train later in the journey, when it was more crowded, and who could not therefore choose a seat, showed the greatest increase in adrenaline (Johansson and Lundberg 1978). These effects were enhanced when measurements were taken during an oil crisis in the 1970s and trains were more crowded. Here again there is some evidence that the effects of travel are greater when there is a perception of lack of control, in this case over seating.

Conclusion

In terms of human evolutionary history, the urban environment is a very new phenomenon and therefore one to which the human body is not adapted. Cities are designed to meet many of our physical needs, but city living is also generally considered to cause psychosocial stress with important consequences for health, particularly cardiovascular disease. Human biologists have an interest in measuring and understanding variation in the physiological systems posited to be responsible for this link between stress and disease. This interest can be advanced best by moving beyond comparisons between rural and urban areas by quantifying aspects of the urban environment considered to cause stress and relating them to physiological measures. Such an approach has the advantage of overcoming the problems caused by generalising about a rural/urban dichotomy across countries where it may have very different meanings, and also in allowing for variation within cities. For example, poor individuals living in cities will be exposed to quite different, and presumably greater, environmental stressors than will the rich, even when they are close neighbors.

The greatest progress along these lines has been made by those interested in work-related stress, perhaps because work is considered to be such an important source of stress. Stress is sometimes defined as an imbalance between perceived demand and ability to meet that demand, and it should not therefore be surprising that the dominant model of work stress is that formulated by Karasek, under which high strain jobs are those where a heavy workload is combined with little opportunity to exert control. There is some evidence that people in such jobs report high levels of stress and suffer from an increased burden of disease, including high blood pressure and overt cardiovascular disease. However, the body of evidence suggesting that they may secrete high levels of stress hormones is much less strong. It appears that demand alone may predict adrenaline levels more successfully than job strain, and there has been little success in relating demand, control or job strain to cortisol levels. Further work is needed to clarify these findings, but it is already clear that it is not possible to make simple predictions about relationships between feelings of stress and so-called stress hormone levels.

Lack of control also appears to play a prominent role in generating feelings of stress outside the workplace. It is a common feature linking stressors such as overcrowded and noisy living conditions and slow and difficult travel. Travel does appear to influence adrenaline levels and blood pressure, but little has been done to investigate the effects of difficult living conditions on these aspects of human biology, although there are suggestions that distress generated by noise can affect growth. These are areas in which human biologists working to understand the human body's response to the novel social and physical environments we have created for ourselves will no doubt make valuable contributions in the future.

References

Aryee, S. (1993). Dual-earner couples in Singapore: an examination of work and nonwork sources of their experienced burnout. *Human Relations*, **46**, 1369–474.

Ashford, S. J., Lee, C. and Bobko, P. (1989). Content, causes, and consequences of job insecurity: a theory-based measure and substantive test. *Academy of Management Journal*, **32**, 803–29.

Barnett, R. C. and Baruch, G. K. (1987). Social roles, gender, and psychological distress. In *Gender and Stress*, eds. R. C. Barnett, L. Biener and G. K. Baruch, pp. 122–43. New York: The Free Press.

Baum, A. and Grunberg, N. (1995). Measurement of stress hormones. In *Measuring Stress: A Guide for Health and Social Scientists*, eds. S. Cohen, R. C. Kessler and L. U. Gordon, pp. 175–92. New York: Oxford University Press.

Boyden, S. (1987). *Western Civilization in Biological Perspective: Patterns in Biohistory*. Oxford: Oxford University Press.

Cameron, A. (1995). Work and change in industrial society: a sociological perspective. In *Work and Health: An Introduction to Occupational Health Care*, ed. M. Bamford, pp. 61–100. London: Chapman and Hall.

Carruthers, M. (1976). Biochemical responses to environmental stress. In *Man in Urban Environments*, eds. G. A. Harrison and J. B. Gibson, pp. 247–73. Oxford: Oxford University Press.

Conway, J. (1993). Ill-health and homelessness: the effects of living in bed-and-breakfast accommodation. In *Unhealthy Housing: Research, Remedies and Reform*, eds. R. Burridge and D. Ormandy, pp. 283–300. London: E. & F. N. Spon.

Department of the Environment (1990). *Report of the Noise Review Working Party 1990*. London: HMSO.

Dimsdale, J. E., Mills, P., Patterson, T., Ziegler, M. and Dillon, E. (1994). Effects of chronic stress on beta-adrenergic receptors in the homeless. *Psychosomatic Medicine*, **56**, 290–5.

Doyal, L. (1995). *What Makes Women Sick: Gender and the Political Economy of Health*. London: Macmillan.

Dressler, W. W. (1995). Modeling biocultural interactions: examples from studies of stress and cardiovascular disease. *Yearbook of Physical Anthropology*, **38**, 27–56.

Dressler, W. W. and Bindon, J. R. (1997). Social status, social context, and arterial blood pressure. *American Journal of Physical Snthropology*, **102**, 55–66.

Dressler, W. W., Dos Santos, J. E. and Viteri, F. E. (1993). Social and cultural influences in the risk of cardiovascular disease in urban Brazil. In *Urban Ecology and Health in the Third World*, eds. L. Schell, M. T. Smith and A. Bilsborough, pp. 10–25. Cambridge: Cambridge University Press.

Ellison, P. T. (1988). Human salivary steroids: methodological considerations and applications in physical anthropology. *Yearbook of Physical Anthropology*, **31**, 115–42.

Evans, G. W. and Carrère, S. (1991). Traffic congestion, perceived control, and psychophysiological stress among urban bus drivers. *Journal of Applied Psychology*, **76**, 658–663.

Fanning, P. M. (1967). Families in flats. *British Medical Journal*, **4**, 382–6.

Frankenhaeuser, M. (1989). A biopsychosocial approach to work life issues. *International Journal of Health Services*, **19**, 747–58.

Frankenhaeuser, M. and Gardell, B. (1976). Underload and overload in working life: outline of a multidisciplinary approach. *Journal of Human Stress*, **2**, 35–46.

Frankenhaeuser, M., Lundberg, U., Fredrikson, M., Melin, B., Tuomisto, M. and Myrsten, A. L. (1989). Stress on and off the job as related to sex and occupational status in white-collar workers. *Journal of Organizational Behavior*, **10**, 321–46.

Freeman, H. (1993). Mental health and high-rise housing. In *Unhealthy Housing: Research, Remedies and Reform*, eds. R. Burridge and D. Ormandy, pp. 168–90. London: E. & F. N. Spon.

Gabe, J. and Williams, P. (1993). Women, crowding and mental health. In *Unhealthy Housing: Research, Remedies and Reform*, eds. R. Burridge and D. Ormandy, pp. 191–208. London: E. & F. N. Spon.

Gardell, B. (1987). Efficiency and health hazards in mechanized work. In *Work Stress*, eds. J. C. Quick, R. Bhagat, J. Dalton and J. D. Quick. New York: Praeger.

Gillis, A. R. (1977). High rise housing and psychological strain. *Journal of Health and Social Behavior*, **18**, 418–31.

Glass, D. C. and Singer, J. E. (1972). *Urban Stress: Experiments on Noise and Social Stressors*. New York: Academic Press.

Harrison, G. A. (1973). The effects of modern living. *Journal of Biosocial Science*, **5**, 217–28.

Harrison, G. A. and Gibson, J. B. (eds.) (1976). *Man in Urban Environments*. Oxford: Oxford University Press.

Harrison, G. A., Palmer, C., Jenner, D. A. and Reynolds, V. (1981). Associations between rates of urinary catecholamine excretion and aspects of lifestyle among adult women in some Oxfordshire villages. *Human Biology*, **53**, 617–33.

Hays, P.A. and Zouari, J. (1995). Stress, coping, and mental health among rural, village, and urban women in Tunisia. *International Journal of Psychology*, **30**, 69–90.

Jahoda, M. (1942). Incentives to work – a study of unemployed adults in a special situation. *Occupational Psychology*, **16**, 20–30.

James, G. D. (1991). Blood pressure response to the daily stressors of urban environments: methodology, basic concepts, and significance. *Yearbook of Physical Anthropology*, **34**, 189–210.

James, G. D., Broege, P. A. and Schlussel, Y. R. (1996). Assessing cardiovascular risk and stress-related blood pressure variability in young women employed in wage jobs. *American Journal of Human Biology*, **8**, 743–9.

James, G. D., Jenner, D. A., Harrison, G. A. and Baker, P. T. (1985). Differences in catecholamine excretion rates, blood pressure and lifestyle among young Western Samoan men. *Human Biology*, **57**, 635–47.

Jenner, D. A., Harrison, G. A. and Prior, I. A. M. (1987). Catecholamine excretion in Tokelauans living in three different environments. *Human Biology*, **59**, 165–72.

Jenner, D. A., Reynolds, V. and Harrison, G. A. (1980). Catecholamine excretion rates and occupation. *Ergonomics*, **23**, 237–46.

Johansson, G. and Lundberg, U. (1978). Psychophysiological aspects of stress and adaptation in technological societies. In *Human Behaviour and Adaptation*, eds. V. Reynolds and N. Blurton Jones. London: Taylor and Francis.

Jonsson, E. and Sorensen, S. (1973). Adaptation to community noise – a case study. *Journal of Sound and Vibration*, **26**, 571–5.

Julius, S. (1996). The defense reaction: a common denominator of coronary risk and blood pressure in neurogenic hypertension? *Clinical and Experimental Hypertension*, **17**, 375–86.

Karasek, R. A. (1979). Job demands, job decision latitude, and mental strain: implications for job redesign. *Administrative Science Quarterly*, **24**, 285–308.

Karasek, R. A., Baker, D., Marxer, F., Ahlbom, A. and Theorell, T. (1981). Job decision latitude, job demands, and cardiovascular disease: a prospective study of Swedish men. *American Journal of Public Health*, **71**, 694–705.

Karasek, R. A., Theorell, T., Schwartz, J. E., Schnall, P. L., Pieper, C. F. and Michela, J. L. (1988). Job characteristics in relation to the prevalence of myocardial infarction in the US Health Examination Survey (HES) and the Health and Nutrition Examination Survey (HANES). *American Journal of Public Health*, **78**, 910–18.

Kirschbaum, C. and Hellhammer, D. (1989). Salivary cortisol in psychobiological research: an overview. *Neuropsychobiology*, **22**, 150–69.

Krantz, D. S. and Falconer, J. J. (1995). Measurement of cardiovascular responses. In *Measuring Stress: A Guide for Health and Social Scientists*, eds. S. Cohen, R. C. Kessler and L. U. Gordon, pp. 175–92. New York: Oxford University Press.

Krupat, E. (1985). *People in Cities: The Urban Environment and its Effects*. Cambridge: Cambridge University Press.

Lin, N. and Lai, G. (1995). Urban stress in China. *Social Science and Medicine*, **41**, 1131–45.

Lundberg, U., Granqvist, M., Hansson, T., Magnusson, M. and Wallin, L. (1989). Psychological and physiological stress responses during repetitive work at an assembly line. *Work and Stress*, **3**, 143–53.

Lundberg, U., Mardberg, B. and Frankenhaueser, M. (1994). The total workload of male and female white collar workers as related to age, occupational level, and number of children. *Scandinavian Journal of Psychology*, **35**, 315–27.

Matthews, K. A. (1989). Interactive effects of behavior and reproductive hormones on sex differences in risk for coronary heart disease. *Health Psychology*, **8**, 373–87.

Mattiasson, I., Lindgarde, F., Nilsson, J. A. and Theorell, T. (1990). Threat of unemployment and cardiovascular risk factors: longitudinal study of quality of sleep and serum cholesterol concentrations in men threatened with redundancy. *British Medical Journal*, **301**, 461–6.

McCann, B. S., Magee, M. S., Broyles, F. C., Vaughan, M., Albers, J. J. and Knopp, R. H. (1995). Acute psychological stress and epinephrine infusion in normolipidemic and hyperlipidemic men: effects of plasma lipid and apoprotein concentrations. *Psychosomatic Medicine*, **57**, 165–76.

McCarthy, P., Byrne, D., Harrison, S. and Keithley, J. (1985). Housing type, housing location and mental health. *Social Psychiatry*, **20**, 125–30.

Mulders, H. P. G., Meijman, T. F., O'Hanlon, J. F. and Mulder, G. (1982). Differential psychophysiological reactivity of city bus drivers. *Ergonomics*, **25**, 1003–11.

Nonogaki, K. and Iguchi, A. (1997). Stress, acute hyperglycemia, and hyperlipidemia: role of the autonomic nervous system and cytokines. *Trends in Endocrinology and Metabolism*, **8**, 192–7.

Panter-Brick, C. and Pollard, T. M. (1999). Hormonal and health correlates of working behavior in subsistence and industrial contexts. In *The Endocrinology of Health and Behavior*, eds. C. Panter-Brick and C. M. Worthman. Cambridge: Cambridge University Press.

Pickering, T. G. (1988). The study of the blood pressure in everyday life. In *Behavioural Medicine in Cardiovascular Disorders*, eds. T. Elbert, W. Langosch, A. Steptoe and D. Vaitl. Chichester: Wiley.

Pickering, T. G., James, G. D., Schnall, P. L., Schlussel, Y. R., Pieper, C. F., Gerin, W. and Karasek, R. A. (1991). Occupational stress and blood pressure: studies in working men and women. In *Women, Work and Health*, eds. M. Frankenhaeuser, U. Lundberg and M. Chesney. New York: Plenum.

Pine, D. S., Cohen, P. and Brook, J. (1996). Emotional problems during youth as predictors of stature during early adulthood: results from a prospective epidemiologic study. *Pediatrics*, **97**, 856–63.

Pollard, T. M. (1995). Use of cortisol as a stress marker: practical and theoretical problems. *American Journal of Human Biology*, **7**, 265–74.

Pollard, T. M. (1997). Physiological consequences of everyday psychosocial stress. *Collegium Antropologicum*, **21**, 17–28.

Pollard, T. M., Ungpakorn, G. and Harrison, G. A. (1992). Some determinants of population variation in cortisol levels in a British urban community. *Journal of Biosocial Science*, **24**, 477–85.

Pollard, T. M., Ungpakorn, G., Harrison, G. A. and Parkes, K. R. (1996). Epinephrine and cortisol responses to work: a test of the models of Frankenhaeuser and Karasek. *Annals of Behavioral Medicine*, **18**, 229–37.

Schell, L. M. (1991). Effects of pollutants on human prenatal and postnatal growth: noise, lead, polychlorobiphenyl compounds, and toxic wastes. *Yearbook of Physical Anthropology*, **34**, 157–88.

Schell, L. M. (1996). Cities and human health. In *Urban Life: Readings in Urban Anthropology*, eds. G. Gmelch and W. P. Zenner, pp. 104–27. Prospect Heights, IL: Waveland Press.

Schnall, P. L., Landsbergis, P. A., Pieper, C. F., Schwartz, J., Dietz, D., Gerin, W., Schlussel, Y., Warren, K. and Pickering, T.G. (1992). The impact of anticipation of job loss on psychological distress and worksite blood pressure. *American Journal of Industrial Medicine*, **21**, 417–32.

Schnall, P. L., Pieper, C., Schwartz, J. E., Karasek, R. A., Schlussel, Y., Devereux, R. B., Ganau, A., Alderman, M., Warren, K. and Pickering, T.G. (1990). The relationship between 'job strain', workplace diastolic blood pressure, and left ventricular mass. *Journal of the American Medical Association*, **263**, 1929–35.

Spring Rice, M. (1939). *Working Class Wives: Their Health and Conditions*. Harmondsworth: Penguin.

Theorell, T., Perski, A., Akerstedt, T., Sigala, F., Ahlberg-Hulten, G., Svensson, J. and Eneroth, P. (1988). Changes in job strain in relation to changes in physiological state. *Scandinavian Journal of Work Environment and Health*, **14**, 189–96.

Tinker, I. (1997). Family survival in an urbanizing world. *Review of Social Economy*, **55**, 251–60.

Toates, F. (1995). *Stress: Conceptual and Biological Aspects*. Chichester: Wiley.

Walmsley, D.J. (1988). *Urban Living: The Individual in the City*. Harlow: Longman.

Walters, V. (1993). Stress, anxiety and depression: women's accounts of their health problems. *Social Science and Medicine*, **36**, 393–402.

Wirth, L. (1938). Urbanism as a way of life. *American Journal of Sociology*, **44**, 1–24.

Xie, J. (1996). Karasek's model in the People's Republic of China: effects of job demands, control, and individual differences. *Academy of Management Journal*, **39**, 1594–618.

13 *Physical activity, lifestyle and health of urban populations*

S. J. ULIJASZEK

Editors' introduction

Ways in which physical activity is influenced by social, cultural, economic and geographical factors in urban settings are discussed in this chapter. Urban populations and the environments they inhabit are highly diverse, this being largely related to position in the economic status system, and reflected in energy expenditure levels. Among urban populations, other factors become increasingly important as the need to work physically hard diminishes with the mechanisation of labor, and the increasing proportion of sedentary paid work. These can be disaggregated into those that operate in nested fashion at the individual, household, neighbourhood and institutional levels. Cultural and demographic factors may have an underlying effect on the way in which the interplay between these factors varies from place to place, although institutional influences may be quite idiosyncratic, and vary enormously within any sociocultural and geographic region. Thus far, there is little data from urban populations on human energetics in an ecological context, and there is a great need to study urban compexity in relation to physical activity. Energetics and physical activity are important to understand in relation to the increasing global prevalence of overweight and obesity (which has taken place predominantly among urban populations), and with respect to ovarian function, which Ellison, in chapter 6, indicates may be influenced by many factors in the urban environment.

Introduction

The structure and compexity of urban environments and the activities that take place in them make the study of physical activity patterns difficult. The type and level of physical activity of populations differ by age, sex, location of residence, occupation or subsistence practice, work and transport technologies, as well as perceptions of potential threat or danger in the environment, and these factors are much more variable in urban environments than in rural ones. Furthermore, there are various ways in which physical activity can be measured, making cross-study comparisons difficult. Perhaps the most consistent measure of physical activity is that made possible by partitioning measures of total energy expenditure into that due to maintenance metabolism, and that due to other factors, including physical activity (Ulijaszek 1992). Although measurements of total energy expenditure have been made since the 1940s in industrialised nations, and since the 1960s among populations in developing countries, there has been no systematic attempt to determine the extent to which energy expenditure in physical activity differs between urban and rural groups in either context. In this chapter, data for energy expenditure and physical activity level (as determined by total daily energy expenditure divided by basal metabolic rate, either measured or predicted) are summarised for a wide range of populations and groups, and urban–rural comparisons made for both developing and already industrialised nations.

The factors that influence physical activity vary across the lifespan, are influenced by cultural attitudes to work, fitness and physique, and mediated at three levels of human organisation: (1) the family or household; (2) community or neighborhood; and (3) population or institution. Urban environments are characterised by strong institutional influences on most aspects of human behaviour, including those behaviours that influence habitual physical activity. Institutions such as nursery schools, day care centres, schools, colleges and workplaces where much of the waking day is spent by much of the population, are a strong influence on habitual activity. Non-institutional influences on physical activity involve the interplay of household and community attitudes to work and activity, and community and neighborhood expectations of participation in social activities which require physical activity. In the second part of this chapter, the ways in which physical activity is influenced by household, neighborhood and institutional effects is examined, and the ways in which this might be peculiarly urban considered.

Physical activity and health

Associations between physical activity and health are well established; disease states associated with low levels of physical activity are summarised in Table 13.1. Various studies show a clear relationship between measures of physical activity and cardiovascular disease risk (e.g. Kannel and Sorlie 1979; Kannel *et al.* 1986; Lindsted *et al.* 1991; Sherman *et al.* 1994; La Croix *et al.* 1996), coronary heart disease (e.g. Morris *et al.* 1953; Khan 1963; Brunner *et al.* 1974; Paffenbarger and Hale 1975; Chave *et al.* 1978; Salonen *et al.* 1982; Pekkanen *et al.* 1987; Slattery *et al.* 1989;

252   S. J. Ulijaszek

Lindsted *et al.* 1991; Seccareccia and Menotti 1992; Rodriguez *et al.* 1994), stroke (e.g. Paffenbarger and Williams 1967; Kannel and Sorlie 1979; Salonen *et al.* 1982; Lapidus and Bengtsson 1986; Menotti *et al.* 1990; Lindsted *et al.* 1991; Abbott *et al.* 1994), hypertension (e.g. Paffenbarger *et al.* 1968; Blair *et al.* 1984), non-insulin-dependent diabetes mellitus (Helrich *et al.* 1991; Manson *et al.* 1992), and hormone dependent cancers in women (e.g. Sternfeld *et al.* 1993; Friendenreich and Rohan 1995). There is also general support for the views that physical activity has a beneficial effect on symptoms of anxiety and depression and can serve to improve mood (United States Department of Health and Human Services 1996), and that low physical activity levels contribute significantly to obesity and overweight (DiPietro 1995; Ching *et al.* 1996).

Negative effects of extreme physical activity include musculoskeletal injuries (United States Department of Health and Human Services 1996), anovulation and amenorrhoea among women (Cumming *et al.* 1994), interpersonal violence among players of team sports (Kraus and Conroy 1984), and increased risk of infection from immunosuppression (Newsholme and Parry-Billings 1994). Moderate physical activity is seen as being most compatible with minimum health risk, at least in the United States (United States Department of Health and Human Services 1996). However, it is not clear to what extent high levels of physical activity among peoples of the developing world are associated with the same negative consequences as for people in the industrialised world. There are likely to be differences in patterns of physical activity between the two settings, whereby the majority of people in developing countries who have high levels of physical activity are more likely to practice those activities at an intensity that can be sustained across the day. In the industrialised nations, people, if active, are more likely to engage in higher intensity activities, often competitive in nature, for shorter periods of time.

Diseases such as cardiovascular disorders, diabetes and stroke are characteristic of modern affluent Western technological communities and have come with changing demography, life expectancy, longevity, exposure to environmental stressors, nutritional status, as well as by aspects of lifestyle which influence physical activity levels (Trowell and Burkitt 1981). Globally, urban populations have experienced a steady rise in the prevalence of overweight and obesity (Byers *et al.* 1994). Increasing obesity has been attributed to low levels of physical activity associated with sedentary lifestyles, as well as the widespread availability of energy dense foods (Durnin 1992), and the money with which to buy them among large sectors of urban populations. Past (overwhealmingly rural) populations may well have been more active than contemporary urban populations. However, part of human evolutionary and ecological success has been due to efficient

Table 13.1. *Diseases associated with low levels of physical activity*

| | Association | | | |
	Strong	Weak	Equivocal	None
Coronary heart disease	+			
Stroke			+	
Hypertension	+			
Colon cancer	+			
Rectal cancer				+
Breast cancer		+		
Prostate cancer				+
Testicular cancer			+	
Non-insulin-dependent diabetes	+			
Osteoporosis			+	
Obesity	+			
Depression & anxiety		+		

Source: Summarised from United States Department of Health and Human Services (1996).

energy capture through different modes of subsistence (Ulijaszek 1995). Across evolutionary time, reduction in physical activity, if associated with increased food capture, must have had survival value, and therefore have been adaptive. This would have ceased to be adaptive under conditions of secure and plentiful food supplies, and when low energy expenditure due to low physical activity was possible among a large proportion of many populations. Human beings are able to regulate appetite and energy intake to energy energy expenditure to moderate levels of precision (Stock and Rothwell 1982). However, this regulation may be easily overwealmed by at least one of two or more mechanisms: (1) possible defects in macronutrient oxidation (Astrup *et al.* 1996) in conjunction with the availability of a high fat intake; and (2) fine lipostatic control of adipose tissue balance may be overridden by behavioural cues to feed on palatable (often high fat) foods (Mela 1996). Low levels of energy expenditure may contribute to either or both mechanisms by setting the physiological energy requirement to a low level and thus making positive energy balance, when it occurs in an individual, larger in degree and resulting in greater weight gain than in an individual with higher activity and higher physiological energy requirement.

The increasing global prevalence of overweight and obesity has taken place predominantly among urban populations where low physical activity lifestyles are the most prevalent. The increasing mechanisation of labour-intensive tasks in the industrialised nations has made low physical activity lifestyles possible, even among categories of workers previously engaged in jobs of high physical intensity.

The role of highly palatable and energy dense foods which are widely affordable in the causation of obesity bears further attention, particularly in respect of the ease with which their over consumption is possible. Comparative studies of primate diets do not support the contention of Rozin (1987) that human food choices may have been shaped in the course of evolution by taste preference for sweet substances and taste aversion to bitter ones (Ulijaszek and Strickland 1993). However, human newborn infants have a positive hedonic response for sweetness, and probably a dislike for bitter and sour tastes, developing a taste for salt between 4–6 months post-natally (Mela and Catt 1996). While it is difficult to establish the evolutionary nature of these innate and early developed preferences, their persistence into adult life encourages the consumption of palatable and usually energy dense foods which can override regulation by appetite alone, and contribute to the risk of overweight and obesity, cardiovascular disease, stroke and non-insulin-dependent diabetes.

Urbanism requires an adequate and safe food supply. Traditionally, preservation of foods has included the use of sugar and salt to inhibit bacterial growth on a wide range to food products. While food preservation techniques have become ever more sophisticated, the availability of high sugar and high salt foods remains high, despite the marketing in recent times of foods including low energy sugar substitutes, and in some cases non-metabolisable fat substitutes. Furthermore, the availability of such foods does nothing to control the population-level tendancy to overeat, since these substitutes do not contribute to satiety in the individual.

The creation of a safe food supply also may have contributed to reduced physical activity among urban women. Prior to regulations ensuring safe food supply, women in well-off homes devoted considerable time to maintenance of domestic cleanliness and the acquisition, preparation and storage of food. In homes where women worked outside the home for long hours, such household management may have been lax, and infant mortality as a consequence of food-related infections, higher. With the provision of a safe food supply, less time and effort needed to be invested by women in such maintenance, and reduced levels of food-related infections are likely to have contributed to the overall decline in infant mortality at this time. Thus, regulation of food supplies not only allowed energy expended in the home economy to decline, but was also a public health measure which had marked impact on the health of urban populations (McMichael, this volume).

Energy expenditure in urban and rural populations

Energetic parsimony is not adaptive for populations in industrialised nations, although the extent of sedentary work has risen since 1980 (Table 13.2). The proportions of the British and United States populations employed in service-oriented work has risen sharply since 1980, as employment in manufacture and heavy industry has declined. In both countries, the very small proportion of the population engaged in agricultural or fisheries work has also declined. In Japan, the trend is very similar, although the decline in employment in manufacture and heavy industry has been slight. The slight increase in the proportion of service-oriented work among males has come from the decline in the proportion of the population engaged in agriculture or fisheries. Among women, the increase in the proportion of women engaged in service-oriented work has come with a reduced proportion of female workers in manufacture. In all three countries, work in manufacture and heavy industry has continued to become more mechanised, and thus requires lower levels of physical activity from workers. In India, only a very small proportion of the adult population is engaged in paid employment. Of those engaged in paid employment, the trend towards increased employment in service industries has come mostly among women, and has come with a decline in the proportion of women engaged in agricultural and manufacturing work.

Although most sedentary groups are likely to be found in urban environments because of the sedentary nature of much paid employment, it is not clear whether this translates into lower overall physical activity, because urban dwellers in sedentary professions are able to spend discretionary time in high physical activity pastimes. In order to understand the extent to which low energy expenditure is an urban phenonmenon, it is important to compare the total energy expenditures and physical activity levels of groups and populations in both rural and urban settings, in developing and industrialised nations. Total daily energy expenditures can give some estimate of the degree of variation across populations practising different modes of subsistence. Table 13.3 shows selected mean total daily energy expenditures of adult males in developing and industrialised countries.

The daily energy expenditure of hunter-gatherers, fishers and hunter-horticulturalists spans a similar range of values to that of agriculturalists in developing countries operating at low levels of technology, with pastoralists spanning a slightly lower range of values. However, there is greater between-population variation in daily energy expenditure within all economic categories than between them. Daily energy expenditures of people in developing countries engaged in cash employment in urban settings in-

Table 13.2. *Proportion of the working population in different categories of employment*

Proportion of total adult population employed	Male		Female	
	1980 (%)	1993 (%)	1980 (%)	1993 (%)
UK (92% of population urban, 1990)				
Agriculture and fisheries	3.5	3.2	1.3	1.0
Manufacture and heavy industry	47.9	36.6	22.5	14.1
Service industries and trades	48.7	60.3	76.0	84.7
USA (74% of population urban, 1990)	1980 (%)	1994 (%)	1980 (%)	1994 (%)
Agriculture and fisheries	5.0	4.0	1.6	1.6
Manufacture and heavy industry	39.8	33.5	18.5	13.4
Service industries and trades	55.2	62.5	79.9	85.0
Japan (77% of population urban, 1990)	1980 (%)	1993 (%)	1980 (%)	1993 (%)
Agriculture and fisheries	8.7	5.4	1.5	0.7
Manufacture and heavy industry	39.9	39.7	32.3	28.4
Service industries and trade	51.3	54.9	66.2	70.8
India (28% of population urban, 1990)	1980 (%)	1989 (%)	1980 (%)	1989 (%)
Agriculture and fisheries	4.2	4.3	17.1	13.2
Manufacture and heavy industry	40.3	38.7	27.0	20.8
Service industries and trade	55.5	57.0	55.9	66.0

Source: Data recalculated from Department of Economic and Social Statistics (1992, 1996).

clude occupational categories with higher values than any group measured practicing any form of traditional subsistence, as well as categories that are within the range of values presented for populations engaged in traditional subsistence. Thus, in the developing world, energy expenditure is higher among poor urban populations engaged in hard physical labor than among rural subsistence populations, but lower among workers engaged in tasks requiring a greater degree of skill and expertise. The greatest range of daily energy expenditure is found in industrialised countries. The upper range of mean values is dominated by forestry workers, who, prior to mechanisation of their work, had the highest values recorded for any occupational category. The range of values represented for the industrialised nations is is higher and broader than for the developing world, urban and rural. Urban populations and the environments they inhabit are highly diverse. This diversity is largely related to position in the economic status system, and is reflected in the energy expenditure levels of those populations.

Physical activity level differences according to gender and urbanism

Although differences in total energy expenditure between developing and industrialised nations appear to be non-existent, any potential differences might be masked by differences in mean body mass; people in countries with high levels of affluence are on average taller and heavier than people in less-developed countries. Following from this, populations in industrialised nations with larger mean body mass also have greater maintenance energy expenditures. Table 13.4 gives mean daily energy expenditures of males and females as multiples of BMR in developing countries. Expressing data in this way controls for differences in body size, and where BMR has been measured as opposed to estimated, for metabolic variation. This is often called the physical activity level (PAL) (James *et al.* 1988) and is a reasonably standardised measure of level of daily exertion. Population values vary enormously, and there is no evidence that women work harder than men in the developing countries, nor that agriculturalists work harder than either hunter-gatherers or hunter-horticulturalists. Indeed, of 15 populations where male–female comparisons of PAL have been made, women work harder than men in only 2: these are the Mvae millet cultivators of Cameroon, and one group of New Guinea highlanders. Men work harder than women in 10 of the 19 populations, while no real difference was seen in 7 of them.

Physical activity level data on occupational groups in the urban developing world is more limited, (Table 13.5), but such data as there is indicates that activity levels are generally in the same range of mean values as rural populations. However type of work or work category, not surprisingly, strongly influences the PAL.

Physical activity level data for industrialised nations needs to take account of the potentially lower activity patterns of the obese members of these societies. Thus, in Table 13.6, PAL values of urban populations are disaggregated into those who were obese, and those who were non-obese. For the non-obese, both males and females, the range of PAL values is similar to that of rural populations in the developing world. Obese males appear to have lower PALs than non-obese groups of males, while there is no such difference in females. Non-obese males have higher PALs than

Table 13.3. *Mean daily energy expenditures of some groups of adult males*

Group or district	Country	Total enegy expenditure (MJ/day)	Reference
Traditional subsistence			
Hunter-gatherers, fishers and hunter-horticulturalists			
Eskimos	Canada	15.36	Godin & Shephard 1973
Ache	Paraguay	13.92	Hill *et al.* 1985
Machiguenga	Peru	13.40	Montgomery & Johnson 1974
Tukanoans	Colombia	12.05	Dufour 1992
Keto	Russia	11.42	Katzmarzyk *et al.* 1994
Yassa	Cameroon	11.41	Pasquet & Koppert 1993
Mvae	Cameroon	9.90	Pasquet & Koppert 1993
!Kung	Botswana	9.11	Lee 1979
Pastoralists and agropastoralists			
Even	Russia	11.91	Katzmarzyk *et al.* 1994
Turkana	Kenya	9.04	Galvin 1985
Nunoa	Peru	8.36	Leonard 1988
Nunoa	Peru	8.31	Thomas 1976
Agriculturalists			
	Western Samoa	16.84	Pelletier 1984
	Western Samoa	16.23	Schendell 1989
	Guatemala	15.50	Viteri *et al.* 1971
	Western Samoa	13.80	Schendell 1989
	Philippines	13.80	de Guzman *et al.* 1974a
Varanin	Iran	13.57[a]	Brun *et al.* 1979
Gurung	Nepal	12.80[b]	Strickland *et al.* 1995
	Myanmar	12.55[b]	Tin-May-Than & Ba-Aye 1985
	Burkina Faso	12.30[c]	Brun *et al.* 1981
Savaii	Western Samoa	12.16	Pearson 1990
Non-Gurung	Nepal	11.30	Strickland *et al.* 1995
Lufa	Papua New Guinea	10.75	Norgan *et al.* 1974
Tamil Nadu	India	10.60[a]	McNeill *et al.* 1988
Genieri	Gambia	9.83[c]	Fox 1953
Kaul	Papua New Guinea	9.82	Norgan *et al.* 1974
Pari	Papua New Guinea	9.33	Hipsley & Kirk 1962
Kaporaka	Papua New Guinea	9.11	Hipsley & Kirk 1962

Country	Occupation	Total energy expenditure (MJ/day)	Reference
Paid labour			
India	Rickshaw pullers	20.42	Bannerjee 1962
	Cottonmill workers	12.76	Bannerjee *et al.* 1959b
	Stone cutters	12.66	Ramanamurthy *et al.* 1962
	Laborers	12.59	Devadas *et al.* 1975
	Laboratory technicians	8.66	Bannerjee *et al.* 1959a
Western Samoa	Laborers, sedentary leisure	16.69	Pelletier 1984
	Sedentary work, active leisure	13.92	

Table 13.3 *(cont.)*

Country	Occupation	Total energy expenditure (MJ/day)	Reference
	Sedentary work and leisure	12.28	
China	Miners, according to employment type:		Ho 1984
	ore porting	15.67	
	ore dressing	15.48	
	hammering	13.41	
	prop setting	13.12	
	drilling	11.21	
China	Shipbuilders, according to employment type:		Ho 1984
	hammerer	15.57	
	planer	12.47	
	carpenter	11.01	
	flame cutting worker	9.88	
	electric welder	9.49	
Philippines	Shoemakers	11.30	de Guzman *et al.* 1974b
	Jepney drivers	10.40	de Guzman *et al.* 1974c
	Textile mill work	10.40	de Guzman *et al.* 1979
	Clerk-typists	9.20	de Guzman *et al.* 1978

Country	Group or occupation	Total energy expenditure (MJ/day)	Reference
Industrialised nations (adults of productive age only)			
Germany	Forestry workers	27.41	Kaminsky 1953
Sweden	Forestry workers	23.84	Lundgren 1946
Holland	Forestry workers	23.00	Streef *et al.* 1959
Japan	Forestry workers	15.48	Kusunoki 1956
UK	Forestry workers	15.36	Durnin & Passmore 1967
UK	Coal miners	15.31	Garry *et al.* 1955
UK	Farm workers	14.85	Durnin & Passmore 1967
Switzerland	Elderly peasants	14.77	Durnin & Passmore 1967
UK	Students	14.64	Passmore *et al.* 1952
UK		14.62	Livingstone *et al.* 1990
Italy	Shipyard Workers:		Norgan & Ferro-Luzzi 1978
	heavy workload	13.84	
	moderate workload	13.13	
	sedentary	11.95	
UK	Steel workers	13.72	Durnin & Passmore 1967
	Building workers	12.55	
	Students	12.26	
	Laboratory technicians	11.88	
	Colliery clerks	11.72	
UK	Students	11.80	Norgan & Durnin 1980
UK	Office clerks	11.72	Garry *et al.* 1955
	Office workers	10.54	Durnin & Passmore 1967
Japan	Assembly workers	9.70	Kashiwazaki *et al.* 1986
	clerical workers	9.51	
USA	Young	9.01	Vaughan *et al.* 1991

a Average of four seasons
b Average of three seasons
c Average of two seasons

Table 13.4. *Physical activity levels for adults engaged in a variety of subsistence practices*

Group or district	Country	Subsistence type	Physical activity level (TEE/BMR)	Reference
Males				
	Guatemala	Maize cultivation	2.32	Viteri et al. 1981
	Western Samoa	Taro cultivation and fishing	2.31	Pelletier 1984
	Philippines	Rice cultivation	2.25	de Guzman et al. 1974
Ache	Paraguay	Hunter-gatherer	2.15	Hill et al. 1985
	Western Samoa	Taro cultivation and fishing	2.12	Schendel 1989
Varanin	Iran	Wheat cultivation	2.10[d]	Brun et al. 1979
Machiguenga	Peru	Hunter-horticulturalist	2.09[b]	Montgomery & Johnson 1974
Ache	Paraguay	Hunter-gatherer	2.08[a]	Hill et al. 1985
Gurung	Nepal	Rice, maize and millet cultivation	2.05[e]	Strickland et al. 1997
	Myanmar	Rice cultivation	2.02[b]	Tin-May-than & Ba-Aye 1985
	Gambia	Rice and peanut cultivation	2.02[b]	Fox 1953
Tamil Nadu	India	Rice cultivation	1.96[d]	McNeill et al. 1988
Sundanese	Indonesia	Rice cultivation	1.96[a]	Suzuki 1988
Non-Gurung	Nepal	rice, maize and millet cultivation	1.91[e]	Stricland et al. 1997
	burkina Faso	Millet cultivation	1.89[b]	Brun et al. 1981
Massa	Cameroon	Millet cultivation, fishing	1.87	Pasquet et al. 1992
Igloolik	Canada	Hunters	1.82	Shephard 1974
	Western Samoa	Taro cultivation and fishing	1.81	Schendell 1989
Iven	Russia	Pastoralism	1.78[a]	Patzmarzyk et al. 1994
Keto	Russia	Fishing	1.72[a]	Katzmarzyk et al. 1994
Yassa	Cameroon	Millet cultivation, fishing	1.71	Pasquet & Koppert 1993
!Kung	Botswana	Hunter-gatherer	1.71	Lee 1979
	Ivory Coast	Mixed cultivation	1.68[a]	Dasgupta 1977
Wopkaimin	Papua New Guinea	Hunter-horticulturalist	1.65[b,c]	Ulijaszek & Brown unpublished
Lufa	Papua New Guinea	Sweet potato cultivation	1.64[a]	Norgan et al. 1974
	Uganda	Maize and plantain cultivation	1.63[a]	Cleave 1970
Savaii	Western Samoa	Cultivation	1.60[a]	Pearson 1990
Mvae	Cameroon	Millet cultivation, hunting	1.60	Pasquet & Koppert 1993
	India	Rice cultivation	1.56	Edmundson & Edmundson 1988

Group or district	Country	Subsistence type	Physical activity level (TEE/BMR)	Reference
Kaul	Papua New Guinea	Taro and plantain cultivation	1.52[a]	Norgan *et al.* 1974
Pari	Papua New Guinea	Hunter-horticulturalist	1.42	Hipsley & Kirk 1962
Nunoa	Peru	Agro-pastoralist	1.32	Thomas 1976
Turkana	Kenya	Pastoralists	1.29	Galvin 1985

Females				
	Gambia	Rural farmers	1.97	Singh *et al.* 1989
	Western Samoa	Rural villagers	1.90[a]	Schendel 1989
Ache	Paraguay	Hunter-gatherer	1.88	Hurtado *et al.* 1985
Tamang	Nepal	Rice, maize, millet & wheat cultivation	1.82[d]	Panter-Brick 1993a
Lufa	Papua New Guinea	Sweet potato cultivation	1.82	Norgan *et al.* 1974
	Burkina Faso	Millet cultivation	1.80	Bleiberg *et al.* 1980
Igloolik	Canada	Hunters	1.79	Shephard 1974
	Western Samoa	Rural villagers	1.79	Schendel 1989
	Gambia	Rural farmers	1.74	Heini *et al.* 1991
Machiguenga	Peru	Hunter-horticulturalist	1.72	Montgomery & Johnson 1974
Mvae	Cameroon	Millet cultivation, hunting	1.72	Pasquet & Koppert 1993
Guangzhou	China	Peasant farmers	1.71[a,b]	Ho 1984
Tukanoa	Colombia	Horticulture	1.71	Dufour 1984
	India	Rice cultivation	1.69[d]	McNeill *et al.* 1988
Yassa	Cameroon	Millet cultivation, fishing	1.67	Pasquet & Koppert 1993
Gurung	Nepal	Rice, maize and millet cultivation	1.67[e]	Strickland *et al.* 1997
Iven	Russia	Pastoralism	1.62[a]	Katzmarzyk *et al.* 1994
	Guatemala	Maize cultivation	1.62	Schutz *et al.* 1980
	Guatemala	Maize cultivation	1.61	Stein *et al.* 1988
Savaii	Western Samoa	Cultivators	1.59[a]	Pearson 1990
Kaul	Papua New Guinea	Taro and plantain cultivation	1.57[a]	Norgan *et al.* 1974
Non Gurung	Nepal	Rice, maize and millet cultivation	1.56[e]	Strickland *et al.* 1997
Keto	Russia	Fishing	1.55[a]	Katzmarzyk *et al.* 1994
	India	Rice cultivation	1.53	Edmundson & Edmundson 1988
!Kung	Botswana	Hunter-gatherer	1.51	Lee 1979
	Ethiopia	Cultivation	1.47	Ferro-Luzzi *et al.* 1990
Turkana	Kenya	Pastoralists	1.37	Galvin 1985
Swazi	Swaziland	Cultivation	1.35	Huss-Ashmore *et al.* 1989

[a]From estimates of BMR from body weight, and total energy expenditure from activity diaries
[b]Average of two seasons
[c]Estimate of total energy expenditure from activity diaries
[d]Average of four seasons
[e]Average of three seasons

Table 13.5. *Physical activity levels (PAL) of urban populations in less-industrialised nations*

Country	Work category	PAL	Reference
Western Samoa	Laborers, sedentary leisure	2.23	Pelletier 1984
	sedentary work, active leisure	1.77	
	urban neighborhood	1.65	Schendel 1989
	sedentary work and leisure	1.57	Pelletier 1984
Chile	Laborers	1.78	Riumallo *et al.* 1989
Philippines	Shoemakers	1.76	de Guzman *et al.* 1987b
	Textile mill workers	1.64	de Guzman *et al.* 1979
	Clerk typists	1.46	de Guzman *et al.* 1978
Samoans in Honolulu		1.48	Pearson 1990
Pima Indians in Gila Community		1.38	Rising *et al.* 1994
American Samoa		1.37	Pearson 1990
Female			
Western Samoa	Urban neighborhood	1.84	Schendel 1989
Philippines	Textile mill workers	1.68	de Guzman *et al.* 1979
	Typists	1.60	de Guzman *et al.* 1978
	Housewives	1.36	de Guzman *et al.* 1987b
American Samoa		1.42	Pearson 1990
Samoans in Honolulu		1.35	Pearson 1990

non-obese females, thus showing a similar pattern of male–female difference in physical activity to rural populations in the developing world.

Human organisation

A range of factors influence physical activity levels. In the rural developing world, the most important is the need for physical labour in the subsistence quest, (Nydon and Thomas 1989), be it cash-related or otherwise. However, with increasing modernisation and the mechanisation of tasks, this becomes less so. Among urban populations, other factors become increasingly important as the need to work physically hard diminishes with the mechanisation of labor, and the increasing proportion of sedentary paid work. These can be disaggregated into those that operate in nested fashion at the individual, household, neighborhood and institutional levels, respectively (Table 13.7).

Cultural and demographic factors may have an underlying effect on the way in which the interplay between these factors varies from place to place,

Table 13.6. *Physical activity levels for adults of productive age, urban populations in industrialised nations*

Group or district	Country	PAL	Reference
Males			
Non-obese			
Chicago	USA	2.21	Schoeller & van Santen 1982
Cambridge	UK	1.82	Diaz *et al.* 1991
Northern Ireland	UK	1.89	Livingstone *et al.* 1991
Boston	USA	1.98	Roberts *et al.* 1991
Berkeley	USA	1.74	Gorsky & Calloway 1983
Boston	USA	1.73	Goran *et al.* 1993
Cambridge	UK	1.71	Stubbs *et al.* 1995
Ste-Foy	Quebec	1.69	Tremblay *et al.* 1992
Wageningen	Netherlands	1.69	Verboeket *et al.* 1993
Wageningen	Netherlands	1.65	Westerterp *et al.* 1992
Wageningen	Netherlands	1.83	Meijer *et al.* 1992
	Italy	1.80	Norgan & Ferro-Luzzi 1978
	USA	1.72	Seale *et al.* 1990
	Italy	1.68	Norgan & Ferro-Luzzi 1978
		1.56	
		1.56	Seale *et al.* 1993
Kamakura City	Japan	1.40	Kashiwazaki *et al.* 1986
Kamakura City	Japan	1.37	
Phoenix	USA	1.23	Vaughan *et al.* 1991
Obese			
Wageningen	Netherlands	1.76	Meijer *et al.* 1992
Wageningen	Netherlands	1.73	Verboeket *et al.* 1993
Wageningen	Netherlands	1.51	Westerterp *et al.* 1991 61
Females			
Non-obese			
Stockholm	Sweden	1.86	Forsum *et al.* 1992
	USA	1.84	Seale *et al.* 1993
Wageningen	Netherlands	1.83	Meijer *et al.* 1992
Stockholm	Sweden	1.78	Sohlstrom 1993
Northern Ireland	UK	1.77	Livingstone *et al.* 1991
Cambridge	UK	1.67	Goldberg *et al.* 1991
Rochester	USA	1.67	Welle *et al.* 1992
Wageningen	Netherlands	1.67	Westerterp *et al.* 1991
Sacramento	USA	1.65	Lovelady *et al.* 1993
Cambridge	UK	1.58	Goldberg *et al.* 1993
Wageningen	Netherlands	1.53	Westerterp *et al.* 1992
Chicago	USA	1.50	Casper *et al.* 1991
Cambridge	UK	1.42	Prentice *et al.* 1986
Obese			
Wageningen	Netherlands	1.84	Meijer *et al.* 1992
Rochester	USA	1.76	Welle *et al.* 1992
Wageningen	Netherlands	1.61	Westerterp *et al.* 1991
Cambridge	UK	1.54	Prentice *et al.* 1986

Table 13.7. *Structural factors influencing physical activity in urban environments*

Cultural	Appropriate physical activity
	Age related declines in exercise in some cultures
	Activity and social status
Institutional	Culture of the work place, school, college, day care center, nursery
Community and neighborhood	Activities organised and affiliative practices which require physical activity
Individual	Choice, free-will, peer group pressure
	Hobbies which are activity-based, or otherwise

although institutional influences may be quite idiosyncratic, and vary enormously within any sociocultural and geographic region. In industrialised nations, the high proportion of young adults living on their own means that household factors influencing physical activity are low; however, preferred levels of physical activity may be related to habitual levels of physical activity in the households from whence they came, those of their parents. Friendships bear strongly on activity levels, and these may be a function of past and present neighborhood and institutional links. The entry of adults into family-building can make activity patterns more strongly determined at household and institutional level than at neighborhood levels, as well as being less influenced by friendships. In Japan, very strong institutional and employment cultures may influence activity patterns more than individual or neighborhood influences in Western industrialised nations, but this remains to be tested.

For individuals in paid employment, at school or in tertiary education, institutional factors control the majority of the waking day when physical activity can be performed. Assuming an 8 hour working day for five days per week, the nature of employment will determine the activity performed during about 35% of the waking day. Slightly lower values apply to children and adolescents at school. One potential source of physical activity is regular travel to and from the workplace, school, college or day-care centre. Table 13.8 shows the PAL on working days of four hypothetical individuals who work at different levels of energy expenditure according to the type of transport used to get to work, assuming that the distance from home to work is a 30 minute walk, or 15 minute cycle ride. The PAL is by definition standardised for body size. In addition, the influence of an hour's sport on PAL of people using motorised transport is calculated. Clearly, the biggest difference in activity level is according to type of occupation, with the physical laborer having almost twice the PAL

Table 13.8. *Hypothetical physical activity levels of working adult males according to employment type and mode of transport to work*

| Type of transport | Mode of transport to work, assuming same distance covered by each type of transport | | | |
	Car	Walk[a]	Cycle[b]	Car, plus 1 hour intensive sport[c]
Physical labor	2.66	2.71	2.74	2.84
Factory worker	2.00	2.05	2.07	2.18
Laboratory worker	1.66	1.71	1.74	1.84
Sedentary worker	1.35	1.40	1.42	1.53

Calculated from values of physical activity ratios (PAR) given in James and Schofield (1990), and based on the assumption that 8 hours of the day is spent sleeping, at $1.0 \times BMR$, 8 hours is spent working at the PAR value associated with that type of work, and part of the remaining 8 hours is spent going to and from work, and the rest in discretionary activities of low intensity, at PAR of $1.4 \times BMR$ for the remainder of this 8 hours.
[a] 30 minutes each way
[b] 15 minutes each way
[c] at PAR of $6 \times BMR$

of the sedentary worker. For the physical laborer and factory worker, intensity of work is such that PAL is very high, irrespective of type of transport taken to work, or whether intensive sport is played. For the laboratory worker, any kind of regular physical activity (walking or cycling to work, an hour's sport daily) raises the PAL slightly, but importantly, from a fairly low level. The most important impact of walking or cycling to work, or of being involved in sport, is for the sedentary worker, who is extremely inactive if he/she takes only motorised transport to work. Any type or level of physical activity outside of the workplace has important consequences for his/her physical activity level. However, none of the activities represented here (walking or cycling to work, one hour of intensive sport) bring the PAL value up to the level of the next least active occupation chosen for this model, the laboratory worker. Thus, physical activity associated with work is important in shaping the activity level of the employee.

In the urban context, distances traveled daily to work and school can be considerable. The proximity to the institution determine the extent to which motorised transport is taken, above walking, cycling, or other modes of personal non-motorised transport. Safety concerns for children mean that while in the past younger children might have walked to school, more children are currently driven in cars.

Community and neighborhood structures are more important than institutional ones in influencing physical activity levels among those not in paid employment. Shared activities of housewives, involving common participation and also serving social function, may influence their level of physical activity. In many urban contexts, it is possible for individuals to chose not to participate in such structures. However, socially oriented physical activity, whether based locally in the community or more distantly, have a strong influence upon total habitual energy expenditure levels, especially among members of urban populations with sedentary occupations.

At the individual level, there is the choice of whether to take exercise or not, and personal interests play an important part in the type of activity undertaken during discretionary time. Cultural factors also play a part in determining the extent to which people will chose physical activity as a means of occupying discretionary time. Safety concerns in the inner city may reduce the extent to which children are able to play in the streets, and may of neccessity make them sedentary, watching television, videos, or playing computer games. The use of television as an electronic babysitter by parents of young children also may predispose children to be less physically active in later life. While a genetic component to energy expended in physical activity has been demonstrated in Canadian twins (Bouchard et al. 1989) a stronger influence is a shared family environment, and/or common tendency for patterns of physical activity and energy expenditure. For example, common family environments for the participation in sport and other outdoor activities are likely to exist, but the extent to which they can be overridden by institutional factors is variable. Activity among children, however, may be susceptible to seasonality and fashion (Bittle 1992), but there is limited data which can be used to address this issue.

Age

The urge to explore is something common to all children, and shows itself with progressive neuromuscular and motor development and maturation (Malina 1992). With crawling, and subsequently walking from 11–14 months of age, the energy expenditure in physical activity increases from its low level in the first 3 months of life. Figure 13.1 shows the proportion of total energy expenditure of maintenance metabolism, growth and physical activity for young Gambian children up to age 18 months. The rural population of the Gambia has very low levels of technology, and levels of physical activity reflect the stimulation-seeking behavior of young children. In industrialised nations, and in urban communities of the

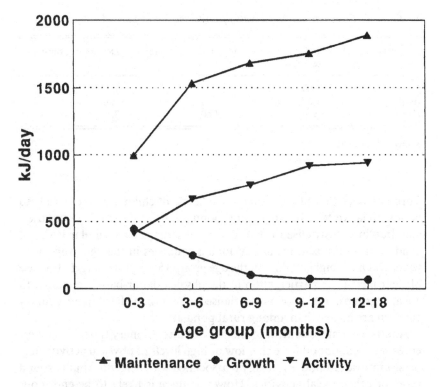

Figure 13.1 Energy expenditure of Gambian children aged o–18 months, partitioned into maintenance metabolism, and energy expended in physical activity, and growth.
Source: Adapted from Vasquez-Velasquez (1988)

developing world, the technological environment offers the possibility of passive stimulation to children from an early age. This is usually in the form of television and video recordings. In the United States in 1990, children aged 2–4 years watched an average of 4 hours of television per day, while 6–11 year olds watched an average of 3.5 hours per day (Nielsen Media Research 1990). Similar values are likely to exist for children in other industrlialised nations. The extent to which passive pursuits result in reduced energy expenditure in physical activity is not clear, however. Dufour (1997) has reviewed associations between physical activity and nutrition in children aged 0 to 10 years, and found that both undernutrition and overnutrition are accompanied by lower levels of physical activity than in controls.

Evidence for lower physical activity levels among children in industrialised and urban populations has been has been sought, and to some extent found. Table 13.9 shows the results of a metanalysis carried out by

Table 13.9. *Physical activity levels of children, sexes pooled*

Age group (years)	Industrialised, urban & rural	Developing, urban	Developing, rural
5–9	1.60	1.56	1.75
10–14	1.60	1.62	1.85
15–19	1.70	1.60	2.13

Source: Torun 1996.

Torun *et al.* (1996), of physical activity levels of children in industrialised nations relative to urban and rural populations of developing countries. Children in industrialised urban and rural settings have similar PALs to children in urban settings in developing countries in the age range 5–14 years, but have higher values in the age group 15–19 years, largely because of their increasing participation in the physical-labor-based economy. In all age groups, PALs of industrialised and urban developing country children are higher than among rural populations.

Adults have more control of their use of discretionary time, but preferences set in childhood for either low or high levels of physical activity may persist into adulthood, especially if associated with behaviors built around sport or other social activities. However, there is likely to be enormous plasticity of activity-related behavior, within the socioeconomic constraints of possibilities for physical activity, such as the availability of sports facilities or of safe places for children to play. In industrialised societies there is often a decline in energy expenditure with increasing age. Although the evidence is limited, the lack of hard and fast rules about retirement from economic activity in the rural developing world probably lead to a more limited reduction in physical activity, more closely related to changes in physical ability as well as physique in later life, as well as physique changes related to ageing.

The ability to work declines with age, for a variety of reasons, many of which are related. High levels of habitual activity are important for the maintenance of physiological performance, and reduced levels of physical ability may arise as a consequence of reduced levels of physical activity due to reduced motor skills or judgment, and accidents arising from these reductions. In traditional rural economies, the potential to maintain reasonable levels of physical activity into older age is high, if physical work is needed for food production. The need to work hard declines with increased commoditisation of life, regardless of whether urban or rural, and the maintenance of levels of physical activity into older age becomes largely a function of peer group activities, and among the non-poverished, also of personal preferences.

Reduced physical activity may also arise from reduced oxidative capacity of mitochondria which occurs as part of the ageing process (Bittles and Sambuy 1986), and from the reduced muscle mass which comes with reduced levels of physical activity. Decreased muscle strength with increasing age also increases the likelihood of falling. Associated with this are the levels of bone mineralisation which determine whether falls result in fractures. In turn, bone mineral mass and density are known to be influenced by muscle strength. Of the variety of factors which can influence the capacity for physical work (including nutrition, habitual levels of activity and disease experience), sociocultural systems which determine the micro-environment, health, nutrition and levels of habitual activity among older people affect neuromuscular function as well as the skeletal system, often resulting in a steady decline in physical work capacity. Women lose bone density earlier and faster than men (Stini 1994), and are more likely to experience reduced activity due to skeletal injury at an earlier age than men. In industrialised societies, older people become much less important in the workforce, and institutional influences that can affect levels of physical activity change, while the desire for involvement in institutions varies from person to person, and may not take place every day.

Summary

Physical activity patterns vary with age, gender, and modes of economic production. Rural–urban differences in physical activity levels reflect these differences. However, the comparisons made in this chapter are at best simplistic, because the studies involving the quantification of physical activity in human populations in urban settings have rarely considered ecological complexity as variables to measure. Thus, it is possible to compare by gender, by occupation, and to a limited extent by age, but not by type and nature of urban environment. One problem with carrying out a metanalysis of energy expenditure data is that while adequate descriptions of rural populations, their environments and economic activities are usually available, such detail is surprisingly missing from most studies of urban populations. One reason for this is that studies of rural populations are often carried out by anthropologists, or nutritionists with some anthropological interest, while studies of urban populations are carried out by nutritionists who perhaps feel that urban populations are well characterised, and nothing further needs to be said, other than some broad statement about socioeconomic status. Often, studies of energy expenditure are carried out to test physiological theory, and consideration of the nature of the environment the subjects live in, is considered of no consequence. In studies of human adaptability, an environmental description is vital (Weiner and Lourie 1981), especially in urban environments, where complexity is greater than in rural environments.

It is possible to make some crude generalisations about physical activity in rural and urban settings, while detailed analysis of physical activity in different types of urban setting are awaited. Broadly, male–female differences in physical activity exist in most societies, regardless of whether they are in urban or rural settings. In most societies, men work harder than women. Furthermore, the rather limited data on energy expenditure among urbanised adults in developing countries suggests that the range of energy expenditure is similar to that of rural populations. However, future comparisons need to consider the extent to which like is compared with like: for example, migrant studies should incorporate measures of physical activity and energy expenditure, to determine whether urban–rural differences are real, or whether they are an artifact of measuring similar populations, but which differ in some subtle way.

Studies of physical activity also need to take into account urban compexity, and how that influences people's behavior. Urban environments are self-made, and people operate within the details of that environment. There is geographical space, safety and fear, community cohesiveness, and the culture of physical activity which has general features, but which also differs from place to place. At present, little is known of the fine weave of ecological complexity as it relates to physical activity in urban environments, despite the fact that many ecological studies among rural populations (e.g. Hill *et al.* 1985; Lee 1979; Morren 1977; Thomas, 1976) have included the measurement of such detail. Human adaptation and adaptability in urban environments represent is poorly understood and thus far little studied; physical activity and energy expenditure need to be examined in detail in this context.

Acknowledgement

I thank Larry Schell for comments on an earlier draft of this paper.

References

Abbott, R. D., Rodriguez, B. L., Burchfiel, C. M. and Curb, J. D. (1994). Physical activity in older middle-aged men and reduced risk of stroke: the Honolulu Heart Program. *American Journal of Epidemiology*, **139**, 881–93.

Astrup, A., Buemann, B., Toubro, S. and Raben, A. (1996). Defects in substrate oxidation involved in the predisposition to obesity. *Proceedings of the Nut-*

rition Society, **55,** 817–28.

Bannerjee, S. (1962). *Studies in Energy Metabolism. Special Report Series, No. 43.* New Delhi: Indian Council of Medical Research.

Bannerjee, S., Sen, R. N. and Acharaya, K. N, (1959a). Energy metabolism in laboratory workers. *Journal of Applied Physiology,* **14,** 625–8.

Bannerjee, S., Acharya, K. N., Chattopadhyay, D. and Sen, R. N. (1959b). Studies on energy metabolism of labourers in a spinning mill. *Indian Journal of Medical Research,* **47,** 657–62.

Bittle, S. (1992). Adherence to physical activity and exercise. In *Physical Activity and Health,* ed. N. G. Norgan, pp. 170–89. Cambridge: Cambridge University Press.

Bittles, A. H. and Sambuy, Y. (1986). Human cell culture systems in the study of ageing. In *The Biology of Human Ageing,* eds. A. H. Bittles and K. J. Collins, pp.49–66. Cambridge: Cambridge University Press.

Blair, S. N. (1994). Physical activity, fitness and coronary heart disease. In *Physical Activity, Fitness, and Health: International Proceedings and Consensus Statement,* eds. C. Bouchard and R. J. Shephard, pp. 579–90. Champaign IL: Human Kinetics.

Bleiberg, F. M., Brun, T. A., and Goihman, S. (1980). Duration of activities and energy expenditure of female farmers in dry and rainy season in Upper-Volta. *British Journal of Nutrition,* **43,** 71–82.

Brun, T., Bleiberg, F., and Goihman, S. (1981). Energy expenditure of male farmers in dry and rainy seasons in Upper-Volta. *British Journal of Nutrition,* **45,** 67–75.

Brun, T. A., Geissler, C. A., Mirbagheri, I., Hormozdiary, H., Bastani, J. and Heydayat, H. (1979). The energy expenditure of Iranian agricultural workers. *American Journal of Clinical Nutrition,* **32,** 2154–61.

Brunner, D., Manelis, G., Modan, M. and Levin, S. (1974). Physical activity at work and the incidence of myocardial infarction, angina pectoris, and death due to ischemic heart disease: an epidemiological study in Israeli collective settlements (kibbutzim). *Journal of Chronic Diseases,* **27,** 211–33.

Byers, T., Wolf, R. and Williamson, D. F. (1994). World-wide increases in body size during the twentieth century: global fattening? In *Nutrition in a Sustainable Environment,* eds. M. L. Wahlqvist, A. S. Truswell, R. Smith and P. J. Nestel, pp. 203–7. London: Smith-Gordon.

Casper, R. C., Schoeller, D. A., Kushner, R., Hnilicka, J. and Trainer Gold, S. (1991). Total daily energy expenditure and activity level in anorexia nervosa. *American Journal of Clinical Nutrition,* **53,** 1143–50.

Chave, S. P. W., Morris, J. N., Moss, S. and Semmence, A. M. (1978). Vigorous exercise in leisure time and the death rate: a study of male civil servants. *Journal of Epidemiology and Community Health,* **32,** 239–43.

Ching, P. L. Y. H., Willett, W. C., Rimm, E. B., Colditz, G. A., Gortmaker, S. L. and Stampfer, M. J. (1996). Activity level and risk of overweight in male health professionals. *American Journal of Public Health,* **86,** 25–30.

Cleave, J. H. (1970). Labour in the development of African agriculture: the evidence of farm surveys. Stanford University: Ph.D. Thesis.

Cumming, D. C., Wheeler, G. D. and Harber, V. J. (1994). Physical activity, nutrition, and reproduction. In *Human Reproductive Ecology. Interactions of Environment, Fertility, and Behavior,* eds. K. L. Campbell and J. W. Wood, pp.

55–76. New York: New York Academy of Sciences.

Dasgupta, B. (1977). *Village Society and Labour Use*. Delhi: Oxford University Press.

Department of Economic and Social Development Statistics. (1992). *United Nations Statistical Yearbook*. New York: United Nations.

Department of Economic and Social Development Statistics (1996). *United Nations Statistical Yearbook*. New York: United Nations.

Devadas, R. P., Anuradha, V. and Rani, A. J. (1975). Energy intake and expenditure of selected manual labourers. *Indian Journal of Nutrition and Dietetics*, 12, 279–84.

Diaz, E., Prentice, A. M., Goldberg, G. R., Murgatroyd, P. R. and Coward, W. A. (1991). Metabolic response to experimental overfeeding in lean and overweight healthy volunteers. *American Journal of Clinical Nutrition*, 56, 641–55.

DiPietro, L. (1995). Physical activity, body weight, and adiposity: an epidemiologic perspective. *Exercise and Sport Sciences Reviews*, 23, 275–303.

Dufour, D. L. (1984). The time and energy expenditure of indigenous women horticulturalists in the northwest Amazon. *American Journal of Physical Anthropology*, 65, 37–46.

Dufour, D. (1992). Nutritional ecology in the tropical rainforests of Amazonia. *American Journal of Human Biology*, 4, 197–207.

Dufour, D. (1997). Nutrition, activity, and health in children. *Annual Reviews in Anthropology*, 26, 541–65.

Durnin, J. V. G. A. (1992). Physical activity levels – past and present. In *Physical Activity and Health*, ed. N. Norgan, pp. 20–7. Cambridge: Cambridge University Press.

Durnin, J. V. G. A. and Passmore, R. (1967). *Energy, Work and Leisure*. London: Heinemann.

Edmundson, W. C. and Edmundson, S. A. (1988). Food intake and work allocation of male and female farmers in an impoverished Indian village. *British Journal of Nutrition*, 60, 433–9.

Forsum, E., Kabir, N., Sadurskis, A. and Westerterp, K. (1992). Total energy expenditure of healthy Swedish women during pregnancy and lactation. *American Journal of Clinical Nutrition*, 56, 334–42.

Fox, R. H. (1953). A study of the energy expenditure of Africans engaged in various activities, with special reference to some environmental and physiological factors which may influence the efficiency of their work. Ph.D. thesis: University of London.

Friedenreich, C. M. and Rohan, T. E. (1995). Physical activity and risk of breast cancer. *European Journal of Cancer Prevention*, 4, 145–51.

Galvin, K. A. (1985). Food procurement, diet, activities and nutrition of Ngisonyoka, Turkana pastoralists in an ecological and social context. Ph.D. thesis: State University of New York, Binghamton.

Garry, R. C., Passmore, R., Warnock, G. M., Durnin, J. V. G. A. (1955). *Expenditure of energy and consumption of food by miners and clerks, Fife, Scotland, 1955*. Medical Research Council Special Report Series No. 289. London: Her Majesty's Stationery Office.

Godin, G. and Shephard, R. J. (1973). Activity patterns of the Canadian Eskimo. In *Polar Human Biology*, eds. O. G. Edholm and E. K. E. Gunderson, pp. 193–215. Chichester: Heinemann Books.

Goldberg, G. R., Prentice, A. M., Coward, W. A., Davies, H. L., Murgatroyd, P. R., Sawyer, M., Ashford, J. and Black, A. E. (1991). Longitudinal assessment of the components of energy balance in well-nourished lactating women. *American Journal of Clinical Nutrition*, **54**, 788–98.

Goldberg, G. R., Prentice, A. M., Coward, W. A., Davies, H. L., Murgatroyd, P. R., Wensing, C., Black, A. E., Ashford, J. and Sawyer, M. (1993). Longitudinal assessment of energy expenditure in pregnancy by the doubly labeled water method. *American Journal of Clinical Nutrition*, **57**, 494–505.

Goran, M. I., Beer, W. H., Wolfe, R. R., Poehlman, E. T. and Young, V. R. (1993). Variation in total energy expenditure in young healthy free-living men. *Metabolism*, **42**, 487–96.

Gorsky, R. D. and Calloway, D. H. (1983). Activity pattern changes with decreases in food energy intake. *Human Biology*, **55**, 577–86.

de Guzman, P. E., Dominguez, S. R., Kalaw, J. M., Basconcillo, R. O. and Santos, V. F. (1974a). A study of the energy expenditure, dietary intake, and pattern of daily activity among various occupational groups. I. Laguna rice farmers. *Philippines Journal of Science*, **103**, 53–65.

de Guzman, P. E., Dominguez, S. R., Kalaw, J. M., Buning, M. N., Basconcillo, R. O., Santos, V. F. (1974b). A study of the energy expenditure, dietary intake and pattern of daily activity among various occupational groups. II. Marikina shoemakers and housewives. *Philippines Journal of Nutrition*, **27**, 21–30.

de Guzman, P. E., Kalaw, J. M., Tan, R. H., Recto, R. C., Basconcillo, R. O., Ferrer, V. T., Tombokon, M. S., Yuchingtat, G. P. and Gaurano, A. L. (1974c). A study of the energy expenditure, dietary intake and pattern of daily activity among various occupational groups. III. Urban jeepney drivers. *Philippines Journal of Nutrition*, **27**, 182–8.

de Guzman, P. E., Cabrera, J. P., Basconcillo, R. O., Gaurano, A. L., Yuchingtat, G. P., Tan, R. M., Kalaw, J. M., Recto, R. C. (1978). A study of the energy expenditure, dietary intake and pattern of daily activity among various occupational groups. V. Clerk-typists. *Philippines Journal of Nutrition*, **31**, 147–56.

de Guzman, P. E., Recto, R. C., Cabrera, J. P., Basconcillo, R. O., Gaurano, A. L., Yuchingtat, G. P., Abanto, Z. U. and Math BS (1979). A study of the energy expenditure, dietary intake and pattern of daily activity among various occupational groups. VI. Textile mill workers. *Philippines Journal of Nutrition*, **32**, 134–48.

Heini, A., Schutz, Y., Diaz, E., Prentice, A. M., Whitehead, R. G. and Jequier, E. (1991). Free-living energy expenditure measured by two independent techniques in pregnant and nonpregnant Gambian women. *American Journal of Physiology*, **261**, E9–17.

Helmrich, S. P., Ragland, D. R., Leung, R. W. and Paffenbarger, R. S. Jr. (1991). Physical activity and reduced occurrence of non-insulin-dependent diabetes mellitus. *New England Journal of Medicine*, **325**, 147–52.

Hill, K. , Kaplan, H. , Hawkes, K. and Hurtado, A. M. (1985). Men's time allocation to subsistence work among the Ache of Eastern Paraguay. *Human Ecology*, **13**, 29–47.

Hipsley, E. H. and Kirk, N. E. (1962). *Studies of dietary intake and the expenditure of energy by New Guineans*. Technical Paper No. 147. Noumea: South Pacific Commision.

Ho, Z. (1984). The energy expenditure of three categories of labourers in Southern China. In *Protein-Energy-Requirement Studies in Developing Countries: Results of International Research*, eds. W. M. Rand, R. Uauy and N. S. Scrimshaw, pp. 193–200. Tokyo: United Nations University.

Hurtado, A. M., Hawkes, K., Hill, K. and Kaplan, H. (1985). Female subsistence strategies among Ache hunter-gatherers of Eastern Paraguay. *Human Ecology*, **13**, 1–28.

Huss-Ashmore, R. A., Goodman, J. I., Sibiya, T. E. and Stein, T. P. (1984). Energy expenditure of young Swazi women as measured by the doubly-labelled water method. *European Journal of Clinical Nutrition*, **43**, 737–48.

James, W. P. T. and Schofield, E. C. (1990). *Human Energy Requirements. A Manual for Planners and Nutritionists*. Oxford: Oxford University Press.

James, W. P. T., Ferro-Luzzi, A. and Waterlow, J. C. (1988). Definition of chronic energy deficiency in adults. *European Journal of Clinical Nutrition*, **42**, 969–81.

Kaminsky, G. (1953). Untersuchungen beim holztransport mit schlitten im winterlichen hochgebirge. *Arbeitsphysiologie*, **15**, 47–56.

Kannel, W. B. and Sorlie, P. (1979). Some health benefits of physical activity: the Framingham study. *Archives of Internal Medicine*, **139**, 857–61.

Kannel, W. B., Belanger, A., D'Agostino, R. and Israel, I. (1986). Physical activity and physical demand on the job and risk of cardiovascular disease and death: the Framingham study. *American Heart Journal*, **112**, 820–5.

Kashiwazaki, H., Inaoka, T., Suzuki, T. and Kondo, Y. (1986). Correlations of pedometer readings with energy expenditure in workers during free-living daily activities. *European Journal of Applied Physiology*, **54**, 585–90.

Katzmarzyk, P. T. Leonard, W. R., Crawford, M. H., Sukernik, R. I. (1994). Resting metabolic rate and daily energy expenditure among two indigenous Siberian populations. *American Journal of Human Biology*, **6**, 719–30.

Kraus, J. F. and Conroy, C. (1984). Mortality and morbidity from injuries in sports and recreation. *Annual Review of Public Health*, **5**, 163–92.

Kusunoki, T. (1956). Forest labours in Japan. *Report of the Institute of Labour*, **49**, 23–5.

LaCroix, A. Z., Leveille, S. G., Hecht, J. A., Grothaus, L. C. and Wagner, E. H. (1996). Does walking decrease the risk of cardiovascular disease hospitalizations and death in older adults? *Journal of the American Geriatrics Society*, **44**, 113–20.

Lapidus, L. and Bengtsson, C. (1986). Socioeconomic factors and physical activity in relation to cardiovascular disease and death: a 12-year follow-up of participants in a population study of women in Gothenburg, Sweden. *British Heart Journal*, **55**, 295–301.

Lee, R. B. (1979). *The !Kung San: Men, Women and Work in a Foraging Society*. Cambridge: Cambridge University Press.

Leonard, W. R. (1988). The impact of seasonality on caloric requirements of human populations. *Human Ecology*, **16**, 343–6.

Lindsted, K. D., Tonstad, S. and Kuzma, J. W. (1991). Self-report of physical activity and patterns of mortality in Seventh-day Adventist men. *Journal of Clinical Epidemiology*, **44**, 355–64.

Livingstone, M. B. E., Stain, J. J., McKenna, P. G., Nevin, G. B., Barker, M. E., Hickey, R. J., Prentice, A. M., Coward, W. A., Ceesay, S. M. and Whitehead, R. G. (1990). Simultaneous measurement of free living energy expenditure by the double-labelled water method and heart rate monitoring. *American Journal of Clinical Nutrition*, **52**, 59–65.

Livingstone, M. B. E., Strain, J. J., Prentice, A. M., Coward, W. A., Nevin, G. B., Barker, M. E., Hickey, R. J., McKenna, P. G. and Whitehead, R. G. (1991). Potential contribution of leisure activity to the energy expenditure patterns of sedentary populations. *British Journal of Nutrition*, **65**, 145–55.

Lovelady, C. A., Meredith, C. N., McCrory, M. A., Nommsen, L. A., Joseph, L. J. and Dewey, K. G. (1993). Energy expenditure in lactating women: a comparison of doubly labeled water and heart-rate-monitoring methods. *American Journal of Clinical Nutrition*, **57**, 512–18.

Lundgren, N. P. V. (1946). Physiological effects of time schedule work on lumberworkers. *Acta Physiologica Scandinavica*, **13**, Supplement 41.

Malina, R. M. (1992). Physical activity and behavioural development during childhood and youth. In *Physical Activity and Health*, ed. N. G. Norgan, pp. 101–20. Cambridge: Cambridge University Press.

Manson, J. E., Nathan, D. M., Krolewski, A. S., Stampfer, M. J., Willett, W. C. and Hennekens, C. H. (1992). A prospective study of exercise and incidence of diabetes among U.S. male physicians. *Journal of the American Medical Association*, **268**, 63–7.

McNeill, G., Payne, P. R., Rivers, J. P. W., Enos, A. M. T., de Britto, J. J. and Mukarji, D. S. (1988). Socio-economic and seasonal patterns of adult energy nutrition in a South Indian village. *Ecology of Food and Nutrition*, **22**, 85–95.

Mela, D. J. (1996). Eating behaviour, food preferences and dietary intake in relation to obesity and bodyweight status. *Proceedings of the Nutrition Society*, **55**, 803–16.

Mela, D. J. and Catt, S. L. (1996). Ontogeny of human taste and smell preferences and their implications for food selection. In *Long-term Consequences of Early Environment. Growth, Development and the Lifespan Developmental Perspective*, eds. C. J. K. Henry and S. J. Ulijaszek, pp. 139–54. Cambridge: Cambridge University Press.

Meijer, G. A. L., Westerterp, K. R., van Hulsel, A. M. P. and ten Hoor, F. (1992). Physical activity and energy expenditure in lean and obese human subjects. *European Journal of Applied Physiology*, **65**, 525–8.

Montgomery, E. and Johnson, A. (1974). Machiguenga energy expenditure. *Ecology of Food and Nutrition*, **6**, 97–105.

Morren, G. E. B. (1977). From hunting to herding: pigs and the control of energy in montane New Guinea. In *Subsistence and Survival. Rural Ecology in the Pacific*, eds. T. P. Bayliss-Smith and R. G. Feachem, pp. 273–315. New York: Academic Press.

Morris, J. N., Heady, J. A., Raffle, P. A. B., Roberts, C. G. and Parks, J. W. (1953). Coronary heart disease and physical activity of work. *Lancet*, **2**, 1111–20.

Nielsen Media Research. (1990). *Nielsen Report on Television*. New York: Nielsen.

Newsholme, E. A. and Parry-Billings, M. (1994). Effects of exercise on the immune system. In *Physical Activity, Fitness, and Health: International Proceedings and Consensus Statement*, eds. C. Bouchard and R.J. Shephard, pp. 451–5. Champaign, IL: Human Kinetics.

Norgan, N. G. (ed.) (1992). *Physical Activity and Health*. Cambridge: Cambridge University Press.

Norgan, N. G. and Durnin, J. V. G. A. (1980). The effects of six weeks of overfeeding on the body weight, body composition and energy metabolism of young men. *American Journal of Clinical Nutrition*, 33, 978–88.

Norgan, N. G. and Ferro-Luzzi, A. (1978). Nutrition, physical activity and physical fitness in contrasting environments. In Nutrition, Physical Fitness, and Health, eds. J. Parizkova and V. A. Rogozkin, pp. 167–93. Baltimore: University Park Press.

Norgan, N. G., Ferro-Luzzi, A., and Durnin, J. V. G. A. (1974). The energy and nutrient intake and the energy expenditure of 204 New Guinean adults. *Philosophical Transactions of the Royal Society of London*, Series B, **268**, 309–48.

Nydon, J. and Thomas, R. B. (1989). Methodological procedures for analysing energy expenditure. In *Research Methods in Nutritional Anthropology*, eds. G. H. Pelto, P. J. Pelto and E. Messer, pp. 57–81. Tokyo: United Nations University.

Paffenbarger, R. S. Jr. and Hale, W. E. (1975). Work activity and coronary heart mortality. *New England Journal of Medicine*, **292**, 545–50.

Paffenbarger, R. S. Jr., Thorne, M. C. and Wing, A. L. (1968). Chronic disease in former college students: early precursors of fatal stroke. *American Journal of Public Health*, **57**, 1290–9.

Paffenbarger, R. S. Jr. and Williams, J. L. (1967). Chronic disease in former college students: early precursors of fatal stroke. *American Journal of Public Health*, **57**, 1290–9.

Panter-Brick, C. (1993). Seasonality of energy expenditure during pregnancy and lactation for rural Nepali women. *American Journal of Clinical Nutrition*, **57**, 620–8.

Pasquet, P. and Koppert, G. J. A. (1993). Activity patterns and energy expenditure in Cameroonian tropical forest populations. In *Tropical Forests, People and Food*, eds. C. M. Hladik, A. Hladik, O. F. Linares, H. Pagezy, A. Semple and M. Hadley, pp. 311–20. Paris: UNESCO.

Pasquet, P., Brigant, L., Froment, A., Koppert, G. A., Bard, D., de Garine, I. and Apfelbaum, M. (1992). Massive overfeeding and energy balance in men: the Guru Walla model. *American Journal of Clinical Nutrition*, **56**, 483–90.

Passmore, R., Thomson, J. G. and Warnock, G. M, (1952). Balance sheet of the estimation of energy intake and energy expenditure as measured by indirect calorimetry. *British Journal of Nutrition*, **6**, 253–64.

Pekkanen, J., Marti, B., Nissinen, A., Tuomilehto, J., Punsar, S. and Karvonen, M. J. (1987). Reduction of premature mortality by high physical activity: a 20-year follow-up of middle-aged Finnish men. *Lancet*, **1**, 1473–7.

Pearson, J. D. (1990). Estimation of energy expenditure in Western Samoa, American Samoa, and Honolulu by recall interviews and direct observation. *American Journal of Human Biology*, **2**, 313–26.

Pelletier, D. L. (1984). Diet, activity, and cardiovascular disease risk factors in Western Samoan men. Ph.D. thesis: Pennsylvania State University, University Park, Pennsylvania.

Prentice, A. M., Black, A. E., Coward, W. A., Davies, H. L., Goldberg, G. R., Ashford, J., Sawyer, M. and Whitehead, R. G. (1986). High levels of energy expenditure in obese women. *British Medical Journal*, **292**, 983–7.

Ramanamurthy, P. S. V. and Dakshayani, R. (1962). Energy intake and expenditure in stone cutters. *Indian Journal of Medical Research*, **50**, 804–9.

Rising, R., Harper, I. T., Fontvielle, A. M., Ferraro, R. T., Spraul, M. and Ravussin, E. (1994). Determinants of total daily energy expenditure: variability in physical activity. *American Journal of Clinical Nutrition*, **59**, 800–4.

Riumallo, J. A., Schoeller, D., Barrera, G., Gattas, V. and Uauy, R. (1989). Energy expenditure in underweight free-living adults: impact of energy supplementation as determined by doubly labeled water and indirect calorimetry. *American Journal of Clinical Nutrition*, **49**, 239–46.

Roberts, S. B., Heyman, M. B., Evans, W. J., Fuss, P., Tsay, R. and Young, V. R. (1991). Dietary energy requirements of young adult men, determined by using the doubly-labeled water method. *American Journal of Clinical Nutrition*, **54**, 499–505.

Rodriguez, B. L., Curb, J. D., Burchfiel, C. M., Abbott, R. D., Petrovitch, H., Masaki, K *et al.* (1994). Physical activity and 23-year incidence of coronary heart disease morbididty and mortality among middle-aged men: the Honolulu Heart Program. *Circulation*, **89**, 2540–4.

Rozin, P. (1987). Psychobiological perspectives on food preferences and avoidances. In *Food and Evolution. Toward a Theory of human Food Habits*, eds. M. Harris and E. B. Ross, pp. 181–206. Philadelphia: Temple University Press.

Salonen, J. T., Puska, P. and Tuomilehto, J. (1982). Physical activity and risk of myocardial infarction, cerebral stroke, and death: a longitudinal study in Eastern Finland. *American Journal of Epidemiology*, **115**, 526–37.

Schendel, D. E. (1989). Sex differences in factors associated with body fatness in Western Samoans. Ph.D. thesis: Pennsylvania State University, University Park, Pennsylvania.

Schoeller, D. A. and van Santen, E. (1982). Measurement of energy expenditure in humans by doubly labeled water method. *Journal of Applied Physiology*, **53**, 955–9.

Schutz, Y., Lechtig, A. and Bradfield, R. B. (1980). Energy expenditures and food intakes of lactating women in Guatemala. *American Journal of Clinical Nutrition*, **33**, 892–903.

Seale, J. L., Conway, J. M. and Canary, J. J. (1993). Seven-day validation of doubly labeled water method using indirect room calorimetry. *Journal of Applied Physiology*, **74**, 402–9.

Seale, J. L., Rumpler, W. V., Conway, J. M. and Miles, C. W. (1990). Comparison of doubly labeled water, intake-balance, and direct- and indirect-calorimetry methods for measuring energy expenditure in adult men. *American Journal of Clinical Nutrition*, **52**, 66–71.

Seccareccia, F. and Menotti, A. (1992). Physical activity, physical fitness, and mortality in a sample of middle-aged men followed up 25 years. *Journal of Sports Medicine and Physical Fitness*, **32**, 206–13.

Shephard, R. J. (1974). Work physiology and activity patterns of circumpolar Eskimos and Ainu. *Human Biology*, **46**, 263–94.

Sherman, S. E., D'Agostino, R. B., Cobb, J. L. and Kannel, W. B. (1994). Physical activity and mortality in women in the Framinham Heart Study. *American Heart Journal*, **128**, 879–84.

Singh, J., Prentice, A. M., Diaz, E., Coward, W. A., Ashford, J., Sawyer, M. and Whitehead, R. G. (1989). Energy expenditure of Gambian women during peak agricultural activity measured by the doubly labelled water methods. *British Journal of Nutrition*, **62**, 315–29.

278   S. J. Ulijaszek

Slattery, M. L., Jacobs, D. R. Jr. and Nichaman, M. Z. (1989). Leisure-time physical activity and coronary heart disease death: the U.S. Railroad Study. *Circulation*, **79**, 304–11.

Solstrom, A. (1993). Body fat during reproduction in a nutritional perspective. Studies in women and rats. Ph.D. thesis: Karolinska Institute, Stockholm.

Stein, T. P., Johnston, F. E. and Greiner, L. (1988). Energy expenditure and socioeconomic status in Guatemala as measured by the doubly labeled water method. *American Journal of Clinical Nutrition*, **47**, 196–200.

Sternfeld, B., Williams, C. S., Quesenberry, C. P., Satariano, W. A. and Sidney, S. (1993). Lifetime physical activity and incidence of breast cancer. *Medicine and Science in Sports and Exercise*, **25** (Supplement 5), S147.

Stini, W. A., Chen, Z. and Stein, P. (1994). Aging, bone loss, and the body mass index in Arizona retirees. *American Journal of Human Biology*, **6**, 43–50.

Stock, M. and Rothwell, N. (1982). *Obesity and Leaness*. London: John Libbey.

Streef, G. M., Gerritsen, A. G., Bol M (1959). Arbeidsfysiologisch onderzoek bij vellingswerk in de bosbouw. Med. Landb. Logesch. Wageningen **59**(14), 1–39.

Strickland, S. S., Tuffrey, V. T. with the assistance of Gurung, G. and Ulijaszek, S. J. (1995). *Form and Function. A Study of Nutrition, Adaptation and Social Inequality in Three Gurung Villages of the Nepal Himalayas*. London: Smith-Gordon and Company.

Stubbs, R. J., Ritz, P., Coward, W. A. and Prentice, A. M. (1995). Covert manipulation of the ratio of dietary fat to carbohydrate and energy density: effect on food intake and energy balance in free-living men eating ad libitum. *American Journal of Clinical Nutrition*, **62**, 330–7.

Suzuki, S. (1988). Villagers' daily life and the environment. In *Health Ecology in Indonesia*, ed. S. Suzuki, pp. 13–22. Tokyo: Gyosei Corporation.

Thomas, R. B. (1976). Energy flow at high altitude. In *Man in the Andes*, eds. P. T. Baker and M. A. Little, pp. 379–404. Stroudsburg, PA: Dowden, Hutchinson and Ross.

Tin-May-Than and Ba-Aye (1985). Energy intake and energy output of Burmese farmers at different seasons. *Human Nutrition: Clinical Nutrition*, **39C**: 7–15.

Torun, B., Davies, P. S. W., Livingstone, M. B. E., Paolisso, M., Sackett, R. and Spurr, G. B. (1996). Energy requirements and dietary energy recommendations for children and adolescents 1 to 18 years old. *European Journal of Clinical Nutrition*, **50**, Supplement 1, S37–81.

Tremblay, A., Despres, J.-P., Theriault, G., Fournier, G. and Bouchard, C. (1992). Overfeeding and energy expenditure in humans. *American Journal of Clinical Nutrition*, **56**, 857–62.

Trowell, H. C. and Burkitt, D. P. (1981). *Western Diseases: their Emergence and Prevention*. London: Edward Arnold.

Ulijaszek, S. J. (1992). Human energetics methods in biological anthropology. *Yearbook of Physical Anthropology*, **35**, 215–42.

Ulijaszek, S. J. (1995). *Human Energetics in Biological Anthropology*. Cambridge: Cambridge University Press.

Ulijaszek, S. J. (1996). Energetics, adaptation, and adaptability. *American Journal of Human Biology*, **8**, 169–82.

Ulijaszek, S. J. and Strickland, S. S. (1993). *Nutritional Anthropology. Prospects and Perspectives*. London: Smith-Gordon.

Vasquez-Velasquez, L. (1988). Energy expenditure and physical activity of malnourished Gambian infants. *Proceedings of the Nutrition Society*, **47**, 233–9.

Vaughan, L., Zurlo, F. and Ravussin, E. (1991). Aging and energy expenditure. *American Journal of Clinical Nutrition*, **53**, 821–5.

Verboeket-van de Venne, W. P. H. G., Westerterp, K. R. and Kester, A. D. M. (1993). Effect of the pattern of food intake on human energy metabolism. *British Journal of Nutrition*, **70**, 103–15.

Viteri, F. E., Torun, B., Garcia, J. C. and Herrera, E. (1971). Determining energy costs of agricultural activities by respirometer and energy balance techniques. *American Journal of Clinical Nutrition*, **24**, 1418–30.

Weiner, J. S. and Lourie, J. A. (eds.) (1981). *Practical Human Biology*, London: Academic Press.

Welle, S., Forbes, G. B., Statt, M., Barnard, R. R. and Amatruda, J. M. (1992). Energy expenditure under free-living conditions in normal-weight and overweight women. *American Journal of Clinical Nutrition*, **55**, 14–21.

Westerterp, K. R., Meijer, G. A. L., Janssen, E. M. E., Saris, W. H. M. and ten Hoor, F. (1992). Long-term effect of physical activity on energy balance and body composition. *British Journal of Nutrition*, **68**, 21–30.

Westertterp, K. R., Miejer, G. A., Saris, W. H., Soeters, P. B., Winants, Y. and ten Hoor, F. (1991). Physical activity and sleeping metabolic rate. *Medical Science and Sports Exercise*, **23**, 166–70.

Widdowson, E. M., Edholm, O. G. and McCance, R. A. (1954). The food intake and energy expenditure of cadets in training. *British Journal of Nutrition*, **8**, 147–55.

World Health Organisation (1985). *Energy and Protein Requirements*. Report of a Joint FAO/WHO/UNU Expert Consultation. World Health Organisation Technical Report Series 724. Rome: World Health Organisation.

14 *HIV transmission in urban environments: London and beyond*

M. PARKER

Editors' introduction

The transmission of the human immunodeficiency virus (HIV) and it's association with tuberculosis (TB) susceptibility has been highlighted by Di Ferdinando in chapter 5. In this chapter, Parker discusses another aspect of HIV transmission, that of sexual networks. Drawing upon fieldwork undertaken in London, Parker's observations show that sexual networks for potential HIV transmission in London cut across social class and across countries: while the majority of contacts were urban British, a significant number of urban United States (New York, San Francisco) and Australian (Sydney, Melbourne) individuals were also part of the network. The importance of the study of urban pathways discussed in the final chapter by Ulijaszek and Schell, and of global urban system given by Clark (chapter 3) is illustrated in this important case study.

Introduction

HIV infections, in common with other sexually transmitted infections, are not randomly distributed. In Europe and North America, for example, the majority of cases occur in large cities among men who have sex with other men. Recent figures published by the Public Health Laboratory Service for the United Kingdom illustrate this point: a total of 30,162 HIV cases were reported in September 1997 and 18,294 or 60.6% of these cases occurred among men who have sex with other men. The majority of HIV infections and AIDS cases in the UK also occurred in London. Indeed, the Thames regions of London accounted for 66% (19,924/30,162) of all HIV infections and 63% (9,265 out of 14,726) of all AIDS cases in the UK (PHLS AIDS and STD Centre 1997).

A large number of social, behavioral and biological factors have been cited to explain the social and geographical variation in the distribution of HIV infections and AIDS cases in countries such as the UK.[1] However,

280

the extent to which any one factor or set of factors sheds light on current or future patterns of transmission is poorly understood. This is not surprising in view of the practical and methodological difficulties of developing a biosocial perspective on HIV and AIDS. Connors and McGrath (1997) have usefully discussed some of these difficulties and this chapter attempts to overcome some of them in the course of investigating the social and behavioral aspects of HIV transmission. In particular, it presents social and sexual network data with a view to assessing the extent to which HIV will remain a predominantly urban phenomenon among men who have sex with other men in London.

Why study sexual networks?

One of the reasons for focusing on the study of sexual networks is that it is frequently asserted by social scientists, epidemiologists and mathematical biologists (such as Anderson *et al.* 1990; Anderson and May 1992; Catania *et al.* 1996; Gupta *et al.* 1989; Klovdahl 1985) that in order to acquire a detailed understanding of current and future patterns of HIV transmission it is necessary to analyse data on the number and rates of sexual partner change; the type of sexual activity; and the links or networks between sexual contacts. To date, most research has been concerned with monitoring the number and rates of sexual partner change and the type of sexual activity. There has been very little empirical research on the network of contacts. This is unfortunate as the structure of networks influences the individual risks for acquiring and transmitting infection and, therefore, the total number of people that will be affected. Thus an individual who is HIV negative and part of a network where the prevalence of HIV is low will have a much lower risk of acquiring HIV infection – even if s/he regularly has unprotected anal sex – than an individual who is HIV negative and has a similar amount of unprotected anal sex with a similar number of contacts, but is part of a network where the prevalence of HIV is high.

To date, the small number of researchers who have investigated sexual networks and the transmission of HIV in the Western world have either analysed egocentric network data[2] and/or undertaken their research in an area where the prevalence of HIV is low (e.g. Coxon 1995; Haraldsdottir *et al.* 1992; Klovdahl *et al.* 1994; Laumann *et al.* 1989; Service and Blower 1995; Woodhouse *et al.* 1994). The network data presented in this chapter is altogether different. It is based upon research with predominantly male participants who have direct and indirect sexual links with each other, and all the participants live in London which, as previously mentioned, is the city where the majority of HIV infections and AIDS cases occur in the United Kingdom. Unlike previous research on sexual networks these data are also interpreted in the light of information and ideas about a wide range of social, economic and political issues affecting the lives of study participants.

Fieldwork site and methods

Fieldwork took place between January 1994 and March 1996. It was carried out in a variety of places including a clinic for sexually transmitted diseases in London, various public venues and the flats and houses of study participants. Two methods were employed to recruit study participants: clinic referrals and snowball sampling. With respect to the latter, participants who had been recruited through the clinic were asked to contact current and past sexual contacts with a view to asking them whether they would be willing for me to contact them, and to participate in the study. Wherever possible, these people were interviewed and they, in turn, were asked to contact past and present sexual partners as well as friends with a view to developing an understanding of their social and sexual networks in London.

Information elicited from all participants

Social, demographic, behavioral and biomedical information was elicited from all participants. This included information about their age, nationality, residence, education, current occupation and current income. It also included information about the participant's sexual history, sexual health and intravenous drug use (see Parker *et al.* 1998 for details). Wherever possible, social, demographic, behavioral and biomedical information was also elicited about each sexual contact mentioned. This included information about each contact's age, nationality, sexual health, ethnicity and the number of sexual contacts' the contacts had had during this 12-month period.

Interviews lasted from one to four hours and the majority of them were tape recorded. Information was elicited by asking open-ended questions rather than following the structured format of a questionnaire, as this seemed the most appropriate way to elicit detailed and accurate information about intimate aspects of their lives. Thus, while I tried to acquire the same information about participants' sexual history, over similar periods of time, the open-ended interviews also enabled detailed discussion, from the participant's perspective, of issues such as, why they have sex with the numbers and types of people they do; why a participant has unprotected sex on one occasion and not another; and how, if at all, their understanding of the risks for acquiring and transmitting HIV infection influenced their actual sexual behavior.

Identifying sexual networks in London

The sexual network presented in this chapter is concerned with sexual links between participants and their contacts for the year August 1994 to July 1995. The identification of this network initially began by investigating the transmission of gonorrhoea but it proved impossible to secure the necessary participation to identify networks facilitating the transmission of gonorrhoea, that is, sexual contacts in the previous three months. It was, however, possible to secure the participation of several people who had had sex with the index patient over the longer time period of 12 months. The identification of this network occurred in the following way: an Australian 20 year old man (participant 1)[3] attended a clinic for sexually transmitted diseases in London to be treated for gonorrhoea in August 1994. He also reported being HIV positive and thought he might have acquired this recent episode of gonorrhoea from one of four possible sexual contacts.

Coincidentally, the four sexual contacts mentioned were also reported to be HIV positive (see Figure 14.1). He was asked to inform all four contacts of his recent infection; to advise them to be screened for gonorrhoea at a clinic for sexually transmitted diseases; and to ask them whether they would be willing to participate in research investigating sexual and social networks.

None of them came to the clinic to be screened for gonorrhoea and they all declined the offer to participate in the research. However, one of these four contacts subsequently attended the clinic to be treated for gonorrhoea in October 1994. He had probably acquired this infection from a different source and after considerable discussion he agreed to participate in the research investigating sexual networks. During the first interview this participant (participant 2) mentioned that he had had four sexual contacts since 31 July 1994 (see Figure 14.2).

The first participant agreed to additional interviews and these took place at two-monthly intervals. This enabled a detailed picture of his sexual links in the years prior to his recent diagnosis for gonorrhoea to be acquired. It also enabled a detailed account to be kept of the 'new' sexual contacts he had had since the previous interview. By August 1995 he reported having had 33 sexual contacts between 1 August 1994 and 31 July 1995 (see Figure 14.3). Eight months and four interviews after first meeting this participant he secured the participation of three friends with whom he had had indirect sexual links as well as one current sexual contact. Interviews were undertaken at their flats and the links established are depicted in Figure 14.4.

During this time, the second participant was interviewed at regular intervals. Eight months and three interviews after our first meeting he secured the participation of his long-term sexual partner. The sexual

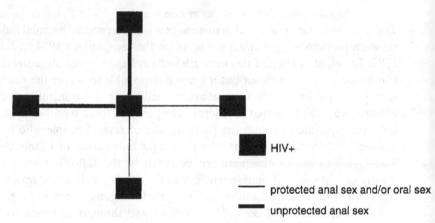

Figure 14.1   The sexual contacts reported by participant 1, Aug. 1994

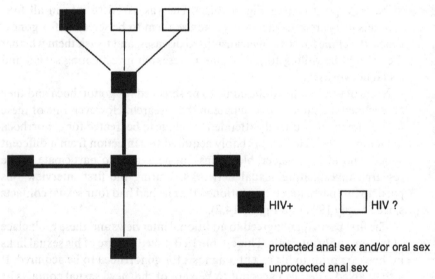

Figure 14.2   Reported sexual links and HIV status, 2 participants, Aug.
1994– Oct. 1994

contacts of participant 2's long-term sexual partner are also depicted in
Figure 14.4. Unfortunately, it was not possible to interview participant 2's
other contacts as they were all resident overseas. However, it was possible
to secure the participation of one of participant 1's sexual contacts through
the clinic referral system. This contact is HIV positive and attended the
clinic to be treated for gonorrhoea in October 1995. In addition, two out of
four of participant 1's friends and contacts persuaded five of their past and

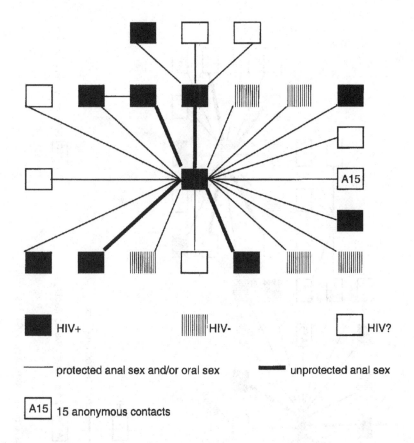

HIV+   ‖‖‖‖‖ HIV-   ☐ HIV?

——— protected anal sex and/or oral sex   ▬▬▬ unprotected anal sex

A15 15 anonymous contacts

Figure 14.3   Reported sexual links and HIV status by 2 participants, Aug.
1994 – July 1995

current sexual contacts to participate in the study. They, in turn, recruited
a further four sexual contacts to the study. During the last month of data
collection, two out of four of these contacts each recruited a further
contact to the study.

The sexual links between all 19 study participants are presented in
Figure 14.5. It took a great deal of time, flexibility and persistence to
develop this network but the study gathered momentum after a slow start.
To some extent, this is reflected in the fact that my identity shifted among
the majority of study participants from "Melissa, the anthropology coun-
sellor" to "Auntie Melissa" and "petal." Indeed, one male prostitute who
was recruited to the study a few months before the end of data collection
turned up to his first interview with a detailed list of the occupations of his

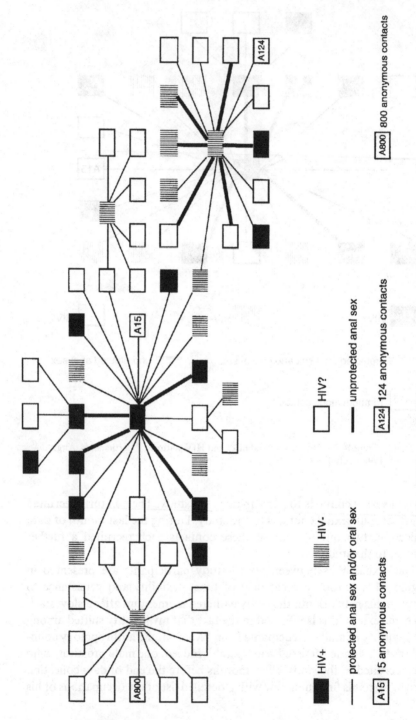

HIV+   HIV-   HIV?

—— protected anal sex and/or oral sex

—— unprotected anal sex

A15   15 anonymous contacts   A124   124 anonymous contacts   A800   800 anonymous contacts

Figure 14.4   Reported sexual links and HIV status, 6 participants, Aug. 1994 – July 1995.

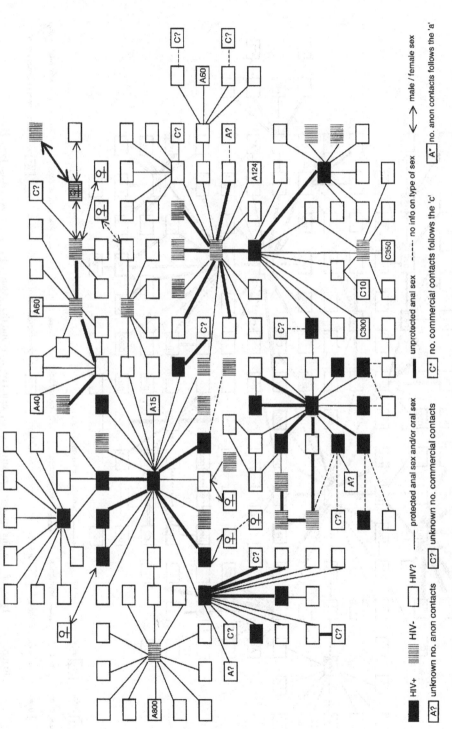

Figure 14.5 Sexual contacts and the sexual contacts' contacts, 19 participants, Aug. 1994–July 1995

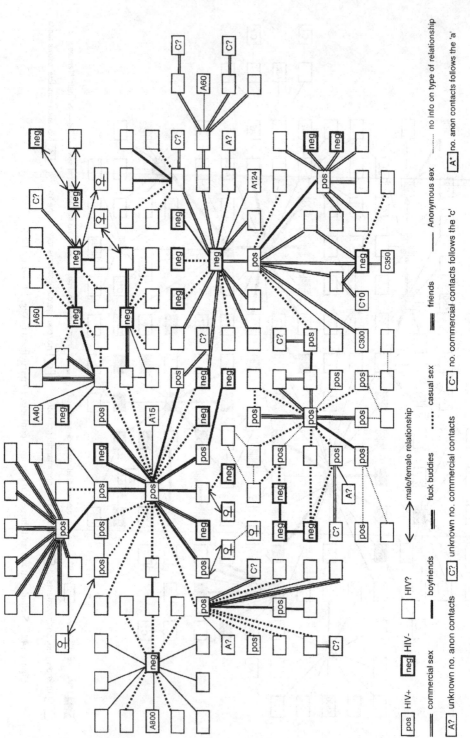

Figure 14.6 Types of relationships between participants, sexual contacts and the sexual contacts' contacts. Aug. 1994–July 1995

commercial contacts saying that he had no doubts about entrusting me with this information as his friend 'Simon' said I was easy to talk to and trustworthy and that was enough for him (see Table 14.1).

Results

By March 1996, 1 woman and 18 men had been recruited to the study and they were all linked to each other through a variety of complex sexual links. A total of 53 interviews were undertaken with these 19 participants and the number of interviews ranged from 1 to 10 per person.

Types of sexual relationships

Figure 14.6 shows the type of sexual relationships linking men rather than the type of sex. Classifying the type of sexual relationships men have with other men is not easy as a wide range of terms are used by participants to describe different types of sexual relationships. These include "boyfriend", "sex friend," "fuck buddy," "a cas [casual] shag," "a quickie," "a bit of a rummage," "trade," "a drunken pick-up," "rent," "a one-off" and so on. For the purposes of this chapter six types of sexual relationship have been identified. These are: "boyfriend," "fuck buddy," "casual," "friend," "anonymous," and "commercial."

Different participants attribute different meanings to these terms but, for the purposes of data analysis, the term "boyfriend" refers to a sexual relationship where the participant has or intends to have protracted sexual contact with the person concerned. The men involved in this type of partnership may or may not live with each other and, over time, they acquire an understanding and/or become a part of each other's social worlds. "Fuck buddies," by contrast, are defined as men who have sex with other men over a protracted period of time (at least six months). The sex is occasional and the men concerned are not part of each other's social worlds and have no intention of becoming a part of each others social worlds.

"Casual" contacts are defined as sexual contacts where the men concerned know or acquire at least some social, demographic or biomedical information about the person concerned (before or after sex). Sex may occur once or several times but the sexual relationship always fizzles out after a few meetings. This type of sexual relationship contrasts with those described by participants as "friends." Here the "friendship" is longstanding (more than six months) and the sex occasional. It is not a central feature of the relationship.

Table 14.1. *The occupations of a male prostitute's commercial contacts*

Judge	Electrical engineer
Barrister	computer software designer
Solicitor	BT engineer
	British Gas engineer
Ambulance driver	Florist
Nurse	Market stall holder
Doctor	Electricity Power Station safety manager
Dentist	
Surgeon	Agronomist
Psychologist	Architect
Psychiatrist	Builder
Social worker	Plumber
Emergency services coordinator	Electrician
Fireman	Glazier
Soldier	Roofer
Sailor (Merchant Navy)	Decorator
Customs and Excise officer	Gardener
	Warehouseman
MP	Footwear designer
Civil Servant	Facilities manager
Librarian	Dealer
Council worker	Mechanic
Refuse collector	Rugby coach
Post Office worker	Hairdresser
Postman	Masseur
Head Master	Security guard
Teacher	
Lecturer	Truck driver
Invigilator/examiner	Taxi driver
Pub landlord	Bus driver
Club manager	Train driver
Hotelier	Chauffeur
Cinema manager	
Album cover designer	Head hunter
Syndicator	Estate agent
	Property developer
Theatre box office clerk	Credit controller
Theatre critic	Insurance salesman
Artist/Sculptor	Financier
Actor	Underwriter
Singer	Accountant
Composer	Company director
Conductor	Director of finance
Musician	PA
Author	Secretary
Publisher	UK Company General manager
Proofreader	
Video tape editor	Student
Assistant film producer	Unemployed
FX Specialist	HIV social worker
Journalist	
Newspaper editor	

"Anonymous" contacts are defined as sexual contacts that have been recruited in public places. The sex usually occurs once and takes place in a public space such as a sauna, the backroom of a pub or club, by a railway line, etc. In addition, the participant acquires little, if any, social, demographic or biomedical information about the contact concerned. Finally, commercial contacts involve the exchange of sex for money. The recruitment of commercial contacts occur through an escort agency, a public place, a "whore house" or a private advertisement.

Two things stand out from the depiction of sexual relationships presented in Figure 14.6: first, all participants reported having two or more different types of sexual relationship in the course of the year. Second, participants reported having a large number of anonymous contacts. That is, 1,378 sexual contacts were mentioned, of which 1,224 (91.5%) were untraceable as they either involved anonymous, one-off sexual encounters in public places (such as saunas, parks, "backrooms," cemeteries and public toilets) or they involved one-off commercial encounters where the prostitute or client concerned did not have information enabling their contact to be traced.

Social, demographic, biomedical and behavioral information about the network

Social, demographic, biomedical and behavioral information elicited from study participants about the remaining network of 154 sexual contacts may be summarised as follows: 147 contacts were men and 7 contacts were women; these contacts ranged in age from 19 to 53 years; the average age was 30 years; and the majority of the participant's sexual contacts (113) were resident in London. A small number of contacts (3) were passing through London on holiday or business while 38 sexual contacts were made by four participants overseas.

With respect to nationality, the majority of the participants' contacts (87) were British but sexual contact was also made with men from a further 21 countries. That is, 19 contacts were from USA and 6 contacts were from Australia; 3 contacts were from Ireland and Brazil respectively; and a further 2 contacts were from Greece, Malaysia, Portugal and Spain respectively. Participants also reported having one contact from Bosnia, Columbia, Croatia, Holland, Israel, Italy, Lebanon, Madagascar, South Africa, Sweden, New Zealand and three contacts were reported to have the dual nationalities of Argentina/Italy, Brazil/Italy, France/Cameroon. Participants did not know the nationality of 32 of their contacts.

Age and nationality aside, participants acquired relatively little information about their contacts. Nevertheless, on the basis of available information, it is apparent that their contacts were engaged in a wide range of activity and employment. That is, 60 contacts were reported to be in full or part-time work and their occupations included the following: solicitor, journalist, university lecturer, priest, merchant banker, green-grocer, air steward, chef, shop assistant, nurse, opera singer, manager of a gay club. A further 10 contacts were students, 19 contacts were either selling sex or selling sex in addition to undertaking other income-generating activities; and 11 contacts were out of work. There was no information about 54 of the 154 contacts.

Participants knew much less about the sexual health of their contacts. Thus 23 men in this network were reported to be HIV positive, 23 people were reported to be HIV negative and there was no information about the HIV status of a further 108 men (excluding anonymous contacts). Contacts and participants were a maximum of five steps away from reported HIV infection and the majority of those who were HIV negative or did not know their status did not realise how close they were to HIV infection.

Discussion

The information presented in the preceding section is detailed and the next part of this chapter discusses some of the most important findings in relation to the following five questions: is there any evidence to suggest there is on-going transmission of HIV among men who have sex with other men in London? What type of sexual links exist between men who reported being HIV positive and men and women who reported being HIV negative? What are the implications of these data for the transmission of HIV in other cities in the United Kingdom as well as large metropolitan centres in the USA, Australia, Europe and South America? How, if at all, can insights emerging from social and sexual network research be harnessed to inform culturally appropriate preventive programs? To what extent are endeavors to design culturally appropriate preventive programs impeded by the fact that political and economic factors continue to promote the transmission of HIV in London?

On-going transmission of HIV in London

First, it is likely that there is considerable on-going transmission of HIV in London as risky sexual behaviors continue to be practiced by those who are HIV positive, HIV negative as well as those who do not know their status. That is, the majority of participants reported having had unprotected anal sex between August 1994 and July 1995 and this occurred in the context of a variety of relationships. Five cases are described to illustrate the range and complexity of contacts involving unprotected anal sex.

Unprotected anal sex and commercial relationships

Case study 1

"John" is a 24-year-old, white South African. He is HIV positive and seeking permanent residence from the Home Office. At our first meeting he said that he ran a small business and that he had had four anonymous contacts in a park close to his flat over the last few months. During the fourth interview he mentioned the fact that he had heard a friend and former sexual contact talking about me in a positive way and that he no longer thought that I was a spy from the Home Office.

Not surprisingly, therefore, he felt much more able to tell me about some of the things that were going on in his life. In particular, he mentioned that he often sold sex at a "whore house" in West London and that he also recruited clients by advertising in the gay press. "John" resented the fact that the Home Office was taking so long to process his application as, among other things, this meant that he was not entitled to a work permit and could not get salaried work. Moreover, he was not entitled to claim any benefits other than a disability allowance which was worth £70 a week. This was not enough to live on and he thus felt that the "easiest" and "obvious" way to resolve pressing economic difficulties was to sell sex. Reflecting on this aspect of his life, he said: "if I infect a lot of people, its not my fault. I tried to get help from a lot of people [lawyers, psychiatrists, doctors]...I didn't get it...what can I do?...I can't help it if I infect others." He also reported having unprotected anal sex in a variety of other contexts. This included unprotected anal sex with a "sort of boyfriend" with whom he had had an "on-off" relationship over the past few years. He assumed his boyfriend knew that he was HIV positive just as he assumed that his boyfriend was HIV positive.

The same informant also described having protected and unprotected anal sex with several "casual" contacts. One of these sexual contacts was a 29-year-old Irish man who lived in London. They met at a bar in Soho and had had sex at his flat later that day. "John" did not tell him that he was HIV positive and he did not ask the Irish man whether he was HIV positive too. Reflecting on the fact that a condom had split while he had had active anal sex with him he said: "maybe I passed the virus to this man and he will pass the virus to his boyfriend [as he knew that he had unprotected anal sex with him] but that's life...its not my fault."

Case study 2

A 42-year-old French man said at his first interview that he was HIV negative. He lives on his own in a one-bedroom basement flat in south London and relies upon income that he inherited from a relative to cover his day-to-day living expenses. He had a six-year relationship with another man which ended a few years ago and since then he has had "anonymous", "casual" and "commercial" relationships. During 1994–95 he reported having oral sex with approximately 110 men at a near-by sauna and a further 14 men at saunas in Chicago and New York, USA. He also reported having unprotected passive anal sex in the context of casual and commercial relationships. He made contact with the majority of these casual contacts at his local sauna in London and then had sex with them at his flat later that day.

The commercial contacts involved purchasing sex at a whore house or responding to adverts in the gay press. He said he bought sex "because its sex and its a lot easier than going to the sauna and, you know, hanging around for four or five or six hours... only to have a quick wank with someone...." There were several occasions during this first interview when he talked openly about his desire to be HIV positive and attributed this to the fact that he finds "the endlessness of being alive very dreary and daunting... too many possibilities and I'm so lazy." And he went on to say:

> maybe this whole thing of wanting to become HIV [positive] is [to do with] being able to control my own dying. I mean I can't engage with the real world terribly well... like in employment and stuff, I've never really worked for anyone... I'm not accustomed to having superiors... and I cannot go out there in the world... the idea of sort of having to do that when my money runs out frightens me... I mean if I had got a [positive] diagnosis last year I know exactly what my ten year plan would have been... it would have been much easier for me in terms of my ability to structure my life.

Unprotected anal sex between men who are HIV positive

Case study 3

"Ryan" is 39-years-old, white Welsh and out of work. He reported being HIV positive and described having protected anal sex with his 25-year-old, black African "boyfriend." His boyfriend "Gerry" is married with two children and has never been tested for HIV. "Ryan" also described having unprotected anal sex with several "casual" contacts between August 1994 and July 1995. This included a 33-year-old, white English man who was reported to be HIV positive and living with a long-term partner who was

HIV negative. They met at a week-end retreat for those interested in complementary therapies. And they were both concerned about having unprotected anal sex with each other as they felt that so long as there was any doubt about the wisdom of being exposed to multiple viral infections then it was better to avoid additional infections. However, on this occasion they thought "what the hell ... and it was very nice too."

During previous interviews at his one-bedroom flat on a council estate in East London "Ryan" always emphasised the fact that he revealed his HIV status before having unprotected anal sex with "casual" contacts. He also commented on the fact that "casual" contacts who reported being HIV positive rarely revealed their HIV status until after they had actually had sex with him. Interviews with two of his "casual" contacts confirmed this observation. They had had unprotected anal sex with "Ryan" but not revealed their HIV status until after the event.

Unprotected anal sex between "boyfriends" who do not both know their HIV status

Case study 4

"Dave" and "Steve" both reported regularly having unprotected anal sex with each other during their interviews. "Dave" is a white English, 27-year-old university graduate. He works for a housing association and earns about £20,000 a year. He was interviewed twice at his one-bedroom flat on a council estate. During these interviews he reported having about twenty "quickies" (i.e. anonymous contacts) between August 1994 and July 1995. These "quickies" generally occurred at various pubs and clubs in London as well as cruising areas such as the side of a railway line in north London, Hampstead Heath and a cemetery close to his home.

It is likely that he underestimated the total number of sexual contacts he had had due, in part, to the fact that he did not experience these one-off sexual encounters as "memorable." Indeed, his response to a question enquiring about the total number of contacts he had had during the year 1994–5 was: "Blimey ... that's like asking me to count drops of rain!" He did, however, talk openly about his concerns with having unprotected anal sex with his boyfriend "Steve." He had had an HIV test a few months before he became involved with "Steve" and the result was negative. But he was now unsure of his status as a friend of "Steve" had told him that "Steve" often had unprotected anal sex with other men.

"Steve" is 27 years old, white English and left school without any qualifications at the age of sixteen. He is from Sunderland, rents a flat with a male, gay-identified friend and earns about £10,000 a year as a musician. Between 1994 and 1995 he reported having "about 60" anonymous contacts in public sex environments as well as unprotected anal sex with a former boyfriend; and sex with two women (involving protected and unprotected vaginal sex respectively). "Steve" has never been tested for HIV as he is "frightened" of getting a positive result. Reflecting on the fact that he regularly has unprotected anal sex with "Dave" he said:

> We were actually using condoms for a short while and one time one of them broke and he came inside me and it just carried on from that point... we kept on saying "this is really naughty... we should not be doing this" ... but we just keep doing it. But most of the time we are safe to a certain extent... I mean he withdraws before he comes or I withdraw just before I come.

The sexual relationship between "Dave" and "Steve" lasted about two years and they split up in 1996 as "Dave" felt that "Steve" was drinking too much, taking too many drugs and having sex with too many men. He was particularly concerned by the fact that "Steve" was having sex with his former boyfriend.

Case study 5

"Mark" is a 23-year-old, white English man from Newcastle. He has sold sex since he was 17 years old and has a boyfriend that he has lived with for the past four years. He also reported having five non-commercial, anonymous contacts at pubs and clubs in London as well as an anonymous contact involving oral sex on a train in Scotland. When I first met "Mark" I assumed that he was HIV negative as I had been told by one of his "regular" clients that he was HIV negative and that he regularly attended a clinic for male prostitutes where he was screened for sexually transmitted infections. During our second interview at his flat in north London it became apparent that he had never been tested for HIV but that he told all his clients that he was HIV negative as he did not wish to lose "trade" or to frighten them unnecessarily. He also mentioned that he had unprotected anal sex with his boyfriend and some of his clients.

The third interview occurred a month later. It had been cancelled several times as he had had a serious bout of 'flu but he clearly wanted to see me as he telephoned me several times at home to arrange and confirm the interview. The sense of urgency in his voice indicated that something was troubling him and it subsequently transpired that he had just had his first HIV test. The result was positive.

"Mark's" boyfriend, "Andy", is a 35-year-old, white English man from the south west of England. He runs the escort agency that "Mark" first worked for when he came to London and he too sells sex. Unlike "Mark", he did not have any non-commercial, anonymous contacts between August 1994 and July 1995. During our first interview he reported testing HIV positive ten years ago and said that he felt his status was beginning to affect his physique. He went on to say that he used to think of himself as a "normal hooker" but due to the deleterious affects of HIV on his body he now finds that he has not got the confidence to take on new clients. To quote:

> I feel like I'm out of shape now... I find it difficult to explain myself [to new clients] on the phone without making myself sound muscular [an allusion to how his body used to be] but then I worry 'cos I know they're going to come and find something different... I feel confident [with regular clients] because I know they know me and I assume they like me... because otherwise they wouldn't come back... I mean they wouldn't come back just for sex...

"Andy" reported having 8–10 commercial contacts. These contacts involved selling oral sex and/or protected anal sex to clients he had known for a minimum of five years. The relatively small number of clients means that he now supplements his income by working as a painter and decorator.

The second interview with "Andy" took place shortly after "Mark" had tested HIV positive. He spoke about the fact that they had had unprotected anal sex throughout their seven-year relationship and that this had been their "secret." Now that "Mark" had sero-converted he felt "guilty" and knew that people would criticise him for "signing his death warrant." But he went on to highlight the complexity of the situation. In particular, he mentioned the fact that "Mark" had always known that he was HIV positive and had consented to having unprotected anal sex with him. But he also questioned the meaning of consent. There was, after all, a considerable age difference between them and he was also troubled by the idea that "Mark" may have sensed how much it meant to him to "strip off the latex" when he was infected with such "a fatal and stigmatising virus."

"Andy" also drew my attention to the fact that when he had first been diagnosed with HIV he had been told that he only had a few years to live, whereas now he was being told that he could expect to live for many more years. Due to shifts in medical opinion he had stopped worrying about how much longer he would live as he could be run over by a bus tomorrow or develop lung cancer and die from that instead. This, he felt, had been "Mark's" attitude too.

Nevertheless, it was clearly important to "Andy" to think that "Mark" had not had unprotected anal sex with anyone else. He felt that "cross infections" would speed up the deterioration in his health and now that "Mark" had sero-converted he was anxious to know whether "Mark" was "telling the truth" about not having had unprotected anal sex with other men. It was, above all, a fraught, emotional and extremely difficult situation for everyone concerned.

The above accounts illustrate the different types of sexual relationships that participants had during 1994–5 and they suggest there is on-going transmission of HIV in London. These findings are particularly interesting when interpreted in the light of a large body of data recently presented by the Unlinked Anonymous HIV Surveys Steering Group (1996). This group monitors the spread of HIV infection in England and Wales and the report suggests there is on-going transmission of HIV, particularly among men who have sex with other men. However, the report does not provide any information about the context in which unprotected anal sex occurs. The case studies presented above are thus particularly useful as they draw attention to the variety of contexts in which unprotected anal sex occurs. They also help to identify pathways of transmission and to investigate the sexual links between men who have sex with other men in London as well as other cities in the UK, Europe and the US. This brings us to the second question addressed in this discussion: what type of sexual links exist between men who reported being HIV positive and men and women who reported or were reported to be HIV negative?

Identifying pathways of transmission: sexual links between men who reported being HIV positive and men or women who were reported to be HIV negative

The sexual network presented in Figure 14.5 includes 19 men who sell sex and at least 10 of these men were reported to be HIV positive. Interviews with 6 of these men and 2 of their clients revealed that a considerable number of men who pay for sex with other men also had non-commercial sex with women. Precise figures are difficult to acquire but, without exception, the men selling sex in this network estimated that 40–50% of their clients fell into this bracket. The following quote from an interview with a male prostitute describes the type of men involved:

> lunch time trade is mainly the married business man...you know, men who either thought they were gay but wanted to do the right thing [by marrying a woman], or men who thought they were gay but wanted kids, or men who thought they were straight, got married and then suddenly had leanings...they're all suited gentlemen,...people who take a short break from work.

The type of sex reported by prostitutes and clients participating in this network study varied a great deal but, without exception, all the male prostitutes said they had sold unprotected anal sex at least once over an

18-month period (August 1994 – January 1996); and they all commented on the fact that many of their friends sold unprotected anal sex too.

One male prostitute, reflecting on the fact that he rarely sold unprotected anal sex said: "very, very rarely do I get screwed . . . umm . . . I screw very few people but . . . umm . . . to my shame there are a few that over the years I've had unprotected sex with." In common with "Mark" (case study 4) he went on to say that he does not tell clients that he is HIV positive as he would lose "trade" and that it would scare them unnecessarily. Moreover, he could not be sure who they would tell and it would increase the chances of the police tracking him down.

Another male prostitute commented on the fact that he sometimes sold unprotected anal sex as it was the most lucrative sex to sell. He went on to say that he knew a number of other male prostitutes who were also willing to sell unprotected anal sex for more money. Together, these data suggest that male prostitutes may be an important bridge between gay-identified and heterosexually-identified populations.

Will HIV infections and AIDS cases continue to occur among men who have sex with other men and reside in London?

It is difficult to assess the extent to which new HIV infections and AIDS cases in the UK will continue to occur among men who have sex with other men and reside in London. There are, however, several indications to suggest that new HIV infections and AIDS cases will continue to be concentrated in London, at least for the next few years. These include the following: London has the largest concentration of gay pubs, clubs and saunas in the UK and while there are other cities in the UK that have gay pubs and clubs (notably Manchester, Glasgow and, to a lesser extent, places such as Bristol, Brighton, Liverpool, Newcastle, Oxford and Sheffield), there is little doubt in the minds of participants that London is the place to be if you want to be at the heart of the "gay scene."

Many of these pubs, clubs and saunas have opened up "backrooms" or "darkrooms" in the last few years. Backrooms vary a great deal from each other but, without exception, they all provide a space for men to have sex with other men in the presence of other men. There is little, if any, light and it is usually not possible to see much more than the outline of the man or men one is having sex with. Several informants felt these places provided an ideal environment for the infection to be transmitted. In particular, they talked about the occurrence of unprotected anal sex, and the fact that some venues were particularly well attended by men who were HIV positive whereas other venues attracted men who were HIV positive and HIV negative (see Parker forthcoming). It is, of course, impossible to enumerate the extent to which these venues facilitate the transmission of HIV as detailed research documenting the age, ethnicity, nationality, residence, sexual health, sexual activity, etc. of men going to these venues has not been undertaken. Preliminary research suggests, however, that many are not widely advertised (due to their illegality) and the majority of men probably reside in London.

Moreover, a report published by Kelley et al. (1996) suggests that the majority of gay-identified men who travel from other parts of the UK to London for an evening out are likely to go to the pubs and clubs in central London, and many of these venues do not have backrooms. In other words, it seems reasonable to speculate that these men will not play a major role in transmitting HIV to other parts of the UK.

It is likely, however, that the future transmission of HIV in London and other parts of the UK will be greatly influenced by global trends in HIV-1 infection. A recent report by the Communicable Disease Surveillance Centre (1996) suggests that the epidemic of HIV-1 in Africa has been most important to England and Wales and data documenting rising levels of infection among intravenous drug users in Europe (Des Jarlais et al. 1996; European Centre for the epidemiological monitoring of AIDS 1995) may be influential in the future. The social and sexual network data presented in this chapter also suggests that British men resident in London have sexual links with men from many other countries in Europe, South America and Africa as well as the USA. It is thus possible that infection could be transmitted from London to some of these countries and vice versa.

Network research as a tool of prevention

It is increasingly suggested that anthropological research investigating sexual networks can usefully inform strategies seeking to prevent HIV transmission (e.g. Klovdahl 1994; Trotter et al. 1995 and Neagius et al. 1994). The rationale for this view can be illustrated with reference to Figure 14.7. This figure has been constructed from interviews with participants A and C and many epidemiologists, public health practitioners and social scientists looking at this figure would say that participants B and C are at considerable risk of acquiring and transmitting infection from A. After all, it would only take a condom to split while B was having anal sex with A and for C to subsequently have unprotected anal sex with B, for C to then be in a position to infect at least two other men.

They would, no doubt, go on to suggest that network research could be used to target individuals at risk of acquiring and transmitting HIV infection as a change in the behaviour of B and C could protect a considerable number of men. This type of reasoning is seductive and, ultimately,

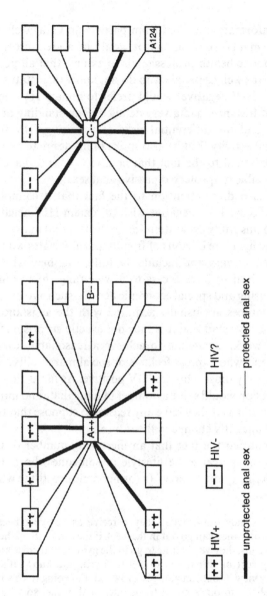

Figure 14.7 Network reseach as a tool for prevention

++ HIV+ |--| HIV-   ☐ HIV?

unprotected anal sex   protected anal sex

misleading. Indeed, information elicited from participant C (as well as a number of other informants) suggests that it would be a great error for epidemiologists and public health professionals to assume that all people who are HIV negative (as well as people who do not know their status but think of themselves as HIV negative) would prefer to remain HIV negative. Participant C, for instance, had a very detailed understanding of the transmission of HIV and the differential risks of acquiring HIV from different types of sexual activity. But he also spoke of his desire to be HIV positive and this contributed to the fact that he often chose to have the riskiest kind of sex possible: unprotected passive anal sex.

Other participants also drew attention to the fact that it cannot be assumed that those who are HIV negative wish to remain HIV negative. Indeed, several men in this study drew attention to the fact that many men, and particularly young men, are envious of friends and associates who are HIV positive. The reasons vary and include the following: men who are HIV positive are perceived to have access to a wide range of resources including access to housing and special economic benefits such as disability allowances. These resources are usually acquired with the assistance of sympathetic and supportive staff and it is thus not exactly surprising that men who are out of work, struggling to achieve a reasonable level of economic well-being and who express feelings of social marginality, depression and loneliness envy those who are HIV positive. As far as they are concerned sero-conversion would solve a number of immediate and important problems – for it is not as if they have any reason to suppose that their circumstances will significantly change in the future.

Participants also reported the fact that an increasing number of men who are in salaried work perceive the lifestyle of someone who is HIV positive to be desirable. To quote "Terry" (a former male prostitute who is HIV positive himself):

> they see their friends out... what they perceive as partying because they're not at work, they can go out in the week if they want to... they're out at the weekends, they've got time to go to the gym... to them it's a big party and they sort of think, I'm in a 9 to 5 job, you know, it's not fair... and they view it that I'm going to grow old, I'm going to be sort of stuck on the shelf, no-one's going to be interested in me, so why not become [HIV] positive?... In their eyes they see it as [there being] more benefits in being positive than negative... I won't get old, I'll get benefits, I can go out and enjoy myself, I don't have to have safe sex anymore...

Similarly, "Peter" (a 40-year-old man who is HIV negative) said:

> I do understand it... I mean, to a lot of people it seems a desirable life style... you don't have to work, you can get invalidity benefit, you can float around and just indulge yourself in a big social whirl... and they

think: "look at him...he's not that ill...[he gets] free cabs and a bus pass"...And then, you know, there's all the sympathy...it's a stupid thing but there's a lot of people that I know who deliberately fly about trying to get it...

"Peter" also drew attention to the glamour of being HIV positive. To quote: "it seems quite glamorous, it does have glamour about it...[I mean] people think I belong...I'm truly gay, I've got HIV". Similarly, a 32-year-old, white English informant (who is HIV positive himself) said: "I wonder if there is a better way of trying to get the message across that catching this thing [HIV] isn't a fashionable thing to do."

Political and economic changes promoting the transmission of HIV

At this point, it is instructive to shift attention away from the study participants and to reflect upon the different ways in which political and economic changes may have promoted the transmission of HIV in London. These links are difficult to establish. However, Wallace's research in New York, USA (1991; 1993) suggests the endeavor is worthwhile. Indeed, his research has shown some of the ways in which public policy has facilitated social disintegration and promoted conditions enabling the transmission of HIV to occur.

Drawing upon information and insights emerging from participant observation fieldwork undertaken at a clinic for sexually transmitted diseases in London (as well as accounts provided by study participants) it is clear that changes in funding imposed upon the National Health Service by the Conservative government during the late 1980s and early 1990s created a situation whereby hospitals were forced to compete with each other for funds and patients.

For instance, many of the London hospitals provide special clinics for those suffering from HIV infections and AIDS. These clinics are based within Departments of Genito-Urinary Medicine (GUM) and the amount of funding they receive is greatly influenced by the number of patients registered at a clinic with HIV infection and AIDS. In view of the limited number of "new" HIV infections (approximately 500–1,000 a year) the pressure is on the financial managers of the clinics to think of ever more imaginative ways to either lure away patients registered at a neighboring clinic or to encourage those "high-risk" patients who do not know their status to have an HIV test.[4] Indeed, it is likely that the growing competition between GUM clinics for limited funds to maintain and develop high standards of care contributes to the desire to be HIV positive.

The standard of care on offer is high. For instance, the HIV clinics currently offer relatively long consultations with doctors as well as a range of other services including acupuncture, nutritional advice and referrals for free psychiatric and psychotherapeutic assistance. The clinics are run in a friendly and informal manner and patients are often on first name terms with receptionists, nurses and/or doctors. In addition, a number of participants referred to the waiting areas as good cruising joints and spoke about the efforts made by clinic staff to organise social events such as Christmas parties to bring together staff and patients, as well as the patient's friends and partners.

In short, the distinct culture of many of the HIV clinics in London-based GUM settings has helped to normalise the experience of being HIV positive. Indeed, several informants felt that, however well intentioned, this was contributing to the disturbing trend whereby one was encouraged to see an HIV diagnosis as "desirable" and an important part of one's identity, entitling one access to a "club with special caring facilities." Others felt that the atmosphere and outlook of staff in GUM clinics encouraged men to view HIV as a "sexy" infection; and they felt this was encapsulated by posters of hunky, muscly men claiming the virtues of being positive about being positive.

There is clearly a need for those engaged in the provision of health services to address these issues and to alter the structure of funding so that clinics do not have to compete with each other in the way they do at the moment for the competition is helping to project an image of HIV infection as enviable and desirable. It is not an easy situation to turn around but it is a matter of some urgency that the issue is addressed. Indeed, endeavors to design culturally appropriate preventive programs will continue to be impeded unless action is taken to redress the errors of the previous Conservative governments.

Conclusion

The social and sexual network data presented in this chapter suggests there is on-going transmission of HIV in London, and detailed case studies have been presented in order to highlight the diverse contexts in which unprotected anal sex occurs. These case studies have also generated information enabling pathways of transmission to be identified. These include the following: first, male prostitutes are sometimes willing to sell unprotected anal sex to other men for more money, and an estimated 40–50% of their clients are also having sex with women. Second, men having anonymous sex with other men in backrooms are also at risk of acquiring and/or transmitting HIV to other men.

These findings are particularly interesting when interpreted in the light of the whole body of social and sexual network data presented in this chapter. These data demonstrate that sexual links exist between British men and women from diverse social and economic backgrounds within London, as well as between British men resident in London and men from large, metropolitan centres in countries such as Australia, Brazil, France, Greece, Holland, Israel, Italy, Lebanon, Malaysia, Spain and the US. The majority of men from these countries were resident in London between August 1994 and July 1995, but many of them continued to have sexual contact with other men resident in their countries of birth.

The information and insights presented in this chapter emerged during a 24-month period of participant observation fieldwork at a clinic for sexually transmitted diseases in London as well as during open-ended, unstructured interviews with predominantly male participants resident in London. The number of study participants was relatively small but, to some extent, this has been offset by the quality and depth of information acquired (see Parker *et al.* 1998 for a detailed discussion of the validity of data presented).

In sum, the data presented in this chapter suggests that the majority of new HIV infections and AIDS cases will continue to occur in London among men who have sex with other men and the rate of transmission will increase. Moreover, there is every indication to suggest that new HIV infections will be acquired in London and subsequently transmitted to men and women residing in metropolitan centres throughout the world. There are, of course, many difficulties with generalising from small data sets but, whatever else, the data presented in this chapter suggests that there is much to be concerned about and public health professionals and policy-makers should take note.

Acknowledgments

Research investigating HIV transmission in London was initiated while working on a collaborative research project focusing on gonorrhoea transmission. This collaborative project was funded by the Wellcome Trust. I would like to thank the Wellcome Trust and my collaborators on the gonorrhoea study (Sophie Day, Azra Ghani, Helen Ward and Jon Weber). I am also grateful to the Health Education Authority for providing a grant enabling me to analyse the data I collected on HIV transmission. Above all, I am grateful to all the study participants who talked so freely and supported the study with such enthusiasm. Special thanks are also extended to Tim Allen and Una Goan for providing encouragement and support throughout the study.

306   M. Parker

Notes

1. Biological factors affecting the transmission of HIV include the infectivity of the viral strain; the infectiousness of the person with HIV; the susceptibility of the uninfected person to infection and the efficiency of the mode of transmission (MacQueen 1994). Behavioral and social factors that are frequently cited to explain the higher prevalence of HIV infections and AIDS cases in London include the following: earlier introduction of the virus; far more sexual mixing between those at "high" risk and "low" risk of infection; higher occurrence of major risk factors for infection (particularly unprotected anal sex among homosexuals and IV drug users); and the fact that cases are continuing to be imported from countries with a higher prevalence of infection (Wadsworth *et al.* 1996).
2. Egocentric network data are derived from individual accounts of sexual behaviour. They involve individuals describing the number of people s/he has had sex with, and the type of sex, as well as any sexual links that s/he knows about between his/her contacts. The term "sexual network" refers to a set of persons in a particular population and the sexual links connecting them. It is a type of social network. Klovdahl usefully distinguishes between egocentric networks and social networks in the following way: "In a population there may be as many personal [egocentric] networks as there are individuals; in the same population all of these personal networks may be connected together to form a single social network" (1985:120).
3. All identifying characteristics of study participants and contacts have been altered to preserve anonymity
4. The following example encapsulates the difficulties faced by staff working at GUM clinics: one of the doctors working in a clinic for sexually transmitted diseases was asked by a senior management figure whether he would be willing to run a special clinic session for men into S&M sex. The suggestion was based on the idea that men into S&M sex would probably be having high risk sex and attracting these men to a clinic catering for their specialist needs may be a way of identifying new cases of HIV infection. But the doctor concerned was uncertain as to whether or not he should run the clinic. He felt he was too well known on the S&M scene and it was potentially awkward for him to see patients that he might have met in very different contexts elsewhere. Other doctors were also concerned about the precedent that such a specialist clinic might set if this particular doctor ran the clinic–namely, that male, gay-identified doctors would feel that they could have greater claim over the gay patients and the "het" (heterosexual) doctors would be left with the "het" patients. The "het" doctors were particularly concerned by this–not because they particularly wanted to run a clinic for men into S&M sex (on the contrary, some of them could think of nothing worse) but because they wanted to have access to the gay HIV positive patients as their research output depended on having access to sufficient numbers of HIV patients. To date, the clinic has not been established but it is an illustration of the thought that is given to attract new patients with undiagnosed HIV infection.

References

Anderson, R. M., Gupta, S. and Ng, W. (1990). The significance of sexual partner contact networks for the transmission dynamics of HIV. *Journal Acquired Immune Deficiency Syndromes*, **3**, 417–29.

Anderson, R. M. and May, R. M. (1992). Understanding the AIDS pandemic. *Scientific American*, **85**, 58–66.

Catania, J. A., Meskowitz, J. T., Ruiz, M. and Cleland, J. (1996). A review of national AIDS-related behavioural surveys. *AIDS*, **10** (suppl. A), S183–S190.

Communicable Disease Surveillance Centre (1996). AIDS and HIV-1 infection in the UK: monthly report. *Communicable Disease Report*, **6**(38), 338.

Connors, M. M. and McGrath, J. W. (1997). The known, unknown and unknowable in AIDS research in anthropology. *Anthropology Newsletter*, **38**(3), 1, 4–5. American Anthropological Association.

Coxon, A. P. M. (1995). Networks and sex: the use of social networks as method and substance in researching gay men's response to HIV and AIDS. In *Conceiving sexuality: approaches to sex research in a postmodern world*, eds. R. G. Parker and J. H. Gagnon. Routledge: New York and London.

Des Jarlais, D. C., Friedman, S. R., Choopanya, K., Vanichseni, S. and Ward, T. P. (1992). International epidemiology of HIV and AIDS among injecting drug users. *AIDS*, **6**, 1053–68.

European Centre for the epidemiological monitoring of AIDS. AIDS surveillance in Europe. Quarterly Report (1995). no. 48 (data to end December 1995).

Gupta, S., Anderson, R. M. and May, R. M. (1989). Networks of sexual contacts: implications for the pattern of spread of HIV. *AIDS*, **3**, 807–17.

Haraldsdottir, S., Gupta, S. and Anderson, R. M. (1992). Preliminary studies of sexual networks in a male homosexual community in Iceland. *Journal of Acquired Immune Deficiency Syndromes*, **5**, 374–81.

Klovdahl, A. S. (1985). Social networks and the spread of infectious diseases: the AIDS example. *Social Science and Medicine*, **21**, 1203–16.

Klovdahl, A. S., Potterat, J. J., Woodhouse, D. E., Muth, D. E., Muth, S. Q. and Darrow, W. W. (1994). Social networks and infectious disease: the Colorado Springs Study. *Social Science and Medicine*, **38**, 79–88.

Laumann, E. O., Gagnon, J. H., Michaels, S., Michael, R. T. and Coleman, J.S. (1989). Monitoring the AIDS epidemic in the United States: a network approach. *Science*, **244**, 1186–89.

Neagius, A., Friedman, S. R., Curtis, R., Des Jarlais, D. C., Furst, R. T., Jose, B., Mota, P., Stepherson, B., Sufian, M., Ward, T. and Wright, J. W. (1994). The relevance of drug injectors' social and risk networks for understanding and preventing HIV infection. *Social Science and Medicine*, **38**, 67–78.

Parker, M., Ward, H. and Day, S. (1998). Sexual networks and the transmission of HIV in London. *Journal of Biosocial Science*, **30**, 63–83.

Parker, M. (forthcoming). Sex in 'backrooms': issues and dilemmas for public health practitioners. London: Health Education Authority.

PHLS AIDS and STD Centre–Communicable Disease Surveillance Centre, and Scottish Centre for Infection and Environmental Health. (1997). Unpublished Quarterly Surveillance Tables No. 37, November.

Service, S. K and Blower, S. M. (1995). HIV transmission in sexual networks: an empirical analysis. *Proceedings of the Royal Society of London*, **260**(1539), 237–44.

Trotter, R. T., Rothenberg, R. B. and Coyle, S. (1995). Drug abuse and HIV prevention research: expanding paradigms and network contributions to risk reduction. *Connections*, **18**(1), 29–45.

Unlinked anonymous surveys steering group (1996). Unlinked anonymous HIV sero-prevalence monitoring programme for England and Wales, data to end of 1995. London: Department of Health, Public Health Laboratory Service, Institute of Child Health.

Wadsworth, J., Hickman, M., Johnson, A. M., Wellings, K. and Field, J. (1996). Goegraphic variation in sexual behaviour in Britain: implications for sexually transmitted disease epidemiology and sexual health promotion. *AIDS*, **10**, 193–9.

Wallace, R. (1991). Social disintegration and the spread of AIDS: thresholds for propogation along 'socio-geographic' networks. *Social Science and Medicine*, **33**, 1155–62.

Wallace, R. (1993). Social disintegration and the spread of AIDS – II. Meltdown of sociogeographic structure in urban minority neighbourhoods. *Social Science and Medicine*, **37**, 887–96.

Woodhouse, D. E., Rothenberg, R. B., Potterat, J. J., Darrow, W. W., Muth, S. Q., Klovdahl, A. S., Zimmerman, H. P., Rogers, H. L., Maldonado, T. S., Muth, J. B. and Reynolds, J. U. (1994). Mapping a social network of heterosexuals at high risk for HIV infection. *AIDS*, **8**, 1331–6.

Part V
The future

15 *The future of urban environments*

S. J. ULIJASZEK AND L. M. SCHELL

Editors' introduction

In the final chapter, possible new ways for the study of urban human biology are discussed. Although various approaches might be possible, two approaches, one of them epidemiological, the other one, anthropological, are considered most pertinent. The use of epidemiological techniques with clearly defined urban variables within an adaptability framework has much mileage, as has the use of ethnography, modernisation studies and studies of urban pathways, for the identification of new, urban characteristics that impact on human biology. Vital to the future of urban human biology is the reformulation of human ecology within an adaptability framework, in which social and organisational constructs are regarded as components of the stress environment.

The human biology of the future must acknowledge the urban existence of human populations, the ways in which they shape their urban environments, and the ways in which their urban environments impact on their health and well-being. Human response to changing environments is a key issue in human adaptability research, and the colonisation of the urban environment is a major adaptive challenge (Boyden 1987; Huss-Ashmore and Thomas, 1997; Schell 1988, 1984), and can be studied at the population and individual levels. Human ecology has contributed significantly to the study of urbanism. Early theories involved the study of the nature and development of community structure, particularly among urban populations (Hawley 1950). Within this structure, relationships between population size and increase, and social organisation and differentiation were sought, and the nature of community structure examined at various levels of organisation, including those of the family, association (profession or school), territorial (urban center or neighborhood), categoric units (those of class or strata), as well as the individual as the fundamental unit of social life. In addition, the types and patterns of

311

relationships between units of organisation were examined (Hawley 1950). The aim of human biology must be in part to reformulate urban human ecology within an adaptability framework, where social and organisational constructs can be considered as components of the stress environment.

In attempting to study urban human adaptability, it is useful to seek fellow-travelers with skills complementary to human biology. Two such fellow-travelers are the disciplines of epidemiology and urban anthropology. Traditionally, epidemiology is defined as the study of the distribution of disease across and within populations in order to understand its causes, and its historical development is related to the health problems created by urbanisation and the sanitary revolution in the nineteenth century (Anderson and May 1992; Lilienfeld and Stolley 1994). Urban anthropology emerged in the 1960s, with studies of rural–urban migration, the ethnography of residential neighborhood life and a strong interest in ethnicity (Sanjek 1990).

Human biology and epidemiology

Modern definitions of epidemiology extend its focus beyond the traditional study of clinically recognised disease entities to include disease-related biological variation in populations (Rose and Barker 1986; Fox et al. 1970; Lilienfeld and Stolley 1994; but for a counter-example see also MacMahon and Pugh 1970). Its origins lie in the study of past epidemics, many of which were urban. Notably, deaths from the plague in fourteenth-century Europe were greatest in urban areas, as were deaths from typhus, cholera and tuberculosis in later centuries; the highly urbanised Aztecs lost about half of their total population to smallpox in 1520 (Anderson and May 1992; Kiple 1993; McNeill 1976). Despite a decline in overall mortality in Europe, the frequency and size of epidemics increased during the eighteenth and nineteenth centuries in large part as a result of urbanisation in increasingly industrialised societies (Anderson and May 1992; Schell 1996). Thus, patterns of disease have been shaped by population size, density, and the social structures developed in urban centers and the behaviors and interactions associated with them. In a complimentary process, the patterns of behavior and the structures of urban social and physical life have been influenced by patterns of disease.

The identification of risk factors for disease and populations at risk of disease is fundamental to epidemiological study (Rose and Barker 1986). Acknowledging the need to increase the complexity of existing epidemiological models to incorporate biocultural interactions, Schell (1992, 1997) put forward the concept of risk focusing. According to Schell, untoward biological outcomes are distributed across society in social and cultural ways, such that there are cultural subsystems in which health risk is focused. This is in contrast to the more traditional ascription of culture as a buffer against stress. Given the complexity of interactions between humans and their urban environments, this approach is especially pertinent to the study of urban human biology. The identification of culture-bound risk factors that can be used in human biological models of stress is an important area to be developed.

The methods of epidemiology rely heavily on the ecological properties of populations, statistics and mathematical modeling (Anderson and May 1992), and population-based surveys must identify target populations at risk of particular diseases. Epidemiology has become ever more sophisticated in its approach to disease patterns, and now deals with a range of issues well beyond the realm of communicable diseases. For much of the twentieth century the most common causes of death in industrialised countries have been chronic, non-infectious ones, e.g. cancers and cardiovascular diseases. Modern epidemiology is applied to determine the relevant risk factors and estimate exposure to them whether they are pathogens of communicable disease or non-infectious environmental agents. This includes the use of many techniques from environmental science and social science. It can, for example, involve estimates of air pollution from outdoor (Suzuki 1990), indoor (Kim 1990) and industrial (Paik 1990) sources, and their effects on health. It could also involve self-reported measurements of the psychosocial environment at work and at home, as well as the new techniques for non-invasive endocrine measurement. Although such work is largely concentrated in the hands of environmental scientists, epidemiologists and psychologists, who examine urban environments from the public health perspective, human biology can consider these potential public health problems in the context of human variation and adaptability. In addition, human biology can employ methods from these fields for the assessment of exposure to risk factors and mathematically model the extent of effects on disease and disease-related variation. Further extensions to this work might include between-population variation in exposure to risk factors for different diseases and related variations, and the study of populations or subpopulations particularly vulnerable to multiple stresses, including lone-parent families, the homeless and drug and alcohol abusers. The development of models to describe and explain biocultural interaction that can be rendered in mathematical terms would be especially useful.

The historical and epidemiological literature abounds with accounts of the effects of infectious diseases on human population size, density and social organisation (Anderson and May 1992). In more recent times, epidemiologists have observed the course of infections and the disease patterns associated with them, created mathematical models of the transmission of infectious agents and of the population biologies of hosts, agents and their environments. Linked aims of human biology as adaptability research include: (1) the identification of the ways in which disease may act as costressor among a number of potential stressors in an environment; (2) explanation of variation within and between populations in response to disease and related biological variation; (3) examination of the extent to which disease constrains the response to other stressors, including other diseases. Studies in human biology should incorporate mathematical modelling techniques, including assessments of relative risk, with particular emphasis on the identification of variables that have particularly urban characteristics. Although existing epidemiological models are able to examine life history and genetic variables of infectious agents in relation to human demographic, genetic and simple behavioral patterns, the use of such models in an adaptability context would require the quantification and integration of more complex human behavioral traits, as well as more detailed quantification of environment.

Human biology and urban anthropology

Urban structures are complex and involve human interactions at different levels of organisation, thus it is important to develop a human biology that incorporates social theories of urbanisation and urbanism. Social and environmental relationships of importance for the study of urban human biology include the following: (1) social, economic and familial relationships between urban and rural populations; (2) urban economic and employment structures; (3) housing and the environment; and (4) social and economic relationships between populations defined in terms of caste, ethnicity, social or economic status. Social relationships between urban and rural populations may influence patterns of migration (Dandekar 1997) and economic well-being of both groups (Gugler 1997a), and thus influence the health patterns of both. Urban economic and employment structures influence the extent of economic disparity, and of unemployment, underemployment and employment which contributes little to social welfare (Gugler 1997b). In addition, it influences gender divisions in the workforce (Nelson 1997; Humphrey 1997) and the extent and nature of child labor (Grootaert and Kanbur 1997). Housing is a fundamental microenvironment in which humans spend a large proportion of their lives. Urban policies concerning water provision, sanitation and waste disposal all influence directly the local environment and the household environment within it (Gilbert 1992). While household habitual practice and behavior can buffer against exposure to stressors in the local environment, there may be limits to this. Furthermore, household behavior and practice may also focus risk upon some households more than others. The use of risk-focusing as a method in the study of urban human biology can include epidemiological techniques incorporating variables derived from ethnographic study, including those which can discriminate between different types of urban–rural relationships, economic and employment structure, housing and environmental factors, as well as ethnicity, class or caste.

There has been little use of ethnographic work in attempting to contextualise stress in urban communities. Culture can contradict biology, in that the biologically least stressful response is not always the observed one, for any number of reasons (Huss-Ashmore and Thomas 1997). In urban communities there are microenvironments that are defined by the socioeconomic status of their inhabitants. These may have their own interpretation of the dominant culture in terms which are structured and framed by the socioeconomic resources and aspirations of the group. These socioeconomically defined subcultures, are, in turn, frameworks for biological and behavioral responses to the urban microenvironment and to the larger urban community and culture. Ideally, such responses are adaptive, but the concept of risk-focusing suggests that they are not necessarily adaptive in every respect, and may lead to more stress and biological disadvantage. Of particular interest is the study of so-called "cycles of deprivation," and what pushes people into them.

More information is needed about the urban environment, the populations that use it, their relationships to each other and to rural and periurban populations. There are numerous problems associated with urban human biology, including the identification of quantifiable cultural traits and markers which can influence human health and well-being. Cultural adaptation is important to understand and measure, but there are no simple ways of so doing. Measures of socioeconomic status are at best crude, and often they dichotomise or make categorical what is usually the product of complex social interactions, any number of which might impact independently on human function. It is important to use tools from the social sciences to identify ways in which socioeconomic status can be disaggregated in biologically more meaningful ways.

Another major challenge is the development of an understanding of how modernisation variables influence human biology globally. The term "modernisation" encompasses all the developments which followed in the wake of industrialisation and mechanisation, and includes the loosening of boundaries between social classes, increased social mobility, as well as the growth of wide-spread education, the evolution of procedures of industrial negotiation, and the development of social welfare systems at the national and regional level. It also includes processes such as the rise of bureaucratic states, the spread of capitalism and the secularisation of culture and education (Smith 1994). These things have happened in most Western nations, as their populations underwent the transition from a predominantly agricultural mode of subsistence, to one of wage-earning within an industrialised and predominantly urban cash-based economy. The experiences of the newer nation-states becoming independent in the twentieth century are quite different from those of Western nations which underwent their industrial revolutions in the eighteenth century. Here, the interaction of less complex energetic, technological and socioeconomic systems characterised by regional production and consumption with contemporary economic systems of industrial technology influenced by the national and international market, social and political factors (McGarvey et al. 1989) has led to different outcomes in relation to urbanisation and urbanism. Thus, modernisation means different things in different places, is associated with urbanism and urbanisation to differing degrees, and represents different social and biological phenomena. As part of its program for the twenty-first century, urban human biology needs to begin to disaggregate the variables which are specifically those of modernisation, discrete from variables which directly reflect urbanism or urbanisation.

There is a further challenge associated with the study of human biology at different levels of social organisation. The types of instruments involved in the measurement of household, community and psychosocial environments are still poorly developed. Households have different structures and meanings even within the same culture, and there are problems with defining households and the transactions that are mediated within them. There is, however, a considerable literature on different aspects of household formation (Brouwer 1988; Ermisch 1988; Morrill and James 1990; Martin 1992), structure (Anderson 1971; Fratkin and Johnson 1990), lifecycle (Rapp 1979; Carter 1984; Wolf 1984), transformation (Wilk and Netting 1984; Johnson 1990; Beittel 1992), as well as economic function (Bender 1967; Roth 1990; Thompson 1992; Wallerstein and Smith 1992). These aspects interact and can influence human biology in different ways. For example, household formation may be related to economic factors which can influence age at marriage and the number of off-spring, while household size and structure may influence their level of economic success relative to other households. Human population biology characteristics are mediated through the household, which may be the primary buffer against, or focus of risk for, many types of ecological stress. Furthermore, households may adapt to changing conditions with varying degrees of success. Better understanding of the processes that take place at this level should improve the understanding of the human biology of urban populations.

Biological processes and outcomes are also mediated at the level of communities, social and occupational groups. Although epidemiologists are able to examine the importance of the latter two units of organisation when they are contiguous with a disease or disease category, they rarely go beyond this to consider the group in relation to broader biological characteristics and other units of organisation. Thus, the study of the behavior and biology of high risk groups is often carried out to understand the prevalence and nature of disease transmission, for example with respect to the transmission of HIV among homosexuals or intravenous drug users. The adaptability of the group itself is not considered. Future studies of urban human biology will need to consider the biological outcomes and implications of social processes that take place at multiple levels of organisation. At a broader level, the study of human adaptability must include the social contexts of human populations, and to appropriate methods from any discipline that will allow the extent and complexity of human variation to be examined in as much detail as possible.

Heterogeneity and population density are markers of urbanism (Gmelch and Zenner 1988) and many studies of urban populations defined on the basis of ethnicity have been carried out. There are also many studies of the social organisation of urban ethnic diversity, and of the relationships between population density, power and the use of urban space (Sanjek 1990). In assessing adaptability of urban populations, it is important to identify discrete populations, their relationships (both biological and cultural) to other populations, and, where possible, to identify aspects of behavior and biology which can be attributed to ethnicity, and which influence adaptability. While human biologists need not be trained as ethnographers, the use of ethnographies to help identify factors potentially involved in human adaptability is important. The use of studies of social organisation, power and use of urban space may help to identify other quantifiable factors associated with stress and illness, as well as health and well-being.

Other urban foci for cultural anthropology which have implications for human biology include: (1) the anthropology of work; (2) gender and sexuality; (3) youth and adolescence; and (4) transitory and tangential groups. The anthropology of work has included studies of involvement in the formal and informal economies, among both adults and children (e.g. Sharff 1981; Schildkrout 1981; Grootaert and Kanbur 1997), and includes work among such groups as female sex workers (Phongpaichit 1982), domestic workers and servants (Cock 1980; Glenn 1986; Hansen 1989) and factory workers (Ong 1987; Westwood 1984). Gender and sexuality have been considered in the context of work (Beneria and Roldan 1987; Hansen 1989; Sharma 1986) and social class (Caplan 1985; Sharma 1986). Youth and adolescence have been considered in terms of risk-taking behavior, while social relations among transitory and tangential groups have been studied among the homeless (Hopper *et al.* 1985), the isolated elderly (Eckert 1980) and personal service workers (Sanjek 1990).

In studying urban human biology, care should be taken to identify new ways of perceiving and interpreting urban complexity, including the social and physical construction of everyday life, many components of which have implications for health and human biology. Hannerz (1980) formulated the theory of human action, which has been used to explore urban pathways of individuals. Hannerz saw individuals as moving through situations, each of which has its own set of behaviors, resources and constraints. For each individual, the path woven through the complex fabric of urban structures and institutions has meaning that often lacks coherence outside of the individual. Furthermore, many of the groups and structures to which the individual belongs often lack social coherence outside of their particular function. For Hannerz, the roles created in situations belong to the human domains of provisioning, household and kinship, leisure, neighboring and transport. Production and social reproduction (Lamphere 1986) have come to replace the domains of provisioning, household and kinship (Sanjek 1990). Individuals follow their particular pathway, moving through situations in which knowledge of, and commitment to others varies enormously; for most members of urban society, such pathways are both predictable and meaningful. Among the variables with implications for urban human biology are economic, medical, educational and other factors which connect the domains of production, social reproduction, neighboring, transport and leisure. The study of such pathways could lead to different understanding of human biology in the urban context, and might reveal new variables which impact on urban human biology and which could be measured at population or group level. One way in which urban pathways might be studied is with the use of classical methods of time allocation measurement (Gross 1988). The tracking of urban pathways in this way might identify situations which are adaptively important. Such methods have been used to examine the time budgets of foraging societies (e.g. McCarthy and McArthur 1960, 1982; Behrens 1981; Ulijaszek and Poraituk 1993), the energetic efficiencies of subsistence strategies (Ulijaszek 1995), the economic value of children (Navera 1978), work carried out by women in the domestic domain (Baumann 1928; Goldschmidt 1982) in developing countries, and in relation to nutritional status (Messer 1989) and energy expenditure (Ulijaszek 1995) of both urban and rural populations. Time allocation studies allow complex behavior and social interaction to be studied (Gross 1988). In urban societies where time is a limiting factor with respect to many aspects of social and productive function, this technique could be powerful in the identification of time-related stressors. Furthermore, time allocation data could be used to examine variation in exposure to potentially stressful factors at the population level.

The use of time allocation data in urban pathway analysis is one possible way forward. There may be others. Future issues in human biology will increasingly be debated on urban terrain, and the challenge is to identify ways in which the high degree of human complexity found in urban environments can be studied without oversimplification.

References

Anderson, M. (1971). *Family Structure in Nineteenth-century Lancashire.* Cambridge: Cambridge University Press.

Anderson, R. M. and May, R. M. (1992). *Infectious Diseases of Humans.* Oxford: Oxford University Press.

Baumann, H. (1928). The division of work according to sex in African hoe culture. *Africa*, **1**, 289–319.

Beittel, M. (1992). The Witwatersrand: black households, white households. In *Creating and Transforming Households*, eds. J. Smith and I. Wallerstein, pp. 197–230. Cambridge: Cambridge University Press.

Bender, D. (1967). A redefinement of the concept of household: families, coresidence, and domestic functions. *American Anthropologist*, **69**, 493–504.

Beneria, L. and Roldan, M. (1987). *The Crossroads of Class and Gender: Industrial Homework, Subcontracting, and Household Dynamics in Mexico City.* Chicago: University of Chicago Press.

Behrens, C. (1981). Time allocation and meat procurement among the Shipibo Indians of eastern Peru. *Human Ecology*, **9**, 189–220.

Boyden, S. (1987). *Western Civilization in Biological Perspective.* Oxford: Clarendon Press.

Brouwer, J. (1988). Application of household models in housing policy. In *Modelling Household Formation and Dissolution*, eds. N. Keilman, A. Kuijsten and A. Vossen, pp. 225–39. Oxford: Clarendon Press.

Caplan, P. (1985). *Class and Gender in India: Women and Their Organisations in a South Indian City.* London: Tavistock Press.

Carter, A. T. (1984). Household histories. In *Households. Comparative and historical studies of the domestic group*, eds. R. McC. Netting, R. R. Wilk and E. J. Arnould, pp. 44–83. Berkeley: University of California Press.

Cock, J. (1980). *Maids and Madams: A Study in the Politics of Exploitation.* Johannesburg: Ravan Press.

Dandekar, H. C. (1997). Changing migration strategies in Deccan Maharashtra, India, 1885–1990. In *Cities in the Developing World. Issues, Theory, and Policy*, ed. J. Gugler, pp. 48–61. Oxford: Oxford University Press.

Eckert, K. (1980). *The Unseen Elderly: A Study of Marginally Subsistent Hotel Dwellers.* San Diego: University of San Diego Press.

Ermisch, J. (1988). An economic perspective on household modelling. In *Modelling Household Formation and Dissolution* , eds. N. Keilman, A. Kuisten and A. Vossen, pp. 23–40. Oxford: Clarendon Press.

Fox, J. P., Hall, C.E. and Elveback, L. R. (1970). *Epidemiology. Man and Disease.* London: The Macmillan Company.

Fratkin, E. and Johnson, P. L. (1990). Empirical approaches to household organisation. *Human Ecology*, **18**, 357–62.

Gilbert, A. (1992). Third world cities: housing, infrastructure and servicing. *Urban Studies*, **29**, 435–60.

Glen, E. (1986). *Issei, Nisei, War Bride: Three Generations of Japanese American Women in Domestic Service.* Philadelphia: Temple University Press.

Gmelch, G. and Zenner, W. (eds.) (1988). *Urban Life: Readings in Urban Anthropology.* Prospect Heights, IL: Waveland Press.

Goldschmidt, C. L. (1982). *Unpaid Work in the Household: A Review of Economic Evaluation Methods.* Geneva: International Labour Office.

Grootaert, C. and Kanbur, R. (1997). Confronting child labour: a gradualist approach. In *Cities in the Developing World. Issues, Theory, and Policy*, ed. J. Gugler, pp. 184–95. Oxford: Oxford University Press.

Gross, D. R. (1988). Time allocation: a tool for the study of cultural behavior. *Annual Reviews in Anthropology*, **15**, 519–58.

Gugler, J. (1997a). Life in a dual system revisited: urban–rural ties in Enugu, Nigeria, 1961–1987. In *Cities in the Developing World. Issues, Theory, and Policy*, ed. J. Gugler, pp. 62–73. Oxford: Oxford University Press.

Gugler, J. (1997b). Overurbanization reconsidered. In *Cities in the Developing World. Issues, Theory, and Policy*, ed. J. Gugler, pp. 114–23. Oxford: Oxford University Press.

Hannerz, U. (1980). *Exploring the City: Inquiries Toward an Urban Anthropology.* New York: Columbia University Press.

Hansen, K. (1989). *Distant Companions: Servants and Employers in Zambia, 1900–1985.* Ithaca: Cornell University Press.

Hawley, A. H. (1950). *Human Ecology. A Theory of Community Structure.* New York: The Ronald Press Company.

Hopper, K., Susser, E. and Conover, S. (1985). Economies of makeshift: deindustrialization and homelessness in New York Ciy. *Urban Anthropology*, **14**, 183–236.

Humphrey, J. (1997). Gender divisions in Brazilian industry. In *Cities in the Developing World. Issues, Theory, and Policy*, ed. J. Gugler, pp. 171–83. Oxford: Oxford University Press.

Huss-Ashmore, R. and Thomas, R. B. (1997). The future of human adaptability research. In *Human Adaptability, Past, Present and Future*, eds. S.J. Ulijaszek and R.A. Huss-Ashmore, pp. 295–319. Oxford: Oxford University Press.

Johnson, P. L. (1990). Changing household composition, labor patterns, and fertility in a highland New Guinea population. *Human Ecology*, **18**, 403–16.

Kiple, K. F. (ed.) (1993). *The Cambridge World History of Human Disease.* New York: Cambridge University Press.

Kim, Y. S., (1990). Exposure to indoor pollutants and its health impact in Seoul residences. *Journal of Human Ergology*, **19**, 107–12.

Lamphere, L. (1986). From working daughters to working mothers: production and reproduction in an industrial community. *American Ethnologist*, **13**,

118–30.

Lilienfeld, A. and Stolley, P. D. (1994). *Foundations of Epidemiology*. 3rd edn. New York: Oxford University Press.

MacMahon, B. and Pugh, T. F. (1970). *Epidemiology. Principles and Methods*. Boston: Little, Brown and Company.

Martin, W. G. (1992). Lesotho: the creation of households. In *Creating and Transforming Households*, eds. J. Smith and I. Wallerstein, pp. 231–49. Cambridge: Cambridge University Press.

McCarthy, F. and McArthur, M. (1960). *The Food Quest and Time Factor in Aboriginal Economic Life*. Melbourne: Melbourne University Press.

McGarvey, S. T., Bindon, J. R., Crews, D. E. and Schendel, D. E. (1989). Modernization and adiposity: causes and consequences. In *Human Population Biology*, eds. M. A. Little and J. D. Haas, pp. 263–79. New York: Oxford University Press.

McNeill, W. H. (1976). *Plagues and Peoples*. Oxford: Blackwell Science Publications.

Messer, E. (1989). The relevance of time allocation analyses for nutritional anthropology. In *Research Methods in Nutritional Anthropology*, eds. G.H. Pelto, P.J. Pelto and E. Messer, pp. 82–125. Tokyo: United Nations University.

Morrill, W. T. and James, A. V. (1990). Modernization and household formation on St. Bart: continuity and change. *Human Ecology*, **18**, 457–74.

Navera, E. R. (1978). The allocation of household time associated with children in rural households in Laguna, Philippines. *Philippine Economic Journal*, **17**, 213–23.

Nelson, N. (1997). How women and men got by and still get by (only not so well): the gender division of labour in a Nairobi shanty-town. In *Cities in the Developing World. Issues, Theory, and Policy*, ed. J. Gugler, pp. 156–70. Oxford: Oxford University Press.

Ong, A. (1987). *Spirits of Resistance and Capitalist Discipline: Factory Women in Malaysia*. Albany, NY: State University of New York Press.

Paik, N. W. (1990). Industrial hygiene in Korea. *Journal of Human Ergology*, **19**, 129–44.

Phongpaichit, P. (1982). *From Peasant Girls to Bangkok Masseuses*. Geneva: International Labour Office.

Rapp, R. (1979). Examining family history: household and family. *Feminist Studies*, **5**, 175–81.

Rose, G. and Barker, D. J. P. (1986). *Epidemiology for the Unitiated*. London: British Medical Journal Publications.

Roth, E. A. (1990). Modeling Rendille household herd composition. *Human Ecology*, **18**, 441–55.

Sanjek, R. (1990). Urban anthropology in the 1980s: a world view. *Annual Reviews in Anthropology*, **19**, 151–86.

Schell, L. M. (1997) Culture as a stressor: a revised model of biocultural interaction. *Amer. J. Phys. Anthropol.*, **102**, 67–77.

Schell, L. M. (1996) Cities and Human Health. In *Urban Life*, 3rd edn., eds. G. Gmelch and W. Zenner, pp. 104–35. Prospect Hts., IL: Waveland Press.

Schell, L. M. (1992). Risk focussing: an example of biocultural interaction. In *Health and Lifestyle Change*, eds. R. Huss-Ashmore, J. Schall and M. Hediger,

322   S. J. Ulijaszek and L. M. Schell

pp. 137–44. MASCA Research Papers in Science and Archaeology, vol. 9. Philadelphia: University Museum of Archaeology and Anthropology.

Schell, L. M. (1986) Community health assessment and physical anthropology: auxological epidemiology. *Human Organization*, 45, 321–7.

Schildkrout, E. (1981). The employment of children in Kano. In *Child Work, Poverty and Underdevelopment*, eds. G. Rogers and G. Standing, pp. 81–112. Geneva: International Labour Office.

Sharff, J. (1981). Free enterprise and the ghetto family. *Psychology Today*, 15, 40–7.

Sharma, U. (1986). *Women's Work, Class, and the Urban Household: A Study of Shimla, North India*. London: Tavistock Press.

Smith, A. D. (1994). The politics of culture: ethnicity and nationalism. In *Companion Encyclopedia of Anthropology*, ed. T. Ingold, pp. 706–33. London: Routledge.

Suzuki, S. (1990). Health effects of lead pollution due to automobile exhaust: findings from field surveys in Japan and Indonesia. *Journal of Human Ergology*, 19, 113–22.

Thompson, L. (1992). Mexico City: the slow rise of wage-centred households. In *Creating and Transforming Households*, eds. J. Smith and I. Wallerstein, pp. 150–69. Cambridge: Cambridge University Press.

Ulijaszek, S. J. (1995). *Human Energetics in Biological Anthropology*. Cambridge: Cambridge University Press.

Ulijaszek, S. J. and Poraituk, S. P. (1993). Making sago: is it worth the effort? In *Tropical Forests, People and Food. Biocultural Interactions and Applications to Development*, eds. C. M. Hladik, A. Hladik, O. F. Linares, H. Pagezy, A. Semple and M. Hadley, pp. 271–80. Paris: UNESCO Publications.

Wallerstein, I. and Smith, J. (1992). Households as an institution of the world-economy. In *Creating and Transforming Households*, eds. J. Smith and I. Wallerstein, pp. 3–23. Cambridge: Cambridge University Press.

Westwood, S. (1984). *All Day, Every Day: Family and Factory in the Making of Women's Lives*. Urbana, IL: University of Illinois Press.

Wilk, R.R. and Netting, R. McC. (1984). Households: changing forms and functions. In *Households. Comparative and historical studies of the domestic group*, eds. R. McC. Netting, R. R. Wilk and E. J. Arnould, pp. 1–28. Berkeley: University of California Press.

Wolf, A. P. (1984). Family life and the life cycle in rural China. In *Households. Comparative and Historical Studies of the Domestic Group*, eds. R. McC. Netting, R. R. Wilk and E. J. Arnould, pp. 279–98. Berkeley: University of California Press.

Index

323